AF368536

Phytic Acid and Mineral Biofortification Strategies

Phytic Acid and Mineral Biofortification Strategies

From Plant Science to Breeding and Biotechnological Approaches

Special Issue Editors

Eleonora Cominelli
Francesca Sparvoli
Roberto Pilu

MDPI • Basel • Beijing • Wuhan • Barcelona • Belgrade

Special Issue Editors

Eleonora Cominelli
Institute of Agricultural Biology
and Biotechnology
Italy

Francesca Sparvoli
Institute of Agricultural Biology
and Biotechnology
Italy

Roberto Pilu
Università degli Studi di Milano
Italy

Editorial Office
MDPI
St. Alban-Anlage 66
4052 Basel, Switzerland

This is a reprint of articles from the Special Issue published online in the open access journal *Plants* (ISSN 2223-7747) from 2019 to 2020 (available at: https://www.mdpi.com/journal/plants/special_issues/phytic_acid).

For citation purposes, cite each article independently as indicated on the article page online and as indicated below:

LastName, A.A.; LastName, B.B.; LastName, C.C. Article Title. *Journal Name* **Year**, *Article Number*, Page Range.

ISBN 978-3-03936-202-8 (Hbk)
ISBN 978-3-03936-203-5 (PDF)

Cover image courtesy of Eleonora Cominelli.

© 2020 by the authors. Articles in this book are Open Access and distributed under the Creative Commons Attribution (CC BY) license, which allows users to download, copy and build upon published articles, as long as the author and publisher are properly credited, which ensures maximum dissemination and a wider impact of our publications.

The book as a whole is distributed by MDPI under the terms and conditions of the Creative Commons license CC BY-NC-ND.

Contents

About the Special Issue Editors . . . . **vii**

Eleonora Cominelli, Roberto Pilu and Francesca Sparvoli
Phytic Acid and Mineral Biofortification Strategies: From Plant Science to Breeding and
Biotechnological Approaches
Reprinted from: *Plants* **2020**, *9*, 553, doi:10.3390/plants9050553 . . . . **1**

Ambuj B. Jha and Thomas D. Warkentin
Biofortification of Pulse Crops: Status and Future Perspectives
Reprinted from: *Plants* **2020**, *9*, 73, doi:10.3390/plants9010073 . . . . **9**

**Shivani Sharma, Gazaldeep Kaur, Anil Kumar, Varsha Meena, Hasthi Ram, Jaspreet Kaur
and Ajay Kumar Pandey**
Gene Expression Pattern of Vacuolar-Iron Transporter-Like (VTL) Genes in Hexaploid Wheat
during Metal Stress
Reprinted from: *Plants* **2020**, *9*, 229, doi:10.3390/plants9020229 . . . . **38**

Paula Pongrac, Iztok Arčon, Hiram Castillo-Michel and Katarina Vogel-Mikuš
Mineral Element Composition in Grain of Awned and Awnletted Wheat (*Triticum aestivum* L.)
Cultivars: Tissue-Specific Iron Speciation and Phytate and Non-Phytate Ligand Ratio
Reprinted from: *Plants* **2020**, *9*, 79, doi:10.3390/plants9010079 . . . . **53**

Victor Raboy
Low phytic acid Crops: Observations Based On Four Decades of Research
Reprinted from: *Plants* **2020**, *9*, 140, doi:10.3390/plants9020140 . . . . **67**

Eleonora Cominelli, Roberto Pilu and Francesca Sparvoli
Phytic Acid and Transporters: What Can We Learn from *low phytic acid* Mutants?
Reprinted from: *Plants* **2020**, *9*, 69, doi:10.3390/plants9010069 . . . . **94**

Gian Attilio Sacchi and Fabio Francesco Nocito
Plant Sulfate Transporters in the Low Phytic Acid Network: Some Educated Guesses
Reprinted from: *Plants* **2019**, *8*, 616, doi:10.3390/plants8120616 . . . . **114**

Giulia Borlini, Cesare Rovera, Michela Landoni, Elena Cassani and Roberto Pilu
lpa1-5525: A New lpa1 Mutant Isolated in a Mutagenized Population by a Novel
Non-Disrupting Screening Method
Reprinted from: *Plants* **2019**, *8*, 209, doi:10.3390/plants8070209 . . . . **124**

Meng Jiang, Yang Liu, Yanhua Liu, Yuanyuan Tan, Jianzhong Huang and Qingyao Shu
Mutation of *Inositol 1,3,4-trisphosphate 5/6-kinase6* Impairs Plant Growth and Phytic Acid
Synthesis in Rice
Reprinted from: *Plants* **2019**, *8*, 114, doi:10.3390/plants8050114 . . . . **138**

Ayaka Fukushima, Ishara Perera, Koki Hosoya, Tatsuki Akabane and Naoki Hirotsu
Genotypic Differences in the Effect of P Fertilization on Phytic Acid Content in Rice Grain
Reprinted from: *Plants* **2020**, *9*, 146, doi:10.3390/plants9020146 . . . . **150**

Nisar Ahmad Taliman, Qin Dong, Kohei Echigo, Victor Raboy and Hirofumi Saneoka
Effect of Phosphorus Fertilization on the Growth, Photosynthesis, Nitrogen Fixation, Mineral
Accumulation, Seed Yield, and Seed Quality of a Soybean Low-Phytate Line
Reprinted from: *Plants* **2019**, *8*, 119, doi:10.3390/plants8050119 . . . . **161**

Catherine Freed, Olusegun Adepoju and Glenda Gillaspy
Can Inositol Pyrophosphates Inform Strategies for Developing Low Phytate Crops?
Reprinted from: *Plants* **2020**, *9*, 115, doi:10.3390/plants9010115 . . . . **174**

About the Special Issue Editors

Eleonora Cominelli has been a researcher at the Institute of Agricultural Biology and Biotechnology of the National Research Council of Italy (CNR) since 2011. Her past research activity has been mainly focused on the functional characterization of transcription factors involved in the regulation of flavonoid biosynthesis and water stress response in plant model species and crops. Her present research activity is mainly focused on seed nutritional quality through the study of low phytic acid (lpa) mutants and genes involved in phytic acid synthesis and transport.

Francesca Sparvoli is a senior scientist at the Institute of Agricultural Biology and Biotechnology of the National Research Council of Italy (CNR), where she has worked since 2000. Her main interests concern the genetic improvement of legume and cereal seeds for nutritional quality traits, with a particular focus on seed biofortification, through the development of low phytic acid (lpa) mutants, and on the exploitation of seed storage protein variants in common beans for the production of novel foods.

Roberto Pilu is an associate professor in Plant Genetics and Breeding at the Department of Agricultural and Environmental Sciences—Production, Landscape, Agroenergy (Di.S.A.A.), of the University of Milan (Italy). After a degree in Biology and a MSc in Applied Genetics at the University of Milan, he completed his PhD in Molecular Biology at the University of Pavia. For more than 25 years, he has been working on topics related to plant genetics and breeding, particularly in maize. In recent years, he has been collaborating with different private companies and developing countries, with the aim to study and improve the genetic materials for special traits. He teaches Essentials of Genetics and Plant Breeding, for different degree courses at the University of Milan.

Editorial

Phytic Acid and Mineral Biofortification Strategies: From Plant Science to Breeding and Biotechnological Approaches

Eleonora Cominelli [1,*], Roberto Pilu [2] and Francesca Sparvoli [1]

[1] Institute of Agricultural Biology and Biotechnology, Consiglio Nazionale delle Ricerche, Via E. Bassini 15, 20133 Milan, Italy; sparvoli@ibba.cnr.it

[2] Department of Agricultural and Environmental Sciences—Production Landscape, Agroenergy, Università degli Studi di Milano, Via G. Celoria 2, 20133 Milan, Italy; salvatore.pilu@unimi.it

* Correspondence: cominelli@ibba.cnr.it; Tel.: +39-02-23699421

Received: 23 March 2020; Accepted: 24 April 2020; Published: 26 April 2020

Abstract: Mineral deficiencies, particularly for iron and zinc, affect over two billion people worldwide, mainly in developing countries where diets are based on the consumption of staple crops. Mineral biofortification includes different approaches aimed to increase mineral concentration and to improve mineral bioavailability in the edible parts of plants, particularly the seeds. A multidisciplinary approach, including agronomic, genetic, physiological, and molecular expertise, is necessary to obtain detailed knowledge of the complex homeostatic mechanisms that tightly regulate seed mineral concentrations and the molecules and mechanisms that determine mineral bioavailability, necessary to reach the biofortification objectives. To increase bioavailability, one strategy is to decrease seed content of phytic acid, a highly electronegative molecule present in the cell that chelates positively charged metal ions, many of which are important for human nutrition. All the contributions of the current Special Issue aim at describing new results, reviewing the literature, and also commenting on some of the economic and sociological aspects concerning biofortification research. A number of contributions are related to the study of mineral transport, seed accumulation, and approaches to increase seed micronutrient concentration. The remaining ones are mainly focused on the study of *low phytic acid* mutants.

Keywords: biofortification; *low phytic acid (lpa)* mutants; metal transporter; mineral deficiencies; phytic acid

---

1. Introduction

In 2015, we edited a *Plants* Special Issue on "Phytic Acid Pathway and Breeding in Plants", with the aim to provide a unique compendium that highlighted new developments in our understanding of how perturbation in phytic acid (PA) synthesis and accumulation contributes to plant function, growth, and response to the environment. After 4 years, we believe that the development of biofortified crops to respond to mineral deficiencies is still challenging. For this reason, we decided to launch a new Special Issue on "Phytic Acid and Mineral Biofortification Strategies: From Plant Science to Breeding and Biotechnological Approaches" in which we collected articles and reviews providing new knowledge and technical advances in the field.

The review from Jha and Warkentin [1] can be considered as an introduction to the topics of the present Special Issue. Although it focuses on the approaches used to biofortify pulses for different key micronutrients, it also provides a general overview of the different strategies to tackle micronutrient deficiency. The authors first describe the requirements and functions of the different key micronutrients in humans and the negative impacts of their deficiency in the diet. Then, different approaches that can be used to improve micronutrient content and absorption from the diet are presented. These approaches include dietary diversification, the use of food supplements, food fortification, and biofortification.

The main objectives for crop seed biofortification are: (i) to increase seed micronutrient concentration; (ii) to decrease seed content of antinutritional factors that reduce micronutrient bioavailability, mainly PA.

To reach the first objective, it is important to acquire basic knowledge of the genes involved in mineral transport, such as the *Vacuolar Iron Transporter-Like* (*VTL*) genes described in the article from Sharma et al. [2], and of how minerals accumulate into seeds, the focus of the article by Pongrac et al. [3]. An overview of breeding approaches used to develop pulses with increased seed mineral concentration is presented in the already cited review [1].

The other contributions of the present Special Issue cover different aspects related to the second objective of crop biofortification. A very fascinating review from Raboy evaluates the possible economic and social impact of *low phytic acid* (*lpa*) mutants [4]. A review and a commentary are focused on *lpa* mutants in transporters [5,6]. Four articles describe the isolation and characterization in different crops of *lpa* mutants [7–10], derived from mutagenized populations [7,9,10] or from the genome editing technology approach [8]. An overview of the role of inositol pyrophosphates and useful recommendations for the development of novel *lpa* mutants is presented by Freed et al. [11].

2. Mineral Transport, Seed Accumulation, and Breeding to Increase Concentration in Seeds

2.1. Mineral Transport and Seed Accumulation

Seed iron is mainly stored in vacuoles. Hence, improving iron uptake into the vacuole is a valuable alternative strategy to increase total iron content. For this reason, the role of vacuolar iron transporters needs to be addressed and exploited. The article by Sharma et al. describes the isolation and preliminary characterization of the family of the *VTL* genes in hexaploid wheat. The authors report data on phylogenetic analysis and on a quantitative expression analysis of *VTL* genes in response to iron surplus and deficiency, under zinc, manganese, and copper deficiency, and under heavy metals treatments [2]. Particularly, 23 wheat *VTL* gene sequences were identified that can be phylogenetically distinguished from the *Vacuolar Iron Transporters* (*VIT*) ones and are grouped as 4 *VTL* genes due to the occurrence of homeologs. The expression data in response to treatments with different concentrations of minerals suggest that these genes have an important role in mineral homeostasis [2].

The knowledge of mechanisms involved in mineral accumulation, in terms of tissue-specificity, speciations, and ligand compounds, is very important to set up a precise biofortification program. In the wheat ear, awns (bristle-like structures extending from lemmas) have transpiration and photosynthetic activity. Hence, their presence could contribute to the translocation of elements taken up by roots on the one hand and/or to the phloem-driven (re)allocation of assimilates on the other hand, thereby affecting mineral element density in the grain. The study by Pongrac et al. presents a comparison of mineral element composition between awned and awnletted (those that have short or no awns) wheat cultivars. Moreover, tissue-specific iron speciation and iron ligands in the cultivars contrasting for seed iron content were also investigated using micro X-ray absorption near edge structure (micro-XANES) [3]. The authors found that among the 20 different cultivars, the awnletted ones showed lower whole-grain concentrations of calcium and manganese, but higher iron concentration, compared to the awned cultivars. Interestingly, no differences were observed either in iron speciation (the percentages of ferric and ferrous iron are similar in the four most contrasting analyzed awned and awnletted cultivars) or in terms of ligands, as on average 53% of the iron is in a phytate form. On the contrary, there was a distinct tissue-specificity in iron speciation and ligands, with the pericarp containing the largest proportion of ferric species with only non-phytate ligands, as also in the nucellar projection. In other tissues, such as the aleurone, scutellum, and embryo, iron was predominantly bound to phytate. The authors conclude that, as iron bioavailability is dependent on iron ligands, its bioavailability in wheat is tissue specific. Further investigation on the genetic and/or metabolic reasons behind the observed differences is needed [3].

2.2. Germplasm Screening: Genetic Variation and Identification of Genomic Regions and Molecular Markers (Quantitative Trait Loci (QTLs), Single Nucleotide Polymorphisms (SNPs))

The increase of seed micronutrient concentration can be achieved through agronomic interventions, genetic engineering, and plant breeding. The first two approaches are briefly summarized in the review by Jha and Warkentin [1]. However, a long section of the review is dedicated to conventional plant breeding approaches that have been used to biofortify pulses, particularly common bean, lentil, chickpea, mungbean, and pea, for some minerals (iron, zinc, and selenium), carotenoids, and folates. The advantages of using conventional plant breeding, compared to the genetic engineering strategy, are the relatively low costs and the high acceptability by consumers. The first step in this kind of approach is the screening for genetic variability for micronutrients' seed concentration. The authors report that a significant effect of the genotype in the determination of micronutrient seed concentration has been shown for the different micronutrients in the various pulses, although the environment (weather and soil factors, such as aeration, water availability, pH, and texture) may have significant effects, for example, in the case of zinc and selenium seed concentrations. Moreover, for the pulses studied, they report the identification of genomic regions (Quantitative Trait Loci, QTLs) and molecular markers, mainly Single Nucleotide Polymorphisms (SNPs), which can be used in marker-assisted selection procedures, aimed at improving micronutrient seed concentration. In some cases, candidate genes involved in the accumulation of the different micronutrients are described. The authors mention the different crops, not only pulses, with increased micronutrient concentration that have been released in recent years in developing countries, mainly thanks to the activity of HarvestPlus, an initiative of the Consultative Group on International Agricultural Research (CGIAR), started in 2003 to enrich various major crops with iron, zinc, and vitamin A. Evidence that the introduction of some of these crops in the diet has helped in overcoming nutrient deficiency is well documented [1].

3. Decreasing Antinutritional Compounds Concentration: Four Decades of Research and Novel Perspectives for *lpa* Mutants

3.1. Some Possibilities to Redeem the so far Neglected lpa Crops?

PA is the most abundant form of phosphorus (P) occurring in seeds. However, it is a "non-available" form of P for monogastric animals devoid of phytase (poultry, swine, fish). Moreover, it is a strong cation chelator, reducing the bioavailability of cations important for nutrition. PA is also a very important signaling molecule involved in different regulatory processes during plant development and responses to different stimuli [12]. Different contributions to the present Special Issue treat aspects related to PA and *lpa* mutants [4–10].

Different *lpa* mutants have been isolated and characterized so far in different species, starting from the first ones, the maize *lpa1-1* and *lpa2-1*, isolated in the early 1990s in the USDA-ARS laboratory of Raboy, but none of them has been commercialized, as the scientist underlines, with a certain regret, in his very interesting commentary [4]. The commentary is strongly felt, as the author is "a person intimately involved in the entire process across a 40-year period," as written in his/her report by one of the anonymous reviewers of the manuscript (the review reports are publicly available at the webpage of the commentary). As still observed by one of the two reviewers, the view presented in this commentary is multidisciplinary, as scientific, economic, and social aspects of the subject are discussed, "encompassing the issues faced by all breeders (and probably agronomists as well) attempting to develop new materials with significant social benefit but difficult to capture short term economic benefit when favored alternatives exist with the reverse tendencies". The author underlines that these mutants have some potential advantages, mainly (i) improving phosphorus management in non-ruminant production, contributing to enhance sustainability and reduce animal waste P, and (ii) increasing mineral bioavailability as a strategy to combat mineral deficiencies, as shown by different studies.

Nevertheless, these mutants have received very little interest. The author thinks that the reasons for this are primarily due to the reduced yield (5–10% decrease) and field performance that characterize some of these mutants, together with the criticism that reducing PA is not wholly advantageous as it may also have positive nutritional benefits (antioxidant and anticancer properties, shown through *in vitro* studies [13], although it has been shown that no phytate is present in human biofluids [14]) that might be lost in the *lpa* mutants. Moreover, another simple explanation may be the tendency to use a conservative approach to crop improvement strategies for crops that provide staple foods to at-risk populations in developing countries. However, no support or time to improve the agronomic performance of these mutants has been provided. Some alternatives that have been preferred in recent years exist. Concerning methods to increase P for feed, it can be directly added or phytase can be used to increase the component of available P. The author says that these methods have been preferred to the use of *lpa* crops, without calculating the possible long-term money-saving deriving from using the *lpa* crops. Moreover, the positive results from animal nutrition studies when animals are fed *lpa* crops (for example: leaner pigs, with enhanced muscle density and less backfat when fed *lpa* maize; eggs with reduced cholesterol from hens fed *lpa* maize) ironically pushed farmers to apply phytase superdoses, also in this case not considering the long-term money-saving if, instead, *lpa* crops might have been used [4]. These savings would have overcome the reduced yield problem with the further advantages of having more nutritious crops. Raboy is also quite critical of the HarvestPlus program and the international agricultural centers participating in the Gates Foundation that concentrated all their efforts only to promote biofortification through breeding crops for elevated micronutrient density and, therefore, pushed the development and promotion of *lpa* crops to the sidelines, although a combination of both approaches would likely give the most promising results.

After the publication of this Special Issue, an article has been published by different authors, including scientists from HarvestPlus and CIAT (a CGIAR institute), comparing the retention of iron and zinc when preparing common household recipes with conventional, biofortified, or *lpa* common beans [15]. The retention of iron was very high and similar using the different bean genotypes, while *lpa* beans exhibited lower retention for zinc. Further studies are needed to understand this difference. However, the authors encourage the development of beans with an increased mineral content combined with a low PA trait, and also with low concentrations of specific polyphenolic compounds, as the research target for the next generation of biofortified beans [15]. This publication can lay the foundation for a brighter future for *lpa* crops.

3.2. lpa Mutants in Different Classes of Transporters: Not Always so Obvious

PA reduction can be achieved with mutations in different types of transporters that control PA transport to the vacuole (MRP-type ABC transporters), or by modifying inorganic P (P_i) availability for PA synthesis through mutations in transporters involved in P_i loading and organ/intracellular distribution (SULFATE TRANSPORTER 3;3 -SULTR3;3- and SULTR3;4, members of the group 3 of putative sulfate transporters) or by P_i acquisition and mobilization during seed development (PHOSPHATE TRANSPORTER 1;4, PHT1;4). The review by Cominelli et al. is focused on the description of genes, proteins, and mutants of these different transporters in cereals and legumes [5]. Particular attention is dedicated to those mutants devoid of negative pleiotropic effects, such as mutants affected by the common bean *MRP1* gene and by the rice and barley *SULTR3;3* and rice *SULTR3;4* genes, suggesting strategies to develop useful *lpa* mutants in other species as well [5].

Sacchi and Nocito in their opinion paper propose a deeper discussion on group 3 putative sulfate transporters and suggest some hypotheses to unveil the links between sulfate and P accumulation in seeds [6]. The fact that genes predicted to encode for sulfate transporters, when mutated, cause a *lpa* phenotype is anything but obvious. Differently from other sulfate transporters, for which capability to transport sulfate has mainly been proven through complementation of yeast mutants, the function of SULTR3s has only been hypothesized based on their sequence homologies. The rice and Arabidopsis SULTR3;4 proteins are able to transport phosphate and not sulfate, as recently shown [16,17], explaining the *lpa* phenotype of the rice mutant. However, the OsSULTR3;3 protein does not show either phosphate or sulfate transport activity. Some hypotheses on the link between sulfate and phosphate homeostasis and the development of the *lpa* phenotype in *sultr3;3* mutants are proposed in the opinion paper [6].

3.3. Response to P Fertilization

Seed PA content is affected by the amount of supplied P in various crops [18–20]. In this Issue, two articles investigate the effects of P fertilization in genotypes differing in their PA content, particularly in a soybean *lpa* mutant compared with two normal-phytate cultivars [10] and in two rice cultivars described for their contrasting grain PA content [9].

In the first article, Taliman et al. [10] compare different parameters, such as dry weight, photosynthetic rate, dinitrogen fixation, mineral accumulation, and grain yield between a soybean *lpa* line and two varieties normally cultivated in Japan (wild type, wt) in response to high and low P fertilization. The authors generally observed increased plant performance and yield at higher P concentration in all three genotypes but very little difference between *lpa* and wt genotypes. Seed yield was higher in the *lpa* line than in the normal-phytate cultivars at both fertilization doses. The results show that the already positive properties of *lpa* seeds in terms of increased mineral cations bioavailability can be also accompanied by good agronomic performance [10].

In the second article, Fukushima et al. [9] analyzed the response to P fertilization of two previously selected rice genotypes differing for their PA content with the World Rice Core Collection 5 (WRC 5) genotype showing the lowest and WRC 6 showing the highest PA content among different accessions [21]. The authors reported that differences in PA content between the two contrasting cultivars were observed only under standard P fertilization conditions, while, if two different doses of P fertilizer were applied at different developmental stages (at seedling or heading stage) an increase in PA content was observed in both genotypes, highly reducing the differences between the two genotypes. Interestingly, the expression level of the *myo-inositol 3-phosphate synthase 1* (*INO1*) gene was suggested to be the genetic basis explaining the natural variation in PA accumulation in rice. Although the DNA sequences of the coding region and a putative promoter region of 1000 bp of the *INO1* gene were identical between WRC 5 and WRC 6, the gene is more expressed in the WRC 6 accession than in the WRC 5 one. Moreover, the *INO1* gene transcript accumulation increased in response to P fertilizer only in the WRC 6 accession. The authors hypothesized the existence of different regulatory mechanisms of PA content besides the DNA mutation in the *INO1* gene [9].

3.4. lpa Mutants: Isolation and Characterization of New Mutants and Description of a Novel Screening Method in Maize

The articles by Jiang et al. and Borlini et al. focus on the isolation and characterization of novel *lpa* mutants in rice and maize, respectively; in the second article, a particularly easy new screening method for maize *lpa* mutants is also described [7,8].

Most of the *lpa* mutants have been isolated by screening mutagenized populations and few examples using transgenic approaches have been also reported [12]. Some maize *lpa* mutant lines were obtained through a genome-editing based method, when this technology was not so popular as today [22] and very recently barley mutant lines have been isolated through the same technology [23]. In the article presented by Jiang et al. [8], the CRISPR/CAS9 method was used to generate four rice mutants in the *OsITPK6* gene, coding for inositol 1,3,4-triphosphate 5/6 kinase. Very recently barley *lpa* allelic variants have been isolated through the same technology. Jiang et al. described that the decrease in PA content and the severity of the negative pleiotropic effects depended on the induced mutations, with the three frameshift mutations resulting in a major reduction in PA content and in a stronger impact on plant germination, growth, reproduction, and abiotic stress tolerance compared to the effects due to the 6-bp in-frame mutation. There is a discrepancy between results obtained from the present study and from a previous one [24], where another mutant affected in the same gene showed a higher decrease in PA content than reported for the mutant in the article by Jiang et al., but normal plant growth. Further studies on other *ositpk6* mutants could clarify this discrepancy and the role of this gene in plant growth and reproduction in addition to its role in PA biosynthesis [8].

Different papers have reported the isolation of *lpa* mutants in different species by the screening of F_2 mutagenized populations through the disruption of the seeds analyzed by Chen's assay [12]. In the paper by Borlini et al. [7], it was proposed to directly identify the putative mutant seeds by a cheap and fast screening method based on the lower density of *lpa1* seeds with respect to the wild type, as reported in previous papers, where among the pleiotropic effects associated with the *lpa* mutation it was also shown that there was a reduction of seed density in maize and rice. This assay was able to identify the *lpa* mutant seeds because the *lpa1* seeds can float in a concentrated sugar solution (density 1.218–1.222 g/cm^3) due to their lower density, unlike the wild sibs that sink [7]. Hence, this method could be used in massive screening of mutagenized populations with the aim to isolate allelic variants at the *lpa* locus.

3.5. Inositol Pyrophosphate: Suggested Strategies for the Development of Novel lpa Mutants

In the cell, a small pool of PA can be further phosphorylated to form inositol pyrophosphates (PP-InsP), containing one or two diphosphate groups (InsP$_7$ and InsP$_8$), through the activity of inositol triphosphate kinase (ITPK) enzymes that phosphorylate PA to InsP$_7$ and the diphosphoinositol-pentakisphosphate kinases (PPIP5Ks) that phosphorylate InsP$_7$ to InsP$_8$. PP-InsP have important roles in energy metabolism, hormone signaling (mainly jasmonate), and P$_i$ sensing. It has been shown that different Arabidopsis *lpa* mutations, affecting PA biosynthetic genes, also cause a reduction in the content of InsP$_8$ and in some cases of InsP$_7$. Starting from this point, Freed et al. recommend the breeders aiming at developing *lpa* mutants to take into account this aspect to avoid negative pleiotropic effects that may reduce pathogen defense, mediated by jasmonate, and affect phosphate homeostasis [11]. To overcome this possibility, one strategy is to develop transgenic *lpa* lines using tissue-specific promoters active only in the seed. However, also in the seed, InsP$_7$ and InsP$_8$ may have important roles in phosphate homeostasis not yet investigated. On the other hand, the Arabidopsis *mrp5* mutant, affecting the PA-MRP vacuolar transporter, shows increased content of both InsP$_7$ and InsP$_8$, representing an interesting target for the development of new *lpa* mutants not compromised in P$_i$ homeostasis and in jasmonate signaling. In this way, the review intends to bridge the gap between the basic science aspects of PP-InsP synthesis and function and the breeding/engineering strategies aimed at developing *lpa* crops [11].

4. Conclusions

Although micronutrient malnutrition is still a challenging problem, progress has been achieved in the development of biofortified crops, either by enhancing the content of key microelements such as iron and zinc or by developing *lpa* mutants with good agronomic performance [1]. Enhancing the content of key microelements has been achieved essentially by breeding, while studies on the elucidation of the mechanisms, and therefore of the genes, involved in micronutrient uptake and efficient storage in the seed are still in progress [1–3].

Advances have been achieved in understanding the function of a number of structural genes involved in PA biosynthesis [7,8,11] and new genes, not obviously correlated to PA biosynthesis and storage, have been discovered to play a role in PA accumulation in the seed [5,6]. Furthermore, there is increasing evidence that many negative pleiotropic effects commonly associated with *lpa* mutants may be overcome by efficient breeding, thus making reasonable and convenient the production of *lpa* mutants [4,9,10]. To reinforce this convenience, there are also social and economic considerations, as clearly explained in the review by Raboy [4].

In conclusion, it is quite clear that, at the moment, the most promising strategy to produce effective biofortified crops is by combining seed PA reduction (*lpa* mutants) with increased seed mineral content, as a number of results provide evidence showing that PA is still the main limiting factor to cations' bioavailability in the diet of humans and monogastric animals [15].

Author Contributions: Conceptualization E.C.; writing E.C., R.P. and F.S.; funding E.C. and F.S. All authors have read and agreed to the published version of the manuscript.

Funding: This research was co-funded by sPATIALS3 project, financed by the European Regional Development Fund under the ROP of the Lombardy Region ERDF 2014–2020—Axis I "Strengthen technological research, development and innovation"—Action 1.b.1.3 "Support for co-operative R&D activities to develop new sustainable technologies, products and services"—Call Hub.

Acknowledgments: We would like to deeply thank all the colleagues who contributed to this Special Issue.

Conflicts of Interest: The authors declare no conflict of interest.

References

1. Jha, A.B.; Warkentin, T.D. Biofortification of pulse crops: Status and future perspectives. *Plants* **2020**, *9*, 73. [CrossRef] [PubMed]
2. Sharma, S.; Kaur, G.; Kumar, A.; Meena, V.; Ram, H.; Kaur, J.; Pandey, A.K. Gene expression pattern of Vacuolar-Iron Transporter-Like (VTL) genes in hexaploid wheat during metal stress. *Plants* **2020**, *9*, 229. [CrossRef] [PubMed]
3. Pongrac, P.; Arčon, I.; Castillo-Michel, H.; Vogel-Mikuš, K. Mineral element composition in grain of awned and awnletted wheat. *Plants* **2020**, *9*, 79. [CrossRef] [PubMed]
4. Raboy, V. Crops: Observations based on four decades of research. *Plants* **2020**, *9*, 140. [CrossRef]
5. Cominelli, E.; Pilu, R.; Sparvoli, F. Phytic acid and transporters: What can we learn from *low phytic acid* mutants. *Plants* **2020**, *9*, 69. [CrossRef]
6. Sacchi, G.A.; Nocito, F.F. Plant sulfate transporters in the low phytic acid network: Some educated guesses. *Plants* **2019**, *8*, 616. [CrossRef]
7. Borlini, G.; Rovera, C.; Landoni, M.; Cassani, E.; Pilu, R. lpa1-5525: A new *lpa1* mutant isolated in a mutagenized population by a novel non-disrupting screening method. *Plants* **2019**, *8*, 209. [CrossRef]
8. Jiang, M.; Liu, Y.; Tan, Y.; Huang, J.; Shu, Q. Mutation of inositol 1,3,4-trisphosphate 5/6-kinase6 impairs plant growth and phytic acid synthesis in rice. *Plants* **2019**, *8*, 114. [CrossRef]
9. Fukushima, A.; Perera, I.; Hosoya, K.; Akabane, T.; Hirotsu, N. Genotypic differences in the effect of P fertilization on phytic acid content in rice grain. *Plants* **2020**, *9*, 146. [CrossRef]
10. Taliman, N.A.; Dong, Q.; Echigo, K.; Raboy, V.; Saneoka, H. Effect of phosphorus fertilization on the growth, photosynthesis, nitrogen fixation, mineral accumulation, seed yield, and seed quality of a soybean low-phytate line. *Plants* **2019**, *8*, 119. [CrossRef]
11. Freed, C.; Adepoju, O.; Gillaspy, G. Can inositol pyrophosphates inform strategies for developing low phytate crops. *Plants* **2020**, *9*, 115. [CrossRef] [PubMed]
12. Sparvoli, F.; Cominelli, E. Seed biofortification and phytic acid reduction: A conflict of interest for the plant. *Plants* **2015**, *4*, 728. [CrossRef] [PubMed]
13. Silva, E.O.; Bracarense, A.P. Phytic acid: From antinutritional to multiple protection factor of organic systems. *J. Food Sci.* **2016**, *81*, R1357–R1362. [CrossRef] [PubMed]
14. Wilson, M.S.; Bulley, S.J.; Pisani, F.; Irvine, R.F.; Saiardi, A. A novel method for the purification of inositol phosphates from biological samples reveals that no phytate is present in human plasma or urine. *Open Biol.* **2015**, *5*, 150014. [CrossRef] [PubMed]
15. Hummel, M.; Talsma, E.F.; Taleon, V.; Londono, L.; Brychkova, G.; Gallego, S.; Raatz, B.; Spillane, C. Iron, zinc and phytic acid retention of biofortified, low phytic acid, and conventional bean varieties when preparing common household recipes. *Nutrients* **2020**, *12*, 658. [CrossRef] [PubMed]
16. Yamaji, N.; Takemoto, Y.; Miyaji, T.; Mitani-Ueno, N.; Yoshida, K.T.; Ma, J.F. Reducing phosphorus accumulation in rice grains with an impaired transporter in the node. *Nature* **2017**, *541*, 92–95. [CrossRef] [PubMed]
17. Ding, G.; Lei, G.J.; Yamaji, N.; Yokosho, K.; Mitani-Ueno, N.; Huang, S.; Ma, J.F. Vascular cambium-localized atspdt mediates xylem-to-phloem transfer of phosphorus for its preferential distribution in Arabidopsis. *Mol. Plant* **2020**, *13*, 99–111. [CrossRef]
18. Buerkert, A.; Haake, C.; Ruckwied, M.; Marschner, H. Phosphorus application affects the nutritional quality of millet grain in the Sahel. *Field Crops Res.* **1998**, *57*, 223–235. [CrossRef]

19. Coelho, C.M.M.; Santos, J.C.P.; Tsai, S.M.; Vitorello, V.A. Seed phytate content and phosphorus uptake and distribution in dry bean genotypes. *Braz. J. Plant Physiol.* **2002**, *14*, 51–58. [CrossRef]

20. Saneoka, H.; Koba, T. Plant growth and phytic acid accumulation in grain as affected by phosphorus application in maize (Zea mays L.). *Grassl. Sci.* **2003**, *48*, 485–489.

21. Perera, I.; Fukushima, A.; Akabane, T.; Horiguchi, G.; Seneweera, S.; Hirotsu, N. Expression regulation of myo-inositol 3-phosphate synthase 1 (INO1) in determination of phytic acid accumulation in rice grain. *Sci. Rep.* **2019**, *9*, 14866. [CrossRef] [PubMed]

22. Shukla, V.K.; Doyon, Y.; Miller, J.C.; DeKelver, R.C.; Moehle, E.A.; Worden, S.E.; Mitchell, J.C.; Arnold, N.L.; Gopalan, S.; Meng, X.; et al. Precise genome modification in the crop species *Zea mays* using zinc-finger nucleases. *Nature* **2009**, *459*, 437–441. [CrossRef] [PubMed]

23. Vlcko, T.; Ohnoutkova, L. Allelic Variants of CRISPR/Cas9 induced mutation in an inositol trisphosphate 5/6 kinase gene manifest different phenotypes in barley. *Plants* **2020**, *9*, 195. [CrossRef] [PubMed]

24. Kim, S.-I.; Tai, T. Identification of novel rice low phytic acid mutations via TILLING by sequencing. *Mol. Breed.* **2014**, *34*, 1717–1729. [CrossRef]

© 2020 by the authors. Licensee MDPI, Basel, Switzerland. This article is an open access article distributed under the terms and conditions of the Creative Commons Attribution (CC BY) license (http://creativecommons.org/licenses/by/4.0/).

 *plants*

Review

Biofortification of Pulse Crops: Status and Future Perspectives

Ambuj B. Jha and Thomas D. Warkentin *

Crop Development Centre/Department of Plant Sciences, University of Saskatchewan, 51 Campus Drive, Saskatoon, SK S7N 5A8, Canada; ambuj.jha@usask.ca
* Correspondence: tom.warkentin@usask.ca; Tel.: +1-306-966-2371; Fax: +1-306-966-5015

Received: 26 November 2019; Accepted: 2 January 2020; Published: 6 January 2020

Abstract: Biofortification through plant breeding is a sustainable approach to improve the nutritional profile of food crops. The majority of the world's population depends on staple food crops; however, most are low in key micronutrients. Biofortification to improve the nutritional profile of pulse crops has increased importance in many breeding programs in the past decade. The key micronutrients targeted have been iron, zinc, selenium, iodine, carotenoids, and folates. In recent years, several biofortified pulse crops including common beans and lentils have been released by HarvestPlus with global partners in developing countries, which has helped in overcoming micronutrient deficiency in the target population. This review will focus on recent research advances and future strategies for the biofortification of pulse crops.

Keywords: biofortification; iron; zinc; selenium; iodine; carotenoid; folate; pulse

---

1. Introduction

Micronutrients, iron (Fe), zinc (Zn), selenium (Se), iodine (I), carotenoids, and folates are essential nutrients required for human growth and development, as these contribute to various metabolic functions in human. The majority of the world's population depends on plant-based foods which are often low in key micronutrients [1], and do not meet the recommended daily allowances (RDA). Micronutrient malnutrition is commonly known as "hidden hunger" and affects one in three people worldwide [2]. Micronutrient deficiencies may lead to serious illnesses such as poor growth, intellectual impairments, perinatal complications, and increased risk of morbidity and mortality [3]. Further, they aggravate infectious and chronic diseases including osteoporosis osteomalacia, thyroid deficiency, colorectal cancer, and cardiovascular diseases and thus greatly impact quality of life [4].

Deficiencies of Fe, Zn, folic acid, and β-carotene are global issues, but they are more predominant in Asian, African, and Latin American countries and affect more than two billion people [4,5]. Micronutrient deficiency and undernourishment of pregnant mothers affects nearly 50% of the world's population, potentially leading to intrauterine growth restriction, low birth weight, protein-energy malnutrition, and chronic energy deficit [6]. Though rates are higher in Africa and Asia, deficiencies of the four common micronutrients, Fe, I, Zn, and vitamin A, alone are responsible for about 12% of deaths globally among children under 5 years of age [6].

Food crops rich in nutrients could address deficiencies of micronutrients and thus provide a sustainable solution to global health issues [7]. Peas (*Pisum sativum* L.), chickpeas (*Cicer arietinum* L.), lentils (*Lens culinaris* Medik.), common beans (*Phaseolus vulgaris* L.), and mungbeans (*Vigna radiate* L.) are major pulse crops grown worldwide [8]. They are great sources of dietary proteins, complex carbohydrates, vitamins, and minerals required for human nutrition [9–15]. Pulse crops are used in traditional diets of people in many parts of the world since they are rich in proteins and amino acid and are slowly digestible carbohydrates [8,9,16]. They are easily available to all groups of people on a regular basis and provide the least expensive source of proteins and micronutrients [17]. Pulse consumption has been increasing owing to their health and environmental benefits [18].

Micronutrient malnutrition has received increased attention in recent decades at a global level and efforts have been made to combat them by various strategies such as increased food production, supplementation, food fortification, and biofortification. Biofortification, enriching the nutritional quality of food crop using either conventional plant breeding or modern biotechnology, is a balanced approach to overcome mineral deficiencies [19–22]. Biofortification through plant breeding to improve the nutritional profile of pulse crops has gained momentum in the past decade. In this regard, several studies in pulse crops have identified genetic variation for the key micronutrients in the available gene pools, with promising breeding lines being used in breeding, and associated genotypic markers for marker assisted selection [11–13,23–30]. This review will focus on recent research advances for the improvement of key micronutrients, Fe, Zn, Se, I, carotenoids, and folates in pulse crops. This review will also discuss challenges and future strategies for the biofortification of pulse crops.

2. Key Micronutrients

2.1. Iron

Iron (Fe) is indispensable for living organisms and vital for various metabolic processes such as electron transport and deoxyribonucleic acid synthesis [31]. In the human body, Fe is required for the synthesis of oxygen transport proteins (hemoglobin and myoglobin) and enzymes involved in electron transfer and oxidation-reductions [32,33]. In hemoglobin, it serves as a transporter of oxygen from the lungs to the tissues. According to the Food and Nutrition Board of the Institute of Medicine, National Academy of Sciences, the RDA of Fe is 8 mg/day for adult males and 18 mg/day for females (https://ods.od.nih.gov/Health_Information/Dietary_Reference_Intakes.aspx). Iron deficiency is considered the most predominant among various micronutrient deficiencies and the major contributor of anemia and affects more than two billion individuals globally [34,35]. It can cause loss of energy, dizziness, and poor pregnancy outcomes such as premature births, low birth weight babies, delayed growth and development in infants, and poor cognitive skills [3,36,37].

2.2. Zinc

Zn is an another important mineral required by humans and is involved in many biological functions, such as improving wound healing by its involvement in membrane signaling systems in cell growth and proliferation [38,39], protecting cells from oxidative damage by quenching reactive oxygen species [40,41], and reducing risk of various cancers including prostate and pancreatic [42]. The RDA for Zn is 11 mg/day for adult males and 8 mg/day for adult females (https://ods.od.nih.gov/Health_Information/Dietary_Reference_Intakes.aspx). Deficiency of Zn has many consequences including a weak immune system, recurrent infections, mental illness, and retarded growth and fertility [43]. It plays an important role in cell division; it thus significantly affects pregnant women.

2.3. Selenium

Se, an essential micronutrient, is required for growth and development and protects the human body against infection, oxidative stress, and progression of cancer [44–47]. The RDI for Se is 55 µg/day for both males and females (https://ods.od.nih.gov/Health_Information/Dietary_Reference_Intakes.aspx). In humans, Se deficiency is associated with several diseases, such as Keshan, Keshin-Beck, and myxedematous cretinism [48].

2.4. Iodine

Iodine is a vital constituent of the thyroid hormones, thyroxine (T4), and triiodothyronine (T3) and essential for normal growth, development, and metabolism. According to the Food and Nutrition Board, Institute of Medicine, the RDI for I is 150 µg/day for both adult males and females (https://ods.od.nih.gov/Health_Information/Dietary_Reference_Intakes.aspx). Deficiency of iodine causes hypothyroidism, goiter, cretinism, mental retardation, and reduced fertility and is accountable for increased prenatal death and infant mortality [49–51]. Deficiency during pregnancy can cause cognitive impairment in the offspring as it is critical for brain development [52,53].

Deficiency of iodine in human populations is different from other micronutrients as it is predominant in developing as well as developed countries [54–56]. This could be due to the low concentration of this mineral in agricultural soils and cereal-based foods [55].

2.5. Carotenoids

Carotenoids are natural pigments produced by plants. Plant-derived foods are sources of carotenoids as humans and animals cannot synthesize carotenoids [57]. Carotenoids act as important antioxidants in the human body and play a key role in various physiological processes. Overall, more than 600 carotenoids are known. Lutein and zeaxanthin prevent age-related macular degeneration [57,58]. Lutein reduces the risk of cataracts and is associated with cardiovascular disease prevention [59,60]. Vitamin A is important for normal vision, bone growth, and cell division in mammals [61]. β-Cryptoxanthin stimulates osteoblastic bone formation and inhibits osteoclastic bone resorption [62], therefore playing an important role in bone formation. Carotenoids have strong cancer-fighting properties [63] and protect cellular organelles from oxidative damage by efficiently scavenging free radicals generated during various metabolic processes [64,65]. Carotenoids are considered as Fe absorption promoters as these improve human Fe bioavailability from plant-based foods [7]. For example, improvement of Fe status was reported in the Venezuelan population after the addition of vitamin A in their food [66,67].

2.6. Folates

Folates are B9 vitamins and act as cofactors in various metabolic functions such as nucleotide biosynthesis and amino acid metabolism in the human body [68,69], and are therefore required for human growth and development. In plants, folates are important for biosynthesis of biomolecules including lignin, alkaloids, and chlorophyll [70]. Humans are dependent on plant and/or animal-based food sources as they cannot synthesize folates [69,71]. Deficiency of folates has been associated with greater risk of various chronic diseases, such as neural tube defects [72], impaired cognitive function [73], Alzheimer's disease [74], cardiovascular diseases [75], and certain types of cancers [76]. Folate-rich diets are highly recommended during pregnancy as these effectively reduce the risk of neural tube defects in newborns [77]. Insufficient folate intake during pregnancy increases the risk of pre-term delivery and fetal growth retardation [78]. Wallock et al. [79] observed a correlation between seminal plasma folate with blood plasma folate; hence, folates are also important for human reproductive health [80].

3. Approaches for Improvement of Nutritional Profile

Dietary diversification, food supplements, food fortification, and biofortification are different approaches used for improvement of the nutritional profile of crops to tackle micronutrient deficiency.

3.1. Dietary Diversification

Dietary diversification is a food-based strategy that involves consuming a wide range of different foods, especially different plant based foods such as vegetables, fruits, and whole grains. Dietary diversification also uses strategies at the household level, such as preparation of food that involves soaking, fermentation, and germination, as these enhance micronutrient content and bioavailability [81]. Fruits and vegetables rich in promoter substances (ascorbate and β-carotene) that increase mineral absorption should be taken along with a reduced intake of foods rich in anti-nutrients (phytic acid and polyphenols), which inhibit mineral absorption. For example, for iron improvement, foods rich in ascorbic acid (Fe absorption promoter) should be consumed [82,83]. Germination and fermentation can improve iron bioavailability, as these methods increase the activity of phytase enzymes that hydrolyze phytic acid in whole grain cereals and legumes [84]. Levels of folates in diets can be improved by the consumption of naturally folate-rich foods [85] or sprouted seed [86].

3.2. Food Supplements

Food supplements are micronutrients consumed in the form of pills, powders, and solutions when diets alone cannot provide an adequate amount of nutrition. Supplementation can be used as a short-term method to improve nutritional health and may be unsustainable for large populations. For example, improvement of folate levels in diets was achieved by the use of folic acid supplements [85,87]. Further, this method had some success with vitamin A and zinc supplementation [88]. Folic acid, iron, and zinc supplements have been helpful for children and pregnant women; however, this method is not cost-effective, especially for low-income consumers [3,89]. Supplementation is a relatively cost-effective method, but may not solve the root cause of micronutrient deficiencies. Supplements for folic acid, zinc, and iron could show different physiological responses and absorption than consuming them in food [3]. Supplementation requires access to medical centers, adequate educational programs, and management of supplies vs. demand, with adequate storage facilities [3,90]. These are manageable in developed countries, but not in rural populations and/or those of developing countries who have little access to these facilities.

3.3. Food Fortification

Fortification is the addition of essential micronutrients including vitamins and minerals to foods to improve their nutritional quality. Several food assistance programs by the World Food Program (WFP) are in place using partially pre-cooked and milled cereals and pulses fortified with micronutrients to overcome nutritional deficiencies and provide health benefits with nominal risk. For food fortification with iron, ferrous sulfate, ferrous fumarate, ferric pyrophosphate, and electrolytic iron powder compounds are commonly used [91]. Similarly, food can be fortified with folic acid to improve levels of folates in diets [85,87]. Salt iodization (fortification with iodine) was successfully achieved to reduce the incidence of goiter [92].

3.4. Biofortification

Biofortification is a process of improvement of nutritional profile of plant-based foods through agronomic interventions, genetic engineering, and conventional plant breeding (Figure 1).

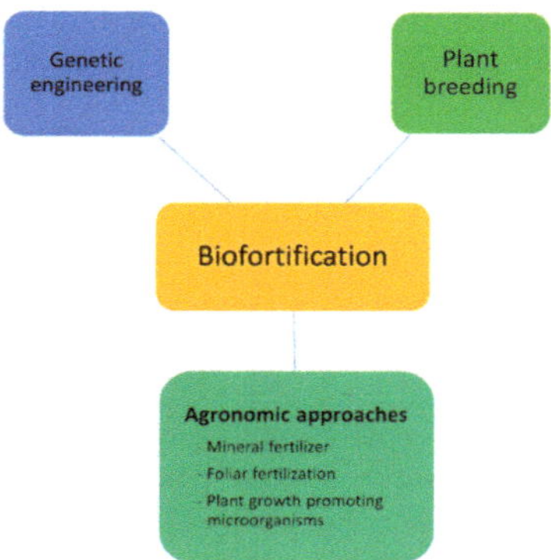

Figure 1. Different approaches of biofortification for improvement of nutritional profile.

3.4.1. Agronomic Approaches

Biofortification through agronomic approaches can be achieved by applying mineral fertilizers to the soil, foliar fertilization [93], and soil inoculation with beneficial microorganisms (http://www.fao.org/agriculture/crops/).

Mineral Fertilizer

Mineral fertilizers are inorganic substances containing essential minerals and can be applied to the soil to improve the micronutrient status of soil and thus plant quality. The phytoavailability of minerals in the soil is often low; thus, to improve the concentration of minerals in the edible plant tissues, the application of mineral fertilizers with improved solubility and mobility of the minerals is required [93]. This method can be used to fortify plants with mineral elements, but not organic nutrients, such as vitamins, which are synthesized by the plant itself. This method was successfully implemented for Se, I, and Zn, as these elements had good mobility in the soil as well as in the plant [93–96]. For example, supplementation of inorganic fertilizers with sodium selenate significantly increased Se concentration in various food items, fruits, vegetables, cereals, meat, dairy products, eggs, and fish in Finland [97,98]. Thus, supplementation of fertilizers with sodium selenate proved to be an effective way to increase Se intake in the human population [99]. Similarly, plants were successfully enriched with I and Zn in China and Thailand using inorganic fertilizers, respectively [100]. However, Fe fertilization was not successful due to a low mobility of Fe in soil [101]. The concentration of Zn was increased in field pea grains by either soil application of Zn fertilizer alone or combined with foliar treatments; thus, these methods could be potentially used for the biofortification of field peas [102].

The fertilization strategy for biofortification typically requires regular applications, which could become harmful for environmental health and may limit the availability of other minerals [93,100,103]. Further, soil composition in the specific geographical location, differences in mineral mobility, and the potential of antinutrient compounds limiting mineral bioavailability are also constraints for successful application of this strategy [104,105].

Foliar Fertilization

Foliar fertilization is the application of fertilizers directly to the leaves. It could be successful when mineral elements are not available immediately in the soil or not readily translocated to edible tissues [93,106]. Pulse crops were biofortified with micronutrients, Fe, Zn, and Se, through foliar application in various studies that resulted in increased levels of these micronutrients in the harvested grain. Márquez-Quiroz et al. [107] reported increased concentration of Fe (29–32%) in seeds of cowpeas. Ali et al. [108] reported increased Fe concentration (46%) in mungbeans upon foliar application of Fe. Similarly, foliar application of Fe and Zn significantly increased the concentration of these minerals along with protein in seeds of cowpeas [109] and chickpeas [110].

Shivay et al. [111] observed a correlation between Zn uptake and the grain yield of chickpeas following foliar application of Zn, and reported that this approach was better than soil application. Similarly, Hidoto et al. [112] evaluated the effects of three Zn fertilization strategies on five varieties of chickpeas and observed that foliar application was an effective method for Zn biofortification with a greater accumulation of Zn in grain compared to soil application and seed priming. Foliar application of Zn fertilizer for Zn biofortification was also reported in common beans [113–115] and field peas [102].

Increased concentration of Se was reported in seeds of peas [116], chickpeas [117], common beans [118], and lentils [119] upon foliar application of Se fertilizers. Further, increased concentration of I was observed in various crops by foliar application, and this could prevent I deficiency in human populations with low dietary I intake [55].

Plant Growth Promoting Microorganisms

Rhizobia, mycorrhizal fungi, actinomycetes, and diazotrophic bacteria are beneficial soil microorganisms associated with plant roots by symbiotic association, and these protect plants by various methods such as promotion of nutrient mineralization and availability and production of plant growth hormones [120]. Though these are naturally present in the soil, their populations can be enhanced by inoculation or agricultural management practices. Various plant growth-promoting (PGP) soil microorganisms including *Enterobacter*, *Bacillus*, and *Pseudomonas* can be exploited to increase the phytoavailability of micronutrients. These are used mostly as seed inoculants and enhance plant growth through the production of growth hormones, antibiotics, chitinases, and siderophores and the induction of systemic resistance and mineralization [121]. PGP microorganisms chelate iron via the production of siderophore compounds, solubilize phosphorus, and inhibit growth of pathogens [122,123], thus playing a significant role in soil fertility and iron fortification. PGP microbes are usually present in soil, compost, and decomposing organic materials and provide an economical and harmless method for increasing crop production and improve environmental and soil health [124].

Numerous studies have shown increased concentrations of Fe, Se, and Zn using microorganism inoculants via mycorrhizal associations [125–127]. Further, enhancement of nitrogen fixation, plant growth, and grain yield have been reported in legumes including chickpeas, soybeans and peas by colonization of *Pseudomonas* sp., *Brevibacterium* sp., *Bacillus* sp., *Enterobacter* sp., and *Acinetobacter* sp. in their roots and nodules [128–132].

In chickpeas, inoculation of PGP actinobacteria increased the concentration of seed minerals including Fe (10–38%) and Zn (13–30%) compared to control (uninoculated) plants [133]. Similarly, arbuscular mycorrhizal fungi field inoculation improved the nutritional profile of chickpea grains by increasing Fe and Zn concentration along with yield and protein content [134]. Khalid et al. [135] reported that application of PGP rhizobacteria with Fe compound ($FeSO_4$) in soil increased iron concentration (up to 81%) in chickpeas compared to a control, and suggested a potential role of microorganisms in the additional uptake of Fe from soil upon supplementation with Fe. Gopalakrishnan et al. [124] inoculated the seeds before sowing and the field plots of chickpeas and pigeonpeas every 15 days until the flowering stage with seven strains of bacteria, *P. plecoglossicida* (SRI-156), *B. antiquum* (SRI-158), *B. altitudinis* (SRI-178), *E. ludwigii* (SRI-211), *E. ludwigii* (SRI-229), *A. tandoii* (SRI-305), and *P. monteilii* (SRI-360), and observed that these bacterial strains significantly improved several growth parameters including nodule, pod number, and grain yield compared to un-inoculated control plots. The harvested grains showed increased concentration of minerals including Fe and Zn. Iron concentration was increased up to 18 and 12%, whereas the concentration of Zn was increased up to 23 and 5% in chickpeas and pigeonpeas, respectively.

3.4.2. Genetic Engineering

Biofortification through genetic engineering is an alternative approach when variation in the desired traits is not available naturally in the available germplasm, a specific micronutrient does not naturally exist in crops, and/or modifications cannot be achieved by conventional breeding [136,137]. This approach was supported by the availability of fully sequenced genomes in various crops in recent years. Along with increasing the concentration of micronutrients, this approach can also be targeted simultaneously for removal of antinutrients or inclusion of promoters that can enhance the bioavailability of micronutrients [93,106,138]. This approach had not only utilized genes associated with various metabolic pathways operated in plants, but also from bacteria and other organisms [139,140]. Development of transgenic crops requires a substantial investment during the initial stage, but this could be a sustainable approach that has the potential to target large populations, especially in developing countries [96,103,141].

Several crops have been successfully modified using a transgenic approach to overcome a micronutrient deficiency. For example, enhanced accumulation (3 to 4 times) of Fe was noted in rice via expression of the iron-storage protein, ferritin [142,143]. In rice, the national levels for Fe and Zn biofortification nutrition targets were attained under field settings in the Philippines and Colombia [144]. The genetically engineered rice (golden rice) was developed to produce β-carotene (pro-vitamin A) to fight against vitamin A deficiency [145]. Recently, transgenic multivitamin corn was produced by the simultaneous modification of three distinct metabolic pathways to increase the levels of three vitamins, i.e., β-carotene (169-fold), ascorbate (6-fold), and folate (2-fold), in the endosperm, and this could pave the way to develop nutritionally complete cereals [146]. Using metabolic engineering, the folate concentration was increased in tomato and rice [85,147]. Storozhenko et al. [148] reported more than 100-fold increase in folate concentration in rice by overexpression of *Arabidopsis thaliana* pterin and para-aminobenzoate genes, precursors of the folate biosynthesis pathway, whereas Hossain et al. [149] reported a two- to four-fold increase in *Arabidopsis* by overexpression of the gene involved in pterin biosynthesis.

To the best of our knowledge, there are no examples of biofortified pulse crops developed through a transgenic approach for Fe, Zn, Se, I, carotenoids, or folates in the available literature. However, a genetic engineering approach has been applied in pulse crops for improvement of other nutritional profile. For example, the concentration of the essential amino acid methionine was significantly increased in transgenic common bean plants (up to 23%) by expression of a methionine-rich storage albumin from the Brazil nut [150], and in the concentration of methionine in transgenic lupins (up to 94%) by expressing a sunflower seed albumin gene [151].

In recent years, targeted gene editing technologies using artificial nucleases, zinc finger nucleases (ZFNs), transcription activator-like effector nucleases (TALENs), and the clustered regularly interspaced short palindromic repeat (CRISPR)/CRISPR-associated protein 9 (Cas9) system (CRISPR/Cas9) have given rise to the possibility to precisely modify genes of interest, and thus have potential application for crop improvement [152,153]. These technologies have been used in various crops including rice [154,155], wheat [156], and tomatoes [157]. Recently, CRISPR/Cas9 and TALENs technologies were used to generate mutant lines for genes involved in small RNA processing of *Glycine max* and *Medicago truncatula* [158]. Similarly, CRISPR/Cas9-mediated genome editing technology was used in cowpeas to successfully disrupt symbiotic nitrogen fixation (SNF) gene activation [159]. These findings pave the way for applicability of use of gene editing technologies for various traits of interest in legumes.

3.4.3. Plant Breeding

Limitations in the long-term effectiveness and sustainability of fertilizer approaches necessitates the development of economical and longstanding strategies for increasing micronutrient density in plants. Genetic engineering technology to produce genetically modified plants with desirable traits has been used in corn, rice, wheat, and soybeans. This can be an effective approach for crop improvement; however, political opposition to GMOs in many countries, a complex legal framework for the acceptance and commercialization of transgenic crops, along with expensive and time-consuming regulatory processes are the major limitations of this method [100,160,161]. For example, golden rice has been available since the early 2000s and has the potential to deliver more than 50% of the estimated average requirement for vitamin A, but unfortunately it has not been commercially introduced in any country to date due to risk factors involved in the regulatory approval processes [162,163]. In developed and developing countries, several groups endorse the arguments made by Greenpeace that approval of golden rice will allow multinational corporations to control developing countries' food supplies and pose risks to human health and the environment (https://gmo.geneticliteracyproject.org/).

Restrictions on the use of genetically modified crops in many countries prompted HarvestPlus to take the initiative to address micronutrient deficiencies through conventional plant breeding [20,22]. Biofortification through plant breeding is a cost-effective and sustainable approach that can improve the health status of low-income people globally [19,21,147]. This approach has been used to control deficiencies of micronutrients including carotenoids, Fe, and Zn [96,164].

HarvestPlus was started in 2003 by the initiative of the Consultative Group on International Agricultural Research (CGIAR) to target several major food crops, including rice, common beans, cassava, maize, sweet potatoes, pearl millet, and wheat in Asia and Africa, to enrich them with three major nutrients, Fe, Zn, and vitamin A through an interdisciplinary and global alliance of scientific institutions and implementing agencies [165,166]. HarvestPlus is part of the CGIAR Research Program on Agriculture for Nutrition and Health (A4NH). The HarvestPlus program is administered by joint venture of the International Center for Tropical Agriculture and the International Food Policy Research Institute and provides global leadership on biofortification evidence and technology. The UK Government, the Bill and Melinda Gates Foundation, the US Government's Feed the Future initiative, the EU Commission, and donors to A4NH are principal investors [163] https://www.harvestplus.org.

Conventional plant breeding approaches can benefit not only large populations but also people living in relatively remote rural areas who have limited access to commercially marketed fortified foods [19,21,22,167]. This approach requires a one-time investment in plant breeding and can be grown and multiplied across years by farmers at virtually zero marginal cost. Recurrent costs are low, and germplasms can be available internationally without any adverse effect on productivity and health; thus, biofortification has widespread public acceptance [20,21,100,168].

Genetic diversity is required in the gene pool to achieve success in biofortification through the plant breeding approach. Several studies have shown substantial variation in the concentration of minerals and vitamins in various crops [93,106]. Parental genotypes with high micronutrient concentration can be identified by screening a wide range of germplasms, and these can be utilized in making crosses, genetic studies, and the development of molecular markers to facilitate marker-assisted selection in breeding. Promising lines can be tested at multiple locations to determine the genotype X environment interaction (G X E) [163]. These can be submitted to national government agencies for testing for agronomic performance and release after robust regional testing across multiple locations over multiple seasons [163]. Using the above-mentioned strategy, various studies in pulse crops have identified substantial variation in the available gene pools and recombinant inbred line populations developed from promising parental genotypes. These were utilized for identifying genomic regions and associated markers for their use in marker-assisted selection. Recent research advances in pulse crops for improvement of key micronutrients, including Fe, Zn, Se, I, carotenoids, and folates, through conventional plant breeding approaches, will be discussed in the following section.

4. Recent Research Advances for Biofortification of Pulse Crops

4.1. Iron

In several studies, a wide range of variation in Fe concentration has been observed in peas, chickpeas, common beans, mungbeans, and lentils (Table 1). Significant genetic variability in Fe concentration was observed in the core collection of common beans [169,170], common bean populations [171–173], and a large collection of lentil accessions [174,175]. Cultivars of lentils (18), peas (17), common beans (10), and chickpeas (8) grown at several locations in Southern Saskatchewan (2005–2006) had Fe concentration of 75.6–100 mg kg^{-1}, 47.7–58.1 mg kg^{-1}, 57.7–80.7 mg kg^{-1}, and 48.6–55.6 mg kg^{-1}, respectively [11]. A 100 g serving of any of these pulse crops provided over 50% of the RDA for Fe. A substantial variation in concentration of Fe was observed in 94 diverse accessions each of chickpeas [12] and peas [13]. Further, Diapari et al. [12] identified several chickpea accessions with high Fe concentration (52–60 mg kg^{-1}) that could be utilized for the development of cultivars with high Fe concentration. Significant variation was observed in Fe concentration (35–87 mg kg^{-1}) in mungbean lines commonly grown in South Asia [176]. In a recent study, Dissanayaka [29] reported significant variation in a pea genome wide association study (GWAS) panel of 177 accessions evaluated at Saskatoon and Rosthern Saskatchewan, Canada, and Fargo, North Dakota, USA.

The environment significantly affected the concentration of minerals in field peas, chickpeas, common beans, and lentils grown at different locations in Saskatchewan, Canada [11]. Diapari et al. [13] and Dissanayaka [29] reported a significant effect of genotypes, year, and location with higher Fe concentration at Rosthern compared with Saskatoon in peas.

Effect of genotypes was also significant in lentils [25,28] and chickpea [12,28]. Along with environment, Fe concentration was also affected by varieties and locations. For example, Ariza-Nieto et al. [177] in common beans and DellaValle et al. [178] in lentils observed differences in Fe concentration at the same location due to varieties, whereas Moraghan et al. [179] reported higher Fe concentration in seeds harvested from acid soil compared to calcareous soil. Diapari et al. [13] reported that the genetic factor was responsible for substantial variation in Fe concentration in pea seeds grown at different locations in Saskatchewan.

Table 1. Genetic variation and identified genomic regions and/or markers for the seed concentration of various micronutrients in pulse crops.

Micronutrient	Pulse Crop	Plant Material	Concentration Range (mg kg^{-1})	Genomic Region/Marker	Reference
Iron	Common bean	Genotype	34–89	7 QTLs	[169]
	Common bean	Genotype	35–92		[170]
	Common bean	DOR364 X G19833	40–85	13 QTLs	[171]
	Common bean	G14519 X G4825	36–97	5 QTLs	[172]
	Common bean	G21242 X G21078	28–95	6 QTLs	[173]
	Common bean	Genotype	30–110		[175]
	Lentil	Genotype	41–109		[174]
	Lentil	ILL 8006 X CDC Milestone	37–176	21 QTLs	[24]
	Lentil	Genotype	41–102	9 SNPs	[25]
	Lentil	Genotype	69–86		[28]
	Chickpea	Genotype	48–57		[28]
	Lentil	Genotype	76–100		[11]
	Pea	Genotype	48–58		[11]
	Common bean	Genotype	58–81		[11]
	Chickpea	Genotype	49–56		[11]
	Chickpea	Genotype	36–86	4 SNPs	[12]
	Chickpea	ICC 4958 X ICC 8261	40–67	6 QTLs	[23]
	Chickpea	Genotype	40–91	10 SNPs	[23]
	Mungbean	Genotype	35–87		[176]
	Pea	Genotype	26–94	9 SNPs	[13]
	Pea	PI 648006 X PI 357292	37–62	5 QTLs	[26]
	Pea	Orb X CDC Striker	26–49	4 QTLs	[27]
	Pea	Carerra X CDC Striker	34–67	6 QTLs	[27]
	Pea	Genotype	29–91	3 SNPs	[29]
Zinc	Common bean	Genotype	21–54	11 QTLs	[169]
	Common bean	Genotype	21–60		[170]
	Common bean	DOR364 X G19833	18–42	13 QTLs	[171]
	Common bean	G14519 X G4825	17–49	8 QTLs	[172]
	Common bean	G21242 X G21078	17–57	3 QTLs	[173]
	Common bean	Genotype	25–60		[175]
	Lentil	Genotype	22–77		[174]
	Lentil	Genotype	23–54	12 SNPs	[25]
	Lentil	Genotype	46–55		[28]
	Chickpea	Genotype	35–43		[28]
	Lentil	Genotype	37–51		[11]
	Pea	Genotype	27–34		[11]
	Common bean	Genotype	25–33		[11]
	Chickpea	Genotype	21–28		[11]
	Chickpea	Genotype	19–62	5 SNPs	[12]
	Chickpea	ICC 4958 X ICC 8261	28–48	5 QTLs	[23]
	Chickpea	Genotype	27–62	10 SNPs	[23]
	Mungbean	Genotype	21–62		[176]
	Pea	Genotype	14–93	2 SNPs	[13]
	Pea	PI 648006 X PI 357292	31–62	5 QTLs	[26]
	Pea	Orb X CDC Striker	25–34	4 QTLs	[27]
	Pea	Carerra X CDC Striker	17–41	6 QTLs	[27]
	Pea	Genotype	13–51	7 SNP	[29]

Table 1. *Cont.*

Micronutrient	Pulse Crop	Plant Material	Concentration Range (mg kg^{-1})	Genomic Region/Marker	Reference
Selenium	Lentil	Genotype	0.3–2.6		[180]
	Lentil	Genotype	0.01–0.3		[181]
	Lentil	Genotype	0.4–0.5		[28]
	Chickpea	Genotype	0.3–0.4		[28]
	Lentil	Genotype	0.9–1.6		[11]
	Pea	Genotype	0.4–0.5		[11]
	Common bean	Genotype	0.4–0.5		[11]
	Chickpea	Genotype	0.6–0.9		[11]
	Mungbean	Genotype	0.2–0.9		[176]
	Pea	Genotype	0.03–1.8		[182]
	Pea	Genotype	0.08–5.5		[13]
	Pea	Orb X CDC Striker	0.3–2.2	3 QTLs	[27]
	Pea	Carerra X CDC Striker	0.1–6.8	6 QTLs	[27]
	Pea	Genotype	0.1– 8.7	44 SNPs	[29]
Carotenoids	Chickpea	Not available		5 QTLs	[183]
	Chickpea	Genotype	311–880		[184]
	Chickpea	Genotype	11–19		[185]
	Chickpea	Genotype	9–31		[186]
	Chickpea	Genotype	22–44		[187]
	Chickpea	CDC Jade X CDC Frontier	15–58	8 QTLs	[188]
	Chickpea	Cory X CDC Jade	2–78	5 QTLs	[188]
	Chickpea	ICC4475 X CDC Jade	22–84	5 QTLs	[188]
	Pea	Genotype	7–23		[185]
	Pea	Genotype	6–27		[186]
Folates	Pea	Genotype	0.6		[189]
	Pea	Genotype	0.3–0.7		[190]
	Common bean	Genotype	1.4–1.6		[190]
	Lentil	Genotype	1.5–2.0		[190]
	Pea	Genotype	0.5		[87]
	Lentil	Genotype	0.7		[87]
	Faba bean	Genotype	1.0		[87]
	Chickpea	Genotype	1.5		[87]
	Chickpea	Genotype	2.7		[191]
	Common bean	Genotype	1.1–1.6		[191]
	Lentil	Genotype	1.1–1.5		[191]
	Pea	Genotype	0.1–0.2		[191]
	Pea	Genotype	0.4–2.0		[192]
	Chickpea	Genotype	0.4–1.2		[192]
	Lentil	Genotype	2.2–2.9		[192]
	Chickpea	Genotype	3.5–5.9		[15]
	Common bean	Genotype	1.6–2.3		[15]
	Lentil	Genotype	1.4–1.8		[15]
	Pea	Genotype	0.2–0.3		[15]
	Chickpea	Genotype	4.0–4.3		[193]
	Common bean	Genotype	2.4–3.0		[193]
	Lentil	Genotype	1.2–1.6		[193]
	Pea	Genotype	0.1–0.2		[193]
	Lentil	Genotype	1.7–5.0		[194]
	Pea	Genotype	0.1–0.6	31 SNPs	[30]

Several SNP markers associated with Fe concentration were identified in peas [13,26,27,29], chickpeas [12,23], and lentils [24,25] that can be used in marker-assisted selection (Table 1). In peas, Diapari et al. [13] reported an association of nine SNPs with Fe concentration in 94 diverse accessions, whereas Dissanayaka [29] reported significant association of three SNP markers with Fe concentration in a pea GWAS panel of 177 accessions. Ma et al. [26] identified five QTLs for Fe concentration in a pea population developed from "Aragorn" (PI 648006) and "Kiflica" (PI 357292). Similarly, Gali et al. [27] observed several QTLs for seed iron concentration on four LGs of pea population PR-02 (Orb X CDC Striker) and six LGs of PR-07 (Carerra X CDC Striker).

In chickpeas, Diapari et al. [12] identified two SNPs on Chromosome (chr) 4 and one each on chr1 and chr6 in 94 diverse accessions, whereas Upadhyaya et al. [23] observed associations of six QTLs for Fe concentration on Chromosomes 1, 3, 4, 5, and 7 in a population developed from a cross between ICC 4958 and ICC 8261. In lentils, Aldemir et al. [24] identified 21 QTLs for Fe concentration on several linkage groups (three each on LG1, LGII, and LGVII, six QTLs on LGIV, four QTLs on LGV, and two QTLs on LGVI), whereas Khazaei et al. [25] reported nine QTLs in a panel of 138 accessions, and two of them were tightly linked to Fe concentration in Chromosomes 5 and 6. Similarly, several QTLs were identified for Fe concentration in common bean populations, DOR364 X G19833 [171], G14519 X G4825 [172], and G21242 X G21078 [173].

4.2. Zinc

Like Fe, a wide range of variation in Zn concentration was observed in peas, chickpeas, common beans, and lentils (Table 1). A significant variation in Zn concentration was noted in 94 diverse chickpea accessions evaluated under field conditions in Saskatchewan, Canada [12]. This study identified three kabuli type accessions, CDC Verano, ILC 2555, and FLIP85-1C (43–48 mg kg^{-1}), and two desi type accessions, FLIP97-677C and FLIP84-48C (42 and 41 mg kg^{-1}), with the greatest Zn concentrations.

A substantial variation in Zn concentration was observed in common beans (24.8–33.3 mg kg^{-1}), peas (27.4–34 mg kg^{-1}), chickpeas (21.1–28.3 mg kg^{-1}), and lentils (36.7–50.6 mg kg^{-1}) grown in two years (2005–2006) at various locations in Saskatchewan [11], and each of these provided over 50% of the RDA in 100 g of dry pulses. Significant variation for Zn was observed in the core collection of common beans (>2400) [169,170], three common bean populations [171–173], lentil accessions (>1600) [174,175], 20 mungbean lines [176], a panel of 94 pea accessions [13], a panel of 177 pea accessions [29], and two pea populations, PR-02 and PR-07 [27].

In various studies, the effect of genotypes, year, and/or location was significant for Zn concentration in peas [13,29], chickpeas [12,28], and lentils [25,28]. Further, Zn concentration was often positively correlated with Fe concentration [12,13,25,26,29].

Several QTLs and/or SNP markers were identified for Zn concentration in common beans, chickpeas, lentils, and peas (Table 1). For example, QTLs were identified for Zn concentration in three common bean populations [171–173]. Five SNPs were identified for Zn concentration in chickpeas, and these were located on Chromosomes 1, 4, and 7 [12]. Further, in chickpeas, Upadhyaya et al. [23] observed an association of eight genomic loci for Zn concentration. In lentils, Khazaei et al. [25] reported twelve SNP markers for Zn concentration in a panel of 138 accessions grown at two locations in Saskatchewan, Canada in 2013–2014.

Using a GWAS, Diapari et al. [13] found two SNPs for Zn on LG III in 94 pea accessions. Ma et al. [26] identified five QTLs for Zn concentration on LGs II, III, V, and VII in a pea population developed from Aragorn X Kiflica. Similarly, Gali et al. [27] identified one QTL each on LG1a and LG3b, two QTLs on LG6 in PR-02, and numerous QTLs on various LGs (1a, 1b, 2b, 3b, 4, and 7a) in the PR-07 pea population. In a recent study, Dissanayaka [29] reported a significant association of seven SNP markers with Zn concentration in a pea GWAS panel of 177 accessions. Further, the SNP marker Sc1512_36017 was co-localized with Sc11336_48840 on LG IIIb in PR-07. In a previous study, Sc11336_48840 was identified as the flanking marker of a QTL for seed Zn concentration [27].

4.3. Selenium

Soil and weather conditions play important roles in Se concentration in harvested pulse crop seeds. Among soil factors, aeration, water availability, pH, and texture are important, as these affect the availability of Se [195]. Diapari et al. [13] observed that variation in Se concentration in peas was mainly due to environment, whereas the effect of genotype was minimal (only 2.7% of the total). Se concentration was higher at the Saskatoon location than Rosthern, and differences in Se concentration were not significant across the 94 genotypes.

Soils of Saskatchewan are generally rich in Se, so pulses grown in this region provide a natural dietary source of this element [180,182,196]. Lentils grown in the Dark Brown and Brown soil zones of Western Canada had a high concentration of Se (425–672 µg kg^{-1}) [180] (Table 1). In comparison, lentils grown in six major lentil-producing countries, Nepal (180 µg kg^{-1}), Southern Australia (148 µg kg^{-1}), Turkey (47 µg kg^{-1}), Morocco (28 µg kg^{-1}), Northwestern USA (26 µg kg^{-1}), and Syria (22 µg kg^{-1}), had a substantially lower Se concentration [181]. A wide range of variation was observed for Se concentration in common beans (381–500 µg kg^{-1}), peas (405–554 µg kg^{-1}), chickpeas (629–864 µg kg^{-1}), and lentils (990–1637 µg kg^{-1}) grown at several locations in Saskatchewan [11]. A 100 g dry weight of any of these pulses could provide 100% of the RDA. Nair et al. [176] observed significant variation for Se concentration (210–910 µg kg^{-1}) in mungbean lines grown in two environments near Hyderabad, India.

Total Se concentration varied from 373 to 519 µg kg^{-1} in 17 field pea cultivars grown at six locations for 2 years in Saskatchewan [182], and this provided 68–94% of the RDA upon the serving of 100 g peas. Evaluation of 80 pea breeding lines obtained from Australia, Czech Republic, Serbia, and the United States had relatively low concentration of Se; however, these lines showed greater concentration of Se when planted in Saskatoon [182]. Similarly, Gali et al. [27] observed a considerable variation in Se concentration in pea populations, PR-02 and PR-07, with a greater range in variation among RILs at the Saskatoon location compared with Rosthern. Compared to Fe and Zn, Dissanayaka [29] observed a substantially different pattern for Se at different locations, with a high coefficient of variation, and the concentration varied from 0.06 to 8.75 ppm. The effects of genotype, genotype × year, and genotype × location were not significant, with the exception of the genotype effect at the trial grown in 2014 at Fargo, North Dakota.

Several QTLs were identified for Se concentration in the PR-02 pea population on LGs 4a, 5a and 7, and in the PR-07 pea population on LG4 and 5b [27] (Table 1). Using a pea GWAS panel of 177 accessions, 44 significant SNP markers were identified for Se concentration, but the majority of the markers were not common among the location-years [29], and this could be due to substantial variation in Se concentration and high coefficient of variation at different locations.

4.4. Iodine

Several studies have reported various methods such as foliar fertilization and application of salt in soil and/or irrigation water for biofortification of crops with iodine; however, little information is available on within-species variation. The consumption of cereal-based foods with low I concentration is the major cause of I deficiency in humans [55,56,197]. Compared to grain, biofortification of leaves and leafy vegetables could be easily achieved due to translocation of the majority of I to xylem tissues, thus the majority of research is focused on I biofortification of vegetables instead of grains [56,198–200].

4.5. Carotenoids

Several studies have reported a carotenoid profile in pulse crops [183,185,186,201–204] (Table 1). Lutein, zeaxanthin, and β-cryptoxanthin were reported in chickpeas [183,202], whereas violaxanthin, lutein, and β-carotene were reported in field peas [201,203]. Thavarajah and Thavarajah [184] reported a high concentration of carotenoids, beta-carotene (166–431 µg/100 g), canthoxanthine (21–68 mg/100 g), and xanthophyll (9–20 mg/100 g) in 10 chickpea genotypes grown in Minot, North Dakota.

Ashokkumar et al. [185] evaluated carotenoids profile in 12 pea and 8 chickpea cultivars grown at multiple locations in Saskatchewan, Canada, using high performance liquid chromatography with a diode array detector. This method is sensitive, reliable, and accurate for the separation and quantification of putative carotenoids. In peas, the concentration of carotenoids was greatest in cotyledon, followed by embryo axis and seed coat. Green cotyledon cultivars (16–21 µg g^{-1}) had generally higher concentrations compared to yellow cotyledon cultivars (7–12 µg g^{-1}). Lutein was the major component (11.45 µg g^{-1}) followed by violaxanthin (0.52 µg g^{-1}), β-carotene (0.47 µg g^{-1}), and zeaxanthin (0.16 µg g^{-1}). In kabuli type chickpea cultivars, carotenoid concentration was highest in the cotyledon, followed by the embryo axis and seed coat, whereas in the desi type, the seed coat, followed by the cotyledon and embryo axis, had the highest carotenoid concentration. Lutein (7.70 μg g^{-1}) was the major component followed by zeaxanthin (5.76 μg g^{-1}), β-carotene (0.40 μg g^{-1}), and violaxanthin (0.05 μg g^{-1}).

In subsequent work, Ashokkumar et al. [186] observed a wide range of variation in concentration of carotenoids in genetically diverse pea (94) and chickpea (121) accessions grown at multiple locations in Saskatchewan, Canada. In the peas, the concentration of lutein was highest (11.2 μg g^{-1}) followed by β-carotene (0.5 μg g^{-1}), zeaxanthin (0.3 μg g^{-1}), and violaxanthin (0.3 μg g^{-1}), whereas in the chickpeas, the concentration of lutein (8.2 μg g^{-1}) was highest followed by zeaxanthin (6.2 μg g^{-1}), b-carotene (0.5 μg g^{-1}), β -cryptoxanthin (0.1 μg g^{-1}), and violaxanthin (0.1 μg g^{-1}). Green cotyledon peas and desi chickpeas had a greater carotenoid concentration than yellow cotyledon peas and kabuli chickpeas, respectively. Pea and chickpea accessions with high carotenoid concentration that can be utilized in future breeding were identified. In five chickpea cultivars with different cotyledon colors, total carotenoid concentration varied from 22 μg g^{-1} (yellow cotyledon kabuli) to 44 μg g^{-1} (green cotyledon desi), with lutein and zeaxanthin as major components [187]. In a recent study, a wide range of total carotenoid concentration was observed in three F_2 populations developed by crossing cultivars with different cotyledon and seed coat colors, CDC Jade X CDC Frontier (14.9–58.1 μg g^{-1}), CDC Cory X CDC Jade (1.9–77.6 μg g^{-1}), and ICC4475 X CDC Jade (21.6–83.7 μg g^{-1}) [188].

Total carotenoids (5.8–26.9 μg g^{-1}) and β-carotene (2.6 μg g^{-1}) were greater in peas [186] compared with potato accessions (1.4–14.3 μg g^{-1}) [205] and golden rice endosperm (1.6 μg g^{-1}) [206]. Similarly, in 121 chickpea accessions, the concentration of total carotenoids (15.0 μg g^{-1}) was three times greater than in 42 banana accessions (4.7 μg g^{-1}) [207] and 37 potato accessions (4.4 μg g^{-1}) [208]. Of all carotenoids identified, lutein was the major compound in chickpeas [183,185,186,202] and peas [185,186,201,203,204]. Further, lutein has positive correlations with chlorophyll concentration in peas [203] and zeaxanthin concentration in chickpeas [183].

Previously, four QTLs for beta-carotene concentration and a single QTL for lutein concentration were detected in chickpeas [183] (Table 1). Using a GWAS, Rezaei et al. [187] identified 32 candidate genes involved in isoprenoid and carotenoid pathways across all eight chromosomes of chickpeas. They observed positive correlation between the expression of genes of carotenoid biosynthesis with various carotenoid components. In a subsequent study, several QTLs were identified for total carotenoids and individual components in three F_2 populations, CDC Jade $\times$ CDC Frontier (8 QTLs on LGs 1, 5, and 8), CDC Cory $\times$ CDC Jade (5 QTLs on LG 8), and ICC4475 $\times$ CDC Jade (5 QTLs on LGs 3 and 8) [188]. Further, several candidate genes associated with carotenoid components were observed with a major gene for cotyledon color on LG 8 in each population.

4.6. Folates

A wide exploration of available genetic resources is necessary to identify the rich source of folates for their potential use for the biofortification of pulse crops. Various methods have been employed to quantify folates from different food sources including pulse crops, and these methods include microbiological assays [209–211], liquid chromatography (LC) coupled with fluorescence detection (FD) [87,192], and mass spectrometry (MS) detection [15,30,191,193,194,212–214].

Previously, researchers have identified a wide range of variation in folates quantified from pulse crops (Table 1). For example, using a microbiological assay, Han and Tyler [190] reported 24.9–64.8 μg/100 g folates in green cotyledon peas and 23.7–55.6 μg/100 g in yellow cotyledon peas grown at multiple locations in Saskatchewan, Canada. Vahteristo et al. [189] and Hefni et al. [87] observed the folate concentration of 59 and 52 μg/100 g using LC-MS in vegetable peas consumed in Finland and dry green peas consumed in Egypt, respectively. Rychlik et al. [191] reported variable folate concentration in different pulse crops—275 μg/100 g (chickpeas), 106–164 μg/100 g (white beans), 110–154 μg/ 100 g (green lentils), and 10–20 μg/100 g (peas) using an LC-MS method, whereas Sen Gupta et al. [192] reported a higher total folate concentration in field peas (41–202 μg/100 g) grown in the USA compared to chickpeas (42–125 μg/100 g) using LC-FD.

Using ultra-performance LC (UPLC), six folate monoglutamates were quantified in four pulse crops, and the total folate concentration was the highest in chickpeas (351–589 µg/100 g), followed by common beans (165–232 µg/100 g), lentils (136–182 µg/100 g), and pea (23–30 µg/100 g) [15]. This method provided high accuracy in quantification of specific folates with the use of isotopically labeled internal standards. Zhang et al. [193] identified eight folate monoglutamates using an optimized one-step extraction approach in peas, chickpeas, beans, and lentils using UPLC-MS. They also observed the highest folate concentration in chickpeas and the lowest concentration in peas. Zhang et al. [194] quantified eight folate monoglutamates in six wild lentil species and one cultivated species using UPLC-MS. They observed that wild lentil species (195–497 µg/100 g) had generally higher folate concentration than cultivated genotypes (174–361 µg/100 g). Most recently, Jha et al. [30] quantified five folate monoglutamates in 85 diverse pea accessions originating from worldwide sources using UPLC-MS, and the results indicated a wide range of variation in the concentration of the sum of folates (14–55 µg/100 g dry seed weight).

Tetrahydrofolate (THF), 5-methyltetrahydrofolate (5-MTHF), and 5-formyltetrahydrofolate (5-FTHF) were the most abundant in common beans, lentils, chickpeas, and peas [15,30,208]. Further, 5-MTHF represented 56% of the total folate concentration in peas [15,30], whereas 5-MTHF and 5-FTHF represented 35–39% and 33–51% of the total folates in common beans, lentils, and chickpeas, respectively [15]. Other studies also reported 5-MTHF as the predominant form of folate in common beans [87,215], lentils [87,191], and chickpeas [87]. 5-MTHF was also the major folate in cereals, vegetables, fruit, bread, milk, and meat products [87,216], and in humans [217]. Scaglione and Panzavolta [218] reported various advantages to using naturally occurring 5-MTHF over synthetic folic acid. For example, it helps in the prevention of the potential negative effects of unconverted folic acid in peripheral circulation. Thus, 5-MTHF may be the most important folate that could be targeted for improvement by breeders. Jha et al. [30] identified several pea accessions with greater folate concentration including MPG87 (42 µg/100 g), Kahuna-NIAB (40 µg/100 g), and OZP0902 (50 µg/100 g).

Using a GWAS, five SNP markers were associated with the sum of folates, and fifteen, eight, and three SNP markers were associated with major individual folates 5-MTHF, 5-FTHF, and THF, respectively [30] (Table 1). Further, SNP markers Sc_6992_86348 and Sc_3060_11265 were validated in an additional 24 accessions, and these markers have potential for marker-assisted selection in pea breeding.

5. Status of Biofortification

In recent years, several crops with increased micronutrient concentration have been introduced in several developing countries, and this has helped in overcoming nutrient deficiency in the target population. For example, the introduction of the orange sweet potato biofortified with β-carotene increased vitamin A intake among children and women in Mozambique [219] and Uganda [220], and maize biofortified with provitamin A increased the concentration of vitamin A in 5–7-year-old children in Zambia who consumed it for three months [221]. Similarly, serum ferritin and total body iron were improved in iron-deficient adolescent boys and girls from Maharashtra, India, who consumed Fe-biofortified pearl millet flat bread for four months [222]. Regarding pulse crops, the consumption of Fe biofortified beans for 4.5 months improved the hemoglobin and total body iron in iron-depleted university women in Rwanda [223].

By the end of 2016, more than 150 biofortified varieties of 10 crops had been released in 30 countries, and these are consumed by more than 20 million people in developing countries [163]. To date, HarvestPlus has released or tested more than 290 varieties of 12 staple food crops including vitamin A orange sweet potato, iron beans, iron pearl millet, vitamin A yellow cassava, vitamin A orange maize, zinc rice, and zinc wheat in 60 countries (www.harvestplus.org). Iron beans are delivered in Rwanda and Democratic Republic of Congo, zinc rice is in Bangladesh, and zinc wheat is in India and Pakistan.

Among pulse crops, HarvestPlus has released 10 Fe-biofortified bean varieties each in Rwanda (RWR 2245, RWR 2154, MAC 42, MAC 44, CAB 2, RWV 1129, RWV 3006, RWV 3316, RWV 3317, and RWV 2887) and the Democratic Republic of Congo (COD MLB 001, COD MLB 032, HM 21-7, RWR 2245, PVA 1438, COD MLV 059, VCB 81013, Nain de Kyondo, Cuarentino, and Namulenga) (www.harvestplus.org). Similarly, several varieties of lentils with high iron and zinc have been released by HarvestPlus and The International Center for Agricultural Research in the Dry Areas (ICARDA) in various countries: seven in Nepal (ILL 7723, Khajurah-1, Khajurah-2, Shital, Sisir Shekhar, Simal), five in Bangladesh (Barimasur-4, Barimasur-5, Barimasur-6, Barimasur-7, and Barimasur-8), two each in India (L4704, Pusa Vaibhav) and Syria (Idlib-2, Idlib-3), and one in Ethiopia (Alemaya). For the effective delivery and production of these crops, HarvestPlus works closely with various public and private organizations [163]. For example, in Rwanda, HarvestPlus with the help of Rwanda Agriculture Board (RAB) facilitated the production of bean seeds through contracted farmers and cooperatives and acquired about 80% of certified seeds during 2011–2015.

Biofortification to enrich nutrient profile of pulse crops is one of the major goals in the pulse crop breeding program at the Crop Development Centre (CDC), University of Saskatchewan, which was established in 1971 with the objectives to improve existing crops and develop new crops (https://agbio. usask.ca/research/centres-and-facilities/crop-development-centre.php#MoreAbouttheCDC). In recent years, several projects have been undertaken to evaluate pulse crops for the profiling of folates, carotenoids, polyphenols, Fe, Zn, and Se.

6. Challenges and Future Strategies for Biofortification

A greater micronutrient density and a high yield are prerequisites for effective biofortification, and these crops must be adopted by farmers and consumed by the target population [21]. Bouis and Saltzman [163] outlined three important challenges for HarvestPlus to reach one billion people by 2030, i.e., building consumer demand, mainstreaming biofortified traits into public and private breeding programs, and integrating biofortification into public and private policies.

Various factors such as genetic diversity in the gene pool, the reduction of antinutrients (especially phytate and polyphenols), and increasing the concentration of promoter substances including certain amino acids (cysteine, lysine, and methionine) and ascorbic acid (vitamin C), which enhance the absorption of essential minerals, and/or high yield, are key for the success of biofortification strategies [93,167].

Narrow genetic variation in the plant gene pool, a long-development time for generating cultivars with a desired trait, and the dependence on the phytoavailability of the mineral nutrients in the soil are limitations for conventional breeding approach [138].

The issue of narrow genetic variation for micronutrient concentration might be overcome by the use of wild germplasm and land races, which may contain a high variation in micronutrient concentration [93,106,138].

For efficient biofortification, the focus should be on increasing the bioavailability of micronutrients simultaneously with increase in their concentration. This can be achieved by increasing the concentration of promoters that stimulate the absorption of minerals and by reducing the concentrations of antinutrients that interfere with absorption [93].

Vitamin E, vitamin D, vitamin C, choline, niacine, and provitamin A are considered promoter substances and stimulate the absorption of Se, Ca P, Fe, Zn, methionine, and tryptophan [224]. In contrast, certain antinutrients including phytate and certain polyphenols reduce the bioavailability of micronutrients in crops [93]. Phytate, a form of phosphorus stored in seed, is not digested by humans or monogastric animals [225]. During digestion, it can bind to iron and zinc and thus restrict their absorption [226]. The concentration of phytate can be controlled by identifying low phytate lines by germplasm screening [227], manipulating the biosynthesis of phytate via mutation of a myo-inositol kinase (MIK) gene [228], and overexpressing phytase, a phytate degrading enzyme [229].

In the recent past, low-phytate lines in pulse crops have been developed and characterized to reduce the concentration of phytate and thus improve mineral absorption [225,230–233]. Warkentin et al. [225] developed low-phytate pea lines, 1-150-81 and 1-2347-144, using chemical mutagenesis of cultivar CDC Bronco, a high-performing pea variety. They observed an approximately 60% reduction in phytate phosphorus in low-phytate lines with an increase in inorganic phosphorus. However, these lines had a slightly lower seed weight and a lower yield compared to CDC Bronco. Nevertheless, these lines are being used to breed for improvement of phosphorus and micronutrient bioavailability, along with high grain yield. Subsequently, Liu et al. [226] evaluated the effects of phytate and seed coat polyphenols on the bioavailability of iron using low-phytate pea lines (1-150-81 and 1-2347-144). The iron bioavailability (FEBIO) was 1.4–1.9 times greater in low-phytate lines compared to normal phytate varieties. Further, pigmented seed coat pea showed a seven times lower FEBIO than non-pigmented seed coats; however, the removal of seeds coats increased the FEBIO up to six times. To understand the genetic basis of the low phytic acid (*lpa*) mutation in the pea, Shunmugam et al. [232] amplified a 1530 bp open reading frame of *myo*-inositol phosphate synthase (MIPS), the rate-limiting step in the phytic acid biosynthesis pathway, from CDC Bronco and two *lpa* pea genotypes, 1-150-81 and 1-2347-144. They did not observe any difference in coding sequence in *MIPS* between CDC Bronco and *lpa* genotypes and noticed that mutation in *MIPS* did not cause the *lpa* trait in pea lines.

Various studies in common beans suggested that *lpa* lines can improve iron bioavailability by reducing the phytic acid level up to 90% [230,234]. Homozygous *lpa* mutant line (*lpa-280-10*) was isolated in common beans from a mutagenized population, and this mutant had 90% less phytic acid and higher free Fe in the seeds compared to the wild type [230]. Further, at the molecular level, it was observed that a recessive mutation was responsible for the *lpa* character. Panzeri et al. [231] mapped the *lpa1*(280-10) mutation and identified and sequenced a candidate gene in common beans for comparison with the soybean genome. They observed that the *lpa1*(280-10) mutation co-segregated with the mutated multidrug resistance-associated protein (MRP) type ATP-binding cassette transporter gene *(Pvmrp1)*, which is orthologous to the *lpa* genes of *Arabidopsis AtMRP5* and maize *ZmMRP4*. They further observed that a defective *Mrp1* gene caused an *lpa1* mutation in common beans that downregulates the phytic acid pathway at the transcriptional level and thus reduced seed myo-inositol. Recently, a new *lpa* line influencing the *PvMRP1* phytic acid transporter was identified in common beans using ethyl methane sulfonate mutagenesis [233]. Further, *PvMRP* promoters were characterized in *Arabidopsis thaliana* and *Medicago truncatula* transgenic plants.

Polyphenols are secondary metabolites including flavonoids and proanthocyanidins [235] and provide protection against various fungal pathogens [236]. They are natural sources of antioxidants in the human diet and are present in fruits, vegetables, cereals, and legumes [237,238]. Previously, all polyphenols were considered as inhibitors of Fe bioavailability in humans. A recent study by Hart et al. [239] reported that four polyphenols inhibited Fe uptake, whereas four other polyphenols promoted Fe uptake upon evaluating the effect of polyphenols present in black bean seed coats on Fe uptake using Caco-2 cells (human cell line). They further concluded that specific polyphenols (promoter of Fe uptake) can be targeted in future breeding for improved Fe bioavailability. Jha et al. [240] detected 30 polyphenols in a recombinant inbred line population developed from crossing pea cultivars CDC Amarillo (white flower) and CDC Dakota (purple flower). Among 30 polyphenols, catechin, 3,4-dihydroxybenzoic acid, and kaempferol 3-glucoside were present in all pea lines, and these were considered promoters of Fe uptake by Hart et al. [239]. Thus, promising accessions having Fe promoter polyphenols can be identified via a wide exploration of germplasms for developing cultivars with additional health benefits.

Postharvest processing can also play an important role in efficient utilization of biofortified crops, as a substantial amount of minerals from the diet can be lost by milling or polishing [241] and cooking. Therefore, efforts should be made to retain the micronutrient concentration in edible seeds, and their absorption by the consumer after processing and cooking [242]. Retention of zinc content after cooking in biofortified rice varieties produced either through traditional breeding or genetic engineering approaches has been discussed in detail by Tsakirpaloglou et al. [243].

Iodization of salt was not enough to overcome I deficiency due to several factors such as the unavailability of iodized salt for all households, the volatilization of I during cooking, and insufficient consumption due to health issues [55,93,200,244]. Hence, for successful biofortification, further research was needed to identify traits that control uptake, mobilization, and retention of I in the plant, and these can be manipulated in plant breeding or using a genetic engineering approach [56].

In pulse crops, growth and productivity are affected by various abiotic and biotic stresses, which can result in significant reduction of grain yield [245–250]. These stresses can significantly alter the nutritional profile of the harvested seeds. As mentioned previously, the targeted micronutrients are either antioxidants or part of enzymes involved in various metabolic processes including electron transfer and oxidation reductions; thus, they protect cells from oxidative damage by quenching reactive oxygen species generated under environmental stresses [32,33,40,41,46,68,69]. Biofortified crops with a greater concentration of micronutrients can better withstand adverse environmental conditions and demonstrate improved adaptation under these conditions.

7. Conclusions

Micronutrients are essential for human growth and development, and their deficiency is a major concern that affects one in three people worldwide. Among various strategies, biofortification through plant breeding is considered the most economical and sustainable approach to tackle micronutrient deficiencies. This approach is universally accepted and has the potential to reach people living in relatively remote rural areas that have limited access to commercially marketed fortified foods. Further, it requires a one-time investment, and seeds can be multiplied across years by farmers at virtually zero marginal cost. In recent years, significant progress has been made with the release of several biofortified crop varieties that are helping to overcome micronutrient deficiencies in the target populations. Pulse crops are an important source of protein and energy, so improvement in their nutritional profile will significantly increase their consumption. Biofortification to improve the nutritional profile of pulse crops has gained momentum in the past decade. However, there are several challenges ahead that need to be addressed if the use of biofortified foods is to be successfully maximized.

Author Contributions: Conceptualization, T.D.W.; Writing—original draft preparation, A.B.J. and T.D.W.; Writing—review and editing, T.D.W. and A.B.J.; Supervision, T.D.W.; Funding acquisition, T.D.W. All authors have read and agreed to the published version of the manuscript.

Funding: This research was funded by the Saskatchewan Ministry of Agriculture and the Saskatchewan Pulse Growers.

Acknowledgments: The authors greatly acknowledge the financial support from the Saskatchewan Ministry of Agriculture and the Saskatchewan Pulse Growers.

Conflicts of Interest: The authors declare no conflict of interest.

References

1. Waters, B.M.; Grusak, M.A. Quantitative trait locus mapping for seed mineral concentrations in two *Arabidopsis thaliana* recombinant inbred populations. *New Phytol.* **2008**, *179*, 1033–1047. [CrossRef] [PubMed]
2. FAO. *The State of Food and Agriculture*; Food and Agriculture Organization: Rome, Italy, 2013.
3. Bailey, R.L.; West, K.P., Jr.; Black, R.E. The epidemiology of global micronutrient deficiencies. *Ann. Nutr. Metab.* **2015**, *66*, 2233. [CrossRef] [PubMed]
4. Tulchinsky, T.H. Micronutrient deficiency conditions: Global health issues. *Public Health Rev.* **2010**, *32*, 243. [CrossRef]
5. Darnton-Hill, I.; Bloem, M.; Chopra, M. Achieving the millennium development goals through mainstreaming nutrition: Speaking with one voice. *Public Health Nutr.* **2006**, *9*, 537–539. [CrossRef]
6. Ahmed, T.; Hossain, M.; Sanin, K.I. Global burden of maternal and child undernutrition and micronutrient deficiencies. *Ann. Nutr. Metab.* **2012**, *61*, 8–17. [CrossRef]

7.   Welch, R.M. Breeding strategies for biofortified staple plant foods to reduce micronutrient malnutrition globally. *J. Nutr.* **2002**, *132*, 495S–499S. [CrossRef]

8.   Duranti, M. Grain legume proteins and nutraceutical properties. *Fitoterapia* **2006**, *77*, 67–82. [CrossRef]

9.   Patterson, C.A.; Maskus, H.; Dupasquier, C. Pulse crops for health. *Cereals Food World* **2009**, *54*, 108–113. [CrossRef]

10. Roy, F.; Boye, J.I.; Simpson, B.K. Bioactive proteins and peptides in pulse crops: Pea, chickpea and lentil. *Food Res. Int.* **2010**, *43*, 432–442. [CrossRef]

11. Ray, H.; Bett, K.E.; Tar'an, B.; Vandenberg, A.; Thavarajah, D.; Warkentin, T. Mineral micronutrient content of cultivars of field pea, chickpea, common bean, and lentil grown in Saskatchewan, Canada. *Crop Sci.* **2014**, *54*, 1698–1708. [CrossRef]

12. Diapari, M.; Sindhu, A.; Bett, K.; Deokar, A.; Warkentin, T.D.; Tar'an, B. Genetic diversity and association mapping of iron and zinc concentrations in chickpea (*Cicer arietinum* L.). *Genome* **2014**, *57*, 459–468. [CrossRef] [PubMed]

13. Diapari, M.; Sindhu, A.; Warkentin, T.D.; Bett, K.; Tar'an, B. Population structure and marker-trait association studies of iron, zinc and selenium concentrations in seed of field pea (*Pisum sativum* L.). *Mol. Breed.* **2015**, *35*, 30. [CrossRef]

14. Jha, A.B.; Tar'an, B.; Diapari, M.; Warkentin, T.D. SNP variation within genes associated with amylose, total starch and crude protein concentration in field pea. *Euphytica* **2015**, *206*, 459–471. [CrossRef]

15. Jha, A.B.; Ashokkumar, K.; Diapari, M.; Ambrose, S.J.; Zhang, H.; Tar'an, B.; Bett, K.E.; Vandenberg, A.; Warkentin, T.D.; Purves, R.W. Genetic diversity of folate profiles in seeds of common bean, lentil, chickpea and pea. *J. Food Compos. Anal.* **2015**, *42*, 134–140. [CrossRef]

16. Messina, M.J. Legumes and soybeans: Overview of their nutritional profiles and health effects. *Am. J. Clin. Nutr.* **1999**, *70*, 439S–450S. [CrossRef] [PubMed]

17. Ghosh, A.; Hasim Reja, M.; Nalia, A.; Kanthal, S.; Maji, S.; Venugopalan, V.; Nath, R. Micronutrient biofortification in pulses: An agricultural approach. *CJAST* **2019**, *35*, 1–12. [CrossRef]

18. Curran, J. The nutritional value and health benefits of pulses in relation to obesity, diabetes, heart disease and cancer. *Br. J. Nutr.* **2012**, *108*, 1–2. [CrossRef]

19. Bouis, H.E. Plant breeding: A new tool for fighting micronutrient malnutrition. *J. Nutr.* **2002**, *132*, 491S–494S. [CrossRef]

20. Nestel, P.; Bouis, H.E.; Meenakshi, J.V.; Pfeiffer, W. Biofortification of staple food crops. *J. Nutr.* **2006**, *136*, 1064–1067. [CrossRef]

21. Bouis, H.E.; Hotz, C.; McClafferty, B.; Meenakshi, J.V.; Pfeiffer, W.H. Biofortification: A new tool to reduce micronutrient malnutrition. *Food Nutr. Bull.* **2011**, *32*, S31–S40. [CrossRef]

22. Saltzman, A.; Birol, E.; Bouis, H.E.; Boy, E.; De Moura, F.F.; Islam, Y.; Pfeiffer, W.H. Biofortification: Progress toward a more nourishing future. *Glob. Food Secur.* **2013**, *2*, 9–17. [CrossRef]

23. Upadhyaya, H.D.; Bajaj, D.; Das, S.; Kumar, V.; Gowda, C.L.L.; Sharma, S.; Tyagi, A.K.; Swarup, K.P. Genetic dissection of seed-iron and zinc concentrations in chickpea. *Sci. Rep.* **2016**, *6*, 24050. [CrossRef] [PubMed]

24. Aldemir, S.; Ateş, D.; Temel, H.Y.; Yağmur, B.; Alsaleh, A.; Kahriman, A.; Özkan, H.; Vandenberg, A.; Tanyolaç, M.B. QTLs for iron concentration in seeds of the cultivated lentil (*Lens culinaris* Medic.) via genotyping by sequencing. *Turk. J. Agric. For.* **2017**, *41*, 243–255. [CrossRef]

25. Khazaei, H.; Podder, R.; Caron, C.T.; Kundu, S.S.; Diapari, M.; Vandenberg, A.; Bett, K.E. Marker-trait association analysis of iron and zinc concentration in lentil (*Lens culinaris* Medik.) seeds. *Plant Genome* **2017**, *10*, 2. [CrossRef]

26. Ma, Y.; Coyne, C.J.; Grusak, M.A.; Mazourek, M.; Cheng, P.; Main, D.; McGee, R.J. Genome-wide SNP identification, linkage map construction and QTL mapping for seed mineral concentrations and contents in pea (*Pisum sativum* L.). *BMC Plant Biol.* **2017**, *17*, 43. [CrossRef]

27. Gali, K.K.; Liu, Y.; Sindhu, A.; Diapari, M.; Shunmugam, A.S.K.; Arganosa, G.; Daba, K.; Caron, C.; Lachagari, R.V.B.; Tar'an, B.; et al. Construction of high-density linkage maps for mapping quantitative trait loci for multiple traits in field pea (*Pisum sativum* L.). *BMC Plant Biol.* **2018**, *18*, 172. [CrossRef]

28. Vandemark, G.J.; Grusak, M.A.; McGee, R.J. Mineral concentrations of chickpea and lentil cultivars and breeding lines grown in the US Pacific Northwest. *Crop J.* **2018**, *6*, 253–262. [CrossRef]

29. Dissanayaka, D. Genome Wide Association Study to Identify Single Nucleotide Polymorphism Markers for Fe, Zn, and Se Concentration in Field Pea Seeds. Master's Thesis, University of Saskatchewan, Saskatoon, SK, Canada, March 2019.

30. Jha, A.B.; Gali, K.K.; Zhang, H.; Purves, R.W.; Vandenberg, A.; Warkentin, T.D. Folate profile diversity and associated SNPs using genome wide association study in pea. *Euphytica* **2020**, accepted.

31. Abbaspour, N.; Hurrell, R.; Kelishadi, R. Review on iron and its importance for human health. *J. Res. Med. Sci.* **2014**, *19*, 164–174.

32. Hurrell, R.F. Bioavailability of iron. *Eur. J. Clin. Nutr.* **1997**, *51*, S4–S8.

33. McDowell, L.R. *Minerals in Animal and Human Nutrition*, 2nd ed.; Elsevier Science: Amsterdam, The Netherlands, 2003; p. 660.

34. World Health Organization. *Iron Deficiency Anaemia: Assessment, Prevention and Control, a Guide for Programme Managers*; WHO: Geneva, Switzerland, 2001.

35. World Health Organization. *Worldwide Prevalence of Anaemia 1993–2005: WHO Global Database on Anaemia*; de Benoist, B., McLean, E., Egli, I., Cogswell, M., Eds.; WHO: Geneva, Switzerland, 2008.

36. Allen, L.H. Anemia and iron deficiency: Effects on pregnancy outcome. *Am. J. Clin. Nutr.* **2000**, *71*, 1280S–1284S. [CrossRef] [PubMed]

37. Lozoff, B.; Clark, K.M.; Jing, Y.; Armony-Sivan, R.; Angelilli, M.L.; Jacobson, S.W. Dose-response relationships between iron deficiency with or without anemia and infant social-emotional behavior. *J. Pediatr.* **2008**, *152*, 696–702. [CrossRef] [PubMed]

38. Prasad, A.S. Zinc deficiency in women, infants, and children. *J. Am. Coll. Nutr.* **1996**, *15*, 113–120. [CrossRef] [PubMed]

39. MacDonald, R.S. The role of zinc in growth and cell proliferation. *J. Nutr.* **2000**, *130*, 1500S–1508S. [CrossRef]

40. Rostan, E.F.; DeBuys, H.V.; Madey, D.L.; Pinnell, S.R. Evidence supporting zinc as an important antioxidant for skin. *Int. J. Dermatol.* **2002**, *41*, 606–611. [CrossRef]

41. Prasad, A.S.; Bao, B.; Beck, F.W.; Kucuk, O.; Sarkar, F.H. Antioxidant effect of zinc in humans. *Free Radic. Biol. Med.* **2004**, *37*, 1182–1190. [CrossRef]

42. Costello, L.C.; Franklin, R.B. Decreased zinc in the development and progression of malignancy: An important common relationship and potential for prevention and treatment of carcinomas. *Expert Opin. Ther. Targets* **2017**, *21*, 51–66. [CrossRef]

43. Roohani, N.; Hurrell, R.; Kelishadi, R.; Schulin, R. Zinc and its importance for human health: An integrative review. *J. Res. Med. Sci.* **2013**, *18*, 144–157.

44. Jansson, B. The role of selenium as a cancer-protecting trace element. *Met. Ions Biol. Syst.* **1980**, *10*, 281–311.

45. Rayman, M.P. Selenium in cancer prevention: A review of the evidence and mechanism of action. *Proc. Nutr. Soc.* **2005**, *64*, 527–542. [CrossRef]

46. Tinggi, U. Selenium: Its role as antioxidant in human health. *Environ. Health Prev. Med.* **2008**, *13*, 102–108. [CrossRef] [PubMed]

47. Zeng, H.; Combs, G.F. Selenium as an anticancer nutrient: Roles in cell proliferation and tumor cell invasion. *J. Nutr. Biochem.* **2008**, *19*, 1–7. [CrossRef] [PubMed]

48. Coppinger, R.J.; Diamond, A.M. Selenium deficiency and human disease. In *Selenium*; Hatfield, D.L., Ed.; Springer: Boston, MA, USA, 2001; pp. 219–233.

49. Delange, F. The disorders induced by iodine deficiency. *Thyroid* **1994**, *4*, 107–128. [CrossRef] [PubMed]

50. World Health Organization. *Iodine Deficiency in Europe: A Continuing Public Health Problem*; Andersson, M., de Benoist, B., Darnton-Hill, I., Eds.; WHO: Geneva, Switzerland, 2007.

51. World Health Organization; United Nations Children's Fund; International Council for the Control of Iodine Deficiency Disorders. Assessment of iodine deficiency disorders and monitoring their elimination. In *A Guide for Programme Managers*, 3rd ed.; World Health Organization: Geneva, Switzerland, 2007.

52. Skeaff, S.A. Iodine deficiency in pregnancy: The effect on neurodevelopment in the child. *Nutrients* **2011**, *3*, 265–273. [CrossRef] [PubMed]

53. Pearce, E.N.; Lazarus, J.H.; Moreno-Reyes, R.; Zimmermann, M.B. Consequences of iodine deficiency and excess in pregnant women: An overview of current knowns and unknowns. *Am. J. Clin. Nutr.* **2016**, *104*, 918S–923S. [CrossRef] [PubMed]

54. Pearce, E.N.; Andersson, M.; Zimmermann, M.B. Global iodine nutrition: Where do we stand in 2013? *Thyroid* **2013**, *23*, 523–528. [CrossRef]

55. Cakmak, I.; Prom-u-thai, C.; Guilherme, L.R.G.; Rashid, A.; Hora, K.; Yazici, A.; Savasli, E.; Kalayci, M.; Tutus, Y.; Phuphong, P.; et al. Iodine biofortification of wheat, rice and maize through fertilizer strategy. *Plant Soil* **2017**, *418*, 319–335. [CrossRef]

56. Gonzali, S.; Kiferle, C.; Perata, P. Iodine biofortification of crops: Agronomic biofortification, metabolic engineering and iodine bioavailability. *Curr. Opin. Biotechnol.* **2017**, *44*, 16–26. [CrossRef] [PubMed]

57. Fraser, P.D.; Bramley, P.M. The biosynthesis and nutritional uses of carotenoids. *Prog. Lipid Res.* **2004**, *43*, 228–265. [CrossRef] [PubMed]

58. Olmedilla, B.; Granado, F.; Blanco, I.; Vaquero, M.; Cajigal, C. Lutein in patients with cataracts and age-related macular degeneration: A long-term supplementation study. *J. Sci. Food Agric.* **2001**, *81*, 904–909. [CrossRef]

59. Moeller, S.M.; Jacques, P.F.; Blumberg, J.B. The potential role of dietary xanthophylls in cataract and age-related macular degeneration. *J. Am. Coll. Nutr.* **2000**, *19*, 522S–527S. [CrossRef]

60. Alves-Rodrigues, A.; Shao, A. The science behind lutein. *Toxicol. Lett.* **2004**, *150*, 57–83. [CrossRef] [PubMed]

61. Stephens, D.; Jackson, P.L.; Gutierrez, Y. Subclinical vitamin A deficiency: A potentially unrecognized problem in the United States. *Pediatr. Nurs.* **1996**, *22*, 377–389. [PubMed]

62. Yamaguchi, M.; Uchiyama, S. beta-Cryptoxanthin stimulates bone formation and inhibits bone resorption in tissue culture in vitro. *Mol. Cell. Biochem.* **2004**, *258*, 137–144. [CrossRef]

63. Tanaka, T.; Shnimizu, M.; Moriwaki, H. Cancer chemoprevention by carotenoids. *Molecules* **2012**, *17*, 3202–3242. [CrossRef] [PubMed]

64. Iannone, A.; Rota, C.; Bergamini, S.; Tomasi, A.; Canfield, L.M. Antioxidant activity of carotenoids: An electron-spin resonance study on beta-carotene and lutein interaction with free radicals generated in a chemical system. *J. Biochem. Mol. Toxicol.* **1998**, *12*, 299–304. [CrossRef]

65. Sujak, A.; Gabrielska, J.; Grudzinski, W.; Borc, R.; Mazurek, P.; Gruszecki, W.I. Lutein and zeaxanthin as protectors of lipid membranes against oxidative damage: The structural aspects. *Arch. Biochem. Biophys.* **1999**, *371*, 301–307. [CrossRef]

66. Garcia-Casal, N.; Layrisse, M.; Solano, L.; Baron, M.A.; Arguello, F.; Liovera, D.; Ramírez, J.; Leets, I.; Tropper, E. Vitamin A and beta carotene can improve nonheme iron absorption from rice, wheat and corn by humans. *J. Nutr.* **1998**, *128*, 646–650. [CrossRef]

67. Garcia-Casal, N.; Leets, I.; Layrisse, M. Beta carotene and inhibitors of iron absorption modify iron uptake by Caco-2 cells. *J. Nutr.* **2000**, *130*, 5–9. [CrossRef]

68. Bailey, L.B.; Gregory, J.F. Folate metabolism and requirements. *J. Nutr.* **1999**, *129*, 779–782. [CrossRef]

69. Scott, J.; R'ebeill'e, F.; Fletcher, J. Folic acid and folate: The feasibility for nutritional enhancement in plant foods. *J. Sci. Food Agric.* **2000**, *80*, 795–824. [CrossRef]

70. Hanson, A.D.; Roje, S. One-carbon metabolism in higher plants. *Annu. Rev. Plant Physiol. Plant Mol. Biol.* **2001**, *52*, 119–137. [CrossRef]

71. Basset, G.J.C.; Quinlivan, E.P.; Gregory, J.F., III; Hanson, A.D. Folate synthesis and metabolism in plants and prospects for biofortification. *Crop Sci.* **2005**, *45*, 449–453. [CrossRef]

72. Geisel, J. Folic acid and neural tube defects in pregnancy—A review. *J. Perinat. Neonat. Nurs.* **2003**, *17*, 268–279. [CrossRef] [PubMed]

73. Ramos, M.I.; Allen, L.H.; Mungas, D.M.; Jagust, W.J.; Haan, M.N.; Green, R.; Miller, J.W. Low folate status is associated with impaired cognitive function and dementia in the Sacramento Area Latino Study on Aging. *Am. J. Clin. Nutr.* **2005**, *82*, 1346–1352. [CrossRef] [PubMed]

74. Seshadri, S.; Beiser, A.; Selhub, J.; Jacques, P.F.; Rosenberg, I.H.; D'Agostino, R.B.; Wilson, P.W.F.; Wolf, P.A. Plasma homocysteine as a risk factor for dementia and Alzheimer's disease. *N. Engl. J. Med.* **2002**, *346*, 476–483. [CrossRef]

75. McCully, K.S. Homocysteine, vitamins, and vascular disease prevention. *Am. J. Clin. Nutr.* **2007**, *86*, 1563S–1568S. [CrossRef]

76. Choi, S.W.; Friso, S. Interactions between folate and aging for carcinogenesis. *Clin. Chem. Lab. Med.* **2005**, *43*, 1151–1157. [CrossRef]

77. Pitkin, R.M. Folate and neural tube defects. *Am. J. Clin. Nutr.* **2007**, *85*, 285S–288S. [CrossRef]

78. Scholl, T.O.; Johnson, W.G. Folic acid: Influence on the outcome of pregnancy. *Am. J. Clin. Nutr.* **2000**, *71*, 1295S–1303S. [CrossRef]

79. Wallock, L.M.; Tamura, T.; Mayr, C.A.; Johnston, K.E.; Ames, B.N.; Jacob, R.A. Low seminal plasma folate concentrations are associated with low sperm density and count in male smokers and nonsmokers. *Fertil. Steril.* **2001**, *75*, 252–259. [CrossRef]

80. Tamura, T.; Picciano, M.F. Folate and human reproduction. *Am. J. Clin. Nutr.* **2006**, *83*, 993–1016. [CrossRef] [PubMed]

81. Gibson, R.S.; Hotz, C. Dietary diversification/modification strategies to enhance micronutrient content and bioavailability of diets in developing countries. *Br. J. Nutr.* **2001**, *85*, S159–S166. [CrossRef] [PubMed]

82. Hurrell, R. How to ensure adequate iron absorption from iron-fortified food. *Nutr. Rev.* **2002**, *60*, S7–S15. [CrossRef]

83. World Health Organization and Food and Agriculture Organization. *Vitamin and Mineral Requirements in Human Nutrition*, 2nd ed.; WHO: Geneva, Switzerland, 2004.

84. Cook, J.D. Diagnosis and management of iron-deficiency anaemia. *Best Pract. Res. Clin. Haematol.* **2005**, *18*, 319–332. [CrossRef]

85. Blancquaert, D.; Storozhenko, S.; Van Daele, J.; Stove, C.; Visser, R.; Lambert, W.; Van Der Straeten, D. Enhancing pterin and paraaminobenzoate content is not sufficient to successfully biofortify potato tubers and *Arabidopsis thaliana* plants with folate. *J. Exp. Bot.* **2013**, *64*, 3899–3909. [CrossRef]

86. Shohag, M.J.I.; Wei, Y.; Yang, X. Changes of folate and other potential health promoting phytochemicals in legume seeds as affected by germination. *J. Agric. Food Chem.* **2012**, *60*, 9137–9143. [CrossRef]

87. Hefni, M.; Öhrvik, V.; Tabekha, M.M.; Witthöft, C. Folate content in foods commonly consumed in Egypt. *Food Chem.* **2010**, *121*, 540–545. [CrossRef]

88. Black, R.E.; Allen, L.H.; Bhutta, Z.A.; Caulfield, L.E.; de Onis, M.; Ezzati, M.; Mathers, C.; Rivera, J. Maternal and child undernutrition: Global and regional exposures and health consequences. *Lancet* **2008**, *371*, 243–260. [CrossRef]

89. Wiltgren, A.R.; Booth, A.O.; Kaur, G.; Cicerale, S.; Lacy, K.E.; Thorpe, M.G.; Keast, R.S.; Riddell, L.J. Micronutrient supplement use and diet quality in university students. *Nutrients* **2015**, *7*, 1094–1107. [CrossRef]

90. Stoltzfus, R.J. Iron interventions for women and children in low-income countries. *J. Nutr.* **2011**, *141*, 756S–762S. [CrossRef] [PubMed]

91. World Health Organization and Food and Agriculture Organization of the United Nations. *Guidelines on Food Fortification with Micronutrients*; Allen, L., de Benoist, B., Dary, O., Hurrell, R., Eds.; WHO: Geneva, Switzerland, 2006.

92. Gómez-Galera, S.; Rojas, E.; Sudhakar, D.; Zhu, C.; Pelacho, A.M.; Capell, T.; Christou, P. Critical evaluation of strategies for mineral fortification of staple food crops. *Transgenic Res.* **2010**, *19*, 165–180. [CrossRef] [PubMed]

93. White, P.J.; Broadley, M.R. Biofortification of crops with seven mineral elements often lacking in human diets-iron, zinc, copper, calcium, magnesium, selenium and iodine. *New Phytol.* **2009**, *182*, 49–84. [CrossRef] [PubMed]

94. Dai, J.L.; Zhu, Y.G.; Zhang, M.; Huang, M.Z. Selecting iodine-enriched vegetables and the residual effect of iodate application to soil. *Biol. Trace Elem. Res.* **2004**, *101*, 265–276. [CrossRef]

95. Hartikainen, H. Biogeochemistry of selenium and its impact on food chain quality and human health. *J. Trace Elem. Med. Biol.* **2005**, *18*, 309–318. [CrossRef]

96. White, P.J.; Broadley, M.R. Biofortifying crops with essential mineral elements. *Trends Plant Sci.* **2005**, *10*, 586–593. [CrossRef]

97. Eurola, M.; Ekholm, P.; Ylinen, M.; Koivistoinen, P.; Varo, P. Effects of selenium fertilization on the selenium content of selected Finnish fruits and vegetables. *Acta Agric. Scand.* **1989**, *39*, 345–350. [CrossRef]

98. Eurola, M.H.; Ekholm, P.I.; Ylinen, M.E.; Koivistoinen, P.E.; Varo, P.T. Selenium in Finnish foods after beginning the use of selenate-supplemented fertilisers. *J. Sci. Food Agric.* **1991**, *56*, 57–70. [CrossRef]

99. Alfthan, G.; Eurola, M.; Ekholm, P. Effects of nationwide addition of selenium to fertilizers on foods, and animal and human health in Finland: From deficiency to optimal selenium status of the population. *J. Trace Elem. Med. Biol.* **2015**, *31*, 42–147. [CrossRef]

100. Winkler, J.T. Biofortification: Improving the nutritional quality of staple crops. In *Access Not Excess*; Pasternak, C., Ed.; Smith-Gordon Publishing: St Ives, UK, 2011; pp. 100–112.

101. Grusak, M.A.; DellaPenna, D. Improving the nutrient composition of plants to enhance human nutrition and health. *Annu. Rev. Plant Physiol. Plant Mol. Biol.* **1999**, *50*, 133–161. [CrossRef]

102. Poblaciones, M.J.; Rengel, Z. Soil and foliar zinc biofortification in field pea (*Pisum sativum* L.). Grain accumulation and bioavailability in raw and cooked grains. *Food Chem.* **2016**, *212*, 427–433. [CrossRef] [PubMed]

103. Hirschi, K.D. Nutrient biofortification of food crops. *Annu. Rev. Nutr.* **2009**, *29*, 401–421. [CrossRef] [PubMed]

104. Frossard, E.; Bucher, M.; Machler, F.; Mozafar, A.; Hurrell, R. Potential for increasing the content and bioavailability of Fe, Zn and Ca in plants for human nutrition. *J. Sci. Food Agric.* **2000**, *80*, 861–879. [CrossRef]

105. Ismail, A.M.; Heuer, S.; Thomson, M.J.; Wissuwa, M. Genetic and genomic approaches to develop rice germplasm for problem soils. *Plant Mol. Biol.* **2007**, *65*, 547–570. [CrossRef] [PubMed]

106. Garg, M.; Sharma, N.; Sharma, S.; Kapoor, P.; Kumar, A.; Chunduri, V.; Arora, P. Biofortified crops generated by breeding, agronomy, and transgenic approaches are improving lives of millions of people around the world. *Front. Nutr.* **2018**, *5*, 12. [CrossRef] [PubMed]

107. Márquez-Quiroz, C.; De-la-Cruz-Lázaro, E.; Osorio-Osorio, R.; Sánchez-Chávez, E. Biofortification of cowpea beans with iron: Iron's influence on mineral content and yield. *J. Soil Sci. Plant Nutr.* **2015**, *15*, 839–847. [CrossRef]

108. Ali, B.; Ali, A.; Tahir, M.; Ali, S. Growth, Seed yield and quality of mungbean as influenced by foliar application of iron sulfate. *Pak. J. Life Soc. Sci.* **2014**, *12*, 20–25.

109. Salih, H.O. Effect of foliar fertilization of Fe, B and Zn on nutrient concentration and seed protein of Cowpea "*Vigna unguiculata*". *IOSR J. Aric. Vet. Sci.* **2013**, *6*, 42–46. [CrossRef]

110. Nandan, B.; Sharma, B.C.; Chand, G.; Bazgalia, K.; Kumar, R.; Banotra, M. Agronomic fortification of Zn and Fe in chickpea an emerging tool for nutritional security—A global perspective. *Acta Sci. Nutr. Health* **2018**, *2*, 12–19.

111. Shivay, Y.S.; Prasad, R.; Pal, M. Effects of source and method of zinc application on yield, zinc biofortification of grain, and Zn uptake and use efficiency in chickpea (*Cicer arietinum* L.). *Commun. Soil Sci. Plant Anal.* **2015**, *46*, 2191–2200. [CrossRef]

112. Hidoto, L.; Worku, W.; Mohammed, H.; Taran, B. Effects of zinc application strategy on zinc content and productivity of chickpea grown under zinc deficient soils. *J. Soil Sci. Plant Nutr.* **2017**, *17*, 112–126. [CrossRef]

113. Ibrahim, E.A.; Ramadan, W.A. Effect of zinc foliar spray alone and combined with humic acid or/and chitosan on growth, nutrient elements content and yield of dry bean (*Phaseolus vulgaris* L.) plants sown at different dates. *Sci. Hortic.* **2015**, *184*, 101–115. [CrossRef]

114. Ram, H.; Rashid, A.; Zhang, W.; Duarte, A.P.; Phattarakul, N.; Simunji, S.; Kalayci, M.; Freitas, R.; Rerkasem, B.; Bal, R.S.; et al. Biofortification of wheat, rice and common bean by applying foliar zinc fertilizer along with pesticides in seven countries. *Plant Soil* **2016**, *403*, 389–401. [CrossRef]

115. Sida-Arreola, J.P.; Sánchez, E.; Ojeda-Barrios, D.L.; Ávila-uezada, G.D.; Flores-Córdova, M.A.; Márquez-Quiroz, C.; Preciado-Rangel, P. Can biofortification of zinc improve the antioxidant capacity and nutritional quality of beans? *Emir. J. Food Agric.* **2017**, *29*, 237–241.

116. Smrkolj, P.; Germ, M.; Kreft, I.; Stibilj, V. Respiratory potential and Se compounds in pea (*Pisum sativum* L.) plants grown from Se-enriched seeds. *J. Exp. Bot.* **2006**, *57*, 3595–3600. [CrossRef]

117. Poblaciones, M.J.; Rodrigo, S.; Santamaria, O.; Chen, Y.; McGrath, S.P. Selenium accumulation and speciation in biofortified chickpea (*Cicer arietinum* L.) under Mediterranean conditions. *J. Sci. Food Agric.* **2014**, *94*, 1101–1106. [CrossRef]

118. Smrkolj, P.; Osvald, M.; Osvald, J.; Stibilj, V. Selenium uptake and species distribution in selenium-enriched bean (*Phaseolus vulgaris* L.) seeds obtained by two different cultivations. *Eur. Food Res. Technol.* **2007**, *225*, 233–237. [CrossRef]

119. Rahman, M.M.; Erskine, W.; Materne, M.A.; McMurray, L.M.; Thavarajah, P.; Thavarajah, D.; Siddique, K.H.M. Enhancing selenium concentration in lentil (*Lens culinaris* subsp. *culinaris*) through foliar application. *J. Agric. Sci.* **2015**, *153*, 656–665.

120. FAO. *The Plant Production and Protection Division (AGP)—Soil Biological Management with Beneficial Microorganisms*; FAO: Rome, Italy, 2019.

121. Mahaffee, W.F.; Kloepper, J.W. Applications of plant growth-promoting rhizobacteria in sustainable agriculture. In *Soil Biota: Management in Sustainable Farming Systems*; Pankhurst, C.E., Doube, B.M., Gupta, V.V.S.R., Grace, P.R., Eds.; CSIRO: Melbourne, Australia, 1994; pp. 23–31.

122. Panhwar, Q.A.; Othman, R.; Rahman, Z.A.; Meon, S.; Ismail, M.R. Isolation and characterization of phosphate-solubilizing bacteria from aerobic rice. *Afr. J. Biotechnol.* **2012**, *11*, 2711–2719.

123. Sreevidya, M.; Gopalakrishnan, S.; Kudapa, H.; Varshney, R.K. Exploring PGP actinomycetes from vermicompost and rhizosphere soil for yield enhancement in chickpea. *Braz. J. Microbiol.* **2016**, *47*, 85–95. [CrossRef]

124. Gopalakrishnan, S.; Vadlamudi, S.; Samineni, S.; Sameer Kumar, C.V. Plant growth-promotion and biofortification of chickpea and pigeonpea through inoculation of biocontrol potential bacteria, isolated from organic soils. *Springerplus* **2016**, *5*, 1882. [CrossRef] [PubMed]

125. Rengel, Z.; Batten, G.D.; Crowley, D.D. Agronomic approaches for improving the micronutrient density in edible portions of field crops. *Field Crop. Res.* **1999**, *60*, 27–40. [CrossRef]

126. Smith, S.E.; Read, D.J. *Mycorrhizal Symbiosis*, 3rd ed.; Elsevier: London, UK, 2007.

127. Cavagnaro, T.R. The role of arbuscular mycorrhizas in improving plant zinc nutrition under low soil zinc concentrations: A review. *Plant Soil* **2008**, *304*, 315–325. [CrossRef]

128. Tokala, R.K.; Strap, J.L.; Jung, C.M.; Crawford, D.L.; Salove, M.H.; Deobald, L.A.; Bailey, J.F.; Morra, M.J. Novel plant-microbe rhizosphere interaction involving Streptomyces lydicus WYEC108 and the pea plant (*Pisum sativum*). *Appl. Environ. Microbiol.* **2002**, *68*, 2161–2171. [CrossRef] [PubMed]

129. Valverde, A.; Burgos, A.; Fiscella, T.; Rivas, R.; Velazquez, E.; Rodrıguez-Barrueco, C.; Cervantes, E.; Chamber, M.; Igual, J.M. Differential effects of co-inoculations with *Pseudomonas jessenii* PS06 (a phosphate-solubilizing bacterium) and *Mesorhizobium ciceri* C-2/2 strains on the growth and seed yield of chickpea under greenhouse and field conditions. *Plant Soil* **2006**, *287*, 43–50. [CrossRef]

130. Minorsky, P.V. On the inside. *Plant Physiol.* **2008**, *146*, 323–324. [CrossRef]

131. Soe, K.M.; Bhromsiri, A.; Karladee, D. Effects of selected endophytic actinomycetes (*Streptomyces* sp.) and *Bradyrhizobia* from Myanmar on growth, nodulation, nitrogen fixation and yield of different soybean varieties. *CMU J. Nat. Sci.* **2010**, *9*, 95–109.

132. Gopalakrishnan, S.; Srinivas, V.; Prakash, B.; Sathya, A.; Vijayabharathi, R. Plant growth-promoting traits of *Pseudomonas geniculata* isolated from chickpea nodules. *3 Biotech* **2015**, *5*, 653–661. [CrossRef]

133. Sathya, A.; Vijayabharati, R.; Srinivas, V.; Gopalakrishnan, S. Plant growth-promoting action-bacteria on chickpea seed mineral density: An upcoming complementary tool for sustainable biofortification strategy. *3 Biotech* **2013**, *6*, 138. [CrossRef]

134. Pellegrino, E.; Bedini, S. Enhancing ecosystem services in sustainable agriculture: Biofertilization and biofortification of chickpea (*Cicer arietinum* L.) by arbuscular mycorrhizal fungi. *Soil Biol. Biochem.* **2014**, *68*, 429–439. [CrossRef]

135. Khalid, S.; Asghar, H.N.; Akhtar, M.J.; Aslam, A.; Zahir, Z.A. Biofortification of iron in chickpea by plant growth promoting rhizobacteria. *Pak. J. Bot.* **2015**, *47*, 1191–1194.

136. Mayer, J.E.; Pfeiffer, W.H.; Bouis, P. Biofortified crops to alleviate micronutrient malnutrition. *Curr. Opin. Plant Biol.* **2008**, *11*, 166–170. [CrossRef] [PubMed]

137. Perez-Massot, E.; Banakar, R.; Gomez-Galera, S.; Zorrilla-Lopez, U.; Sanahuja, G.; Arjo, G.; Miralpeix, B.; Vamvaka, E.; Farré, G.; Rivera, S.M.; et al. The contribution of transgenic plants to better health through improved nutrition: Opportunities and constraints. *Genes Nutr.* **2013**, *8*, 29–41. [CrossRef] [PubMed]

138. Carvalho, S.M.P.; Vasconcelos, M.W. Producing more with less: Strategies and novel technologies for plant-based food biofortification. *Food Res. Int.* **2013**, *54*, 961–971. [CrossRef]

139. Christou, P.; Twyman, R.M. The potential of genetically enhanced plants to address food insecurity. *Nutr. Res. Rev.* **2004**, *17*, 23–42. [CrossRef] [PubMed]

140. Newell-McGloughlin, M. Nutritionally improved agricultural crops. *Plant Physiol.* **2008**, *147*, 939–953. [CrossRef]

141. Hefferon, K.L. Can biofortified crops help attain food security? *Curr. Mol. Biol. Rep.* **2016**, *2*, 180–185. [CrossRef]

142. Goto, F.; Yoshihara, T.; Saiki, H. Iron accumulation and enhanced growth in transgenic lettuce plants expressing the iron-binding protein ferritin. *Theor. Appl. Genet.* **2000**, *100*, 658–664. [CrossRef]

143. Vasconcelos, M.; Datta, K.; Oliva, N.; Khalekuzzaman, M.; Torrizo, L.; Krishnan, S.; Oliveira, M.; Goto, F.; Datta, S.K. Enhanced iron and zinc accumulation in transgenic rice with the ferritin gene. *Plant Sci.* **2003**, *164*, 371–378. [CrossRef]

144. Trijatmiko, K.R.; Dueñas, C.; Tsakirpaloglou, N.; Torrizo, L.; Arines, F.M.; Adeva, C.; Balindong, J.; Oliva, N.; Sapasap, M.V.; Borrero, J.; et al. Biofortified indica rice attains iron and zinc nutrition dietary targets in the field. *Sci. Rep.* **2016**, *6*, 19792. [CrossRef]

145. Paine, J.A.; Shipton, C.A.; Chaggar, S.; Howells, R.M.; Kennedy, M.J.; Vernon, G.; Wright, S.Y.; Hinchliffe, E.; Adams, J.L.; Silverstone, A.L.; et al. Improving the nutritional value of Golden Rice through increased pro-vitamin A content. *Nat. Biotechnol.* **2005**, *23*, 482–487. [CrossRef] [PubMed]

146. Naqvi, S.; Zhu, C.; Farre, G.; Ramessar, K.; Bassie, L.; Breitenbach, J.; Perez Conesa, D.; Ros, G.; Sandmann, G.; Capell, T.; et al. Transgenic multivitamin corn through biofortification of endosperm with three vitamins representing three distinct metabolic pathways. *Proc. Natl. Acad. Sci. USA* **2009**, *106*, 7762–7767. [CrossRef] [PubMed]

147. Blancquaert, D.; De Steur, H.; Gellynck, X.; Van Der Straeten, D. Present and future of folate biofortification of crop plants. *J. Exp. Bot.* **2014**, *65*, 895–906. [CrossRef] [PubMed]

148. Storozhenko, S.; De Brouwer, V.; Volckaert, M.; Navarrete, O.; Blancquaert, D.; Zhang, G.F.; Lambert, W.; Van Der Straeten, D. Folate fortification of rice by metabolic engineering. *Nat. Biotechnol.* **2007**, *25*, 1277–1279. [CrossRef] [PubMed]

149. Hossain, T.; Rosenberg, I.; Selhub, J.; Kishore, G.; Beachy, R.; Schubert, K. Enhancement of folate in plants through metabolic engineering. *Proc. Natl. Acad. Sci. USA* **2004**, *101*, 5158–5163. [CrossRef] [PubMed]

150. Aragao, F.J.L.; Barros, L.M.G.; De Sousa, M.V.; Grossi de Sa, M.F.; Almeida, E.R.P.; Gander, E.S.; Rech, E.L. Expression of a methionine-rich storage albumin from the Brazil nut (*Bertholletia excelsa* H.B.K., Lecythidaceae) in transgenic bean plants (*Phaseolus vulgaris* L., Fabaceae). *Genet. Mol. Biol.* **1999**, *22*, 445–449. [CrossRef]

151. Molvig, L.; Tabe, L.M.; Eggum, B.O.; Moore, A.E.; Craig, S.; Spencer, D.; Higgins, T.J. Enhanced methionine levels and increased nutritive value of seeds of transgenic lupins (*Lupinus angustifolius* L.) expressing a sunflower seed albumin gene. *Proc. Natl. Acad. Sci. USA* **1997**, *94*, 8393–8398. [CrossRef]

152. Bortesi, L.; Fischer, R. The CRISPR/Cas9 system for plant genome editing and beyond. *Biotechnol. Adv.* **2015**, *33*, 41–52. [CrossRef]

153. Jaganathan, D.; Ramasamy, K.; Sellamuthu, G.; Jayabalan, S.; Venkataraman, G. CRISPR for crop improvement: An update review. *Front. Plant Sci.* **2018**, *9*, 985. [CrossRef]

154. Li, T.; Liu, B.; Spalding, M.H.; Weeks, D.P.; Yang, B. High-efficiency TALEN-based gene editing produces disease-resistant rice. *Nat. Biotechnol.* **2012**, *30*, 390–392. [CrossRef]

155. Zhang, H.; Zhang, J.; Wei, P.; Zhang, B.; Gou, F.; Feng, Z.; Mao, Y.; Yang, L.; Zhang, H.; Xu, N.; et al. The CRISPR/Cas9 system produces specific and homozygous targeted gene editing in rice in one generation. *Plant Biotechnol. J.* **2014**, *12*, 797–807. [CrossRef] [PubMed]

156. Wang, Y.; Cheng, X.; Shan, Q.; Zhang, Y.; Liu, J.; Gao, C.; Qiu, J.L. Simultaneous editing of three homoeoalleles in hexaploid bread wheat confers heritable resistance to powdery mildew. *Nat. Biotechnol.* **2014**, *32*, 947–951. [CrossRef] [PubMed]

157. Brooks, C.; Nekrasov, V.; Lippman, Z.B.; Van Eck, J. Efficient gene editing in tomato in the first generation using the clustered regularly interspaced short palindromic repeats/CRISPR-associated9 system. *Plant Physiol.* **2014**, *166*, 1292–1297. [CrossRef] [PubMed]

158. Curtin, S.J.; Xiong, Y.; Michno, J.M.; Campbell, B.W.; Stec, A.O.; Čermák, T.; Starker, C.; Voytas, D.F.; Eamens, A.L.; Stupar, R.M. CRISPR/Cas9 and TALENs generate heritable mutations for genes involved in small RNA processing of *Glycine max* and *Medicago truncatula*. *Plant Biotechnol. J.* **2018**, *16*, 1125–1137. [CrossRef]

159. Ji, J.; Zhang, C.; Sun, Z.; Wang, L.; Duanmu, D.; Fan, Q. Genome editing in cowpea *Vigna unguiculata* using CRISPR-Cas9. *Int. J. Mol. Sci.* **2019**, *20*, 2471. [CrossRef]

160. Inaba, M.; Macer, D. Policy, regulation and attitudes towards agricultural biotechnology in Japan. *J. Int. Biotechnol. Laws* **2004**, *1*, 45–53. [CrossRef]

161. Watanabe, K.N.; Sassa, Y.; Suda, E.; Chen, C.H.; Inaba, M.; Kikuchi, A. Global political, economic, social and technological issues on transgenic crops-review. *Plant Biotechnol. J.* **2005**, *22*, 515–522. [CrossRef]

162. Wesseler, J.; Zilberman, D. The economic power of the Golden Rice opposition. *Environ. Dev. Econ.* **2014**, *19*, 724–742. [CrossRef]

163. Bouis, H.E.; Saltzman, A. Improving nutrition through biofortification: A review of evidence from HarvestPlus, 2003 through 2016. *Glob. Food Sec.* **2017**, *12*, 49–58. [CrossRef]

164. Welch, R.M.; Graham, R.D. Agriculture: The real nexus for enhancing bioavailable micronutrients in food crops. *J. Trace Elem. Med. Biol.* **2005**, *18*, 299–307. [CrossRef]

165. Pfeiffer, W.H.; McClafferty, B. HarvestPlus: Breeding crops for better nutrition. *Crop Sci.* **2007**, *47*, S88–S100. [CrossRef]

166. Bouis, H.E.; Welch, R.M. Biofortification—A sustainable agricultural strategy for reducing micronutrient malnutrition in the global south. *Crop Sci.* **2010**, *50*, S20–S32. [CrossRef]

167. Bouis, H.E. Micronutrient fortification of plants through plant breeding: Can it improve nutrition in man at low cost? *Proc. Nutr. Soc.* **2003**, *62*, 403–411. [CrossRef] [PubMed]

168. Bouis, H.E. Enrichment of food staples through plant breeding: A new strategy for fighting micronutrient malnutrition. *Nutrition* **2000**, *16*, 701–704. [CrossRef]

169. Beebe, S.; Gonzalez, A.V.; Rengifo, J. Research on trace minerals in the common bean. *Food Nutr. Bull.* **2000**, *21*, 387–391. [CrossRef]

170. Islam, F.M.A.; Basford, K.E.; Jara, C.; Redden, R.J.; Beebe, S.E. Seed compositional and disease resistance differences among gene pools in cultivated common bean. *Genet. Resour. Crop Evol.* **2002**, *49*, 285–293. [CrossRef]

171. Blair, M.W.; Astudillo, C.; Grusak, M.; Graham, R.; Beebe, S. Inheritance of seed iron and zinc content in common bean (*Phaseolus vulgaris* L.). *Mol. Breed.* **2009**, *23*, 197–207. [CrossRef]

172. Blair, M.W.; Medina, J.I.; Astudillo, C.; Rengifo, J.; Beebe, S.E.; Machado, G.; Graham, R. QTL for seed iron and zinc concentrations in a recombinant inbred line population of Mesoamerican common beans (*Phaseolus vulgaris* L.). *Theor. Appl. Genet.* **2010**, *121*, 1059–1070. [CrossRef]

173. Blair, M.W.; Astudillo, C.; Rengifo, J.; Beebe, S.E.; Graham, R. QTL for seed iron and zinc concentrations in a recombinant inbred line population of Andean common beans (*Phaseolus vulgaris* L.). *Theor. Appl. Genet.* **2011**, *122*, 511–521. [CrossRef]

174. Sarker, A.; El-Askhar, F.; Uddin, M.J.; Million, E.; Yadav, N.K.; Dahan, R.; Wolfgang, P. Lentil improvement for nutritional security in the developing world. Presented at the ASA-CSSA-SSSA International Annual Meeting, New Orleans, LA, USA, 4–8 November 2007.

175. HarvestPlus. Biofortification Progress Briefs: Iron and Zinc Lentils. 2014. Available online: www.HarvestPlus.org (accessed on 15 September 2019).

176. Nair, R.M.; Thavarajah, D.; Thavarajah, P.; Giri, R.R.; Ledesma, D.; Yang, R.Y.; Hanson, P.; Easdown, W.; Hughes, J.D.A.; Keatinge, J.D.H. Mineral and phenolic concentrations of mungbean [*Vigna radiata* (L.) R. Wilczek var. radiata] grown in semi-arid tropical India. *J. Food Compos. Anal.* **2015**, *39*, 23–32. [CrossRef]

177. Ariza-Nieto, M.; Blair, M.W.; Welch, R.M.; Glahn, R.P. Screening of bioavailability patterns in eight bean (*Phaseolus vulgaris* L.) genotypes using the Caco-2 cell in vitro model. *J. Agric. Food Chem.* **2007**, *55*, 7950–7956. [CrossRef] [PubMed]

178. DellaValle, D.M.; Vandenberg, A.; Glahn, R.P. Seed coat removal improves iron bioavailability in cooked lentils: Studies using an in vitro digestion/Caco-2 cell culture model. *J. Agric. Food Chem.* **2013**, *61*, 8084–8089. [CrossRef] [PubMed]

179. Moraghan, J.T.; Padilla, J.; Etchevers, J.D.; Grafton, K.; Acosta-Gallegos, J.A. Iron accumulation in seed of common bean. *Plant Soil* **2002**, *246*, 175–183. [CrossRef]

180. Thavarajah, D.; Ruszkowski, J.; Vandenberg, A. High potential for selenium biofortification of lentils (*Lens culinaris* L.). *J. Agric. Food Chem.* **2008**, *56*, 10747–10753. [CrossRef] [PubMed]

181. Thavarajah, D.; Thavarajah, P.; Sarker, A.; Materne, M.; Vandemark, G.; Shrestha, R.; Idrissi, O.; Hacikamiloglu, O.; Bucak, B.; Vandenberg, A. A global survey of effects of genotype and environment on selenium concentration in lentils (*Lens culinaris* L.): Implications for nutritional fortification strategies. *Food Chem.* **2011**, *125*, 72–76. [CrossRef]

182. Thavarajah, D.; Warkentin, T.; Vandenberg, A. Natural enrichment of selenium in Saskatchewan field peas (*Pisum sativum* L.). *Can. J. Plant Sci.* **2010**, *90*, 383–389. [CrossRef]

183. Abbo, S.; Molina, C.; Jungmann, R.; Grusak, M.A.; Berkovitch, Z.; Reifen, R.; Kahl, G.; Winter, P.; Reifen, R. QTL governing carotenoid concentration and weight in seeds of chickpea (*Cicer arietinum* L.). *Theor. Appl. Genet.* **2005**, *111*, 185–195. [CrossRef]

184. Thavarajah, D.; Thavarajah, P. Evaluation of chickpea (*Cicer arietinum* L.) micronutrient composition: Biofortification opportunities to combat global micronutrient malnutrition. *Food Res. Int.* **2012**, *49*, 99–104. [CrossRef]

185. Ashokkumar, K.; Tar'an, B.; Diapari, M.; Arganosa, G.; Warkentin, T.D. Effect of cultivar and environment on carotenoid profile of pea and chickpea. *Crop Sci.* **2014**, *54*, 2225–2235. [CrossRef]

186. Ashokkumar, K.; Diapari, M.; Jha, A.B.; Tar'an, B.; Arganosa, G.; Warkentin, T.D. Genetic diversity of nutritionally important carotenoids in 94 pea and 121 chickpea accessions. *J. Food Compos. Anal.* **2015**, *43*, 49–60. [CrossRef]

187. Rezaei, M.K.; Deokar, A.; Tar'an, B. Identification and expression analysis of candidate genes involved in carotenoid biosynthesis in chickpea seeds. *Front. Plant Sci.* **2016**, *7*, 1867. [CrossRef] [PubMed]

188. Rezaei, M.K.; Deokar, A.A.; Arganosa, G.; Roorkiwal, M.; Pandey, S.K.; Warkentin, T.D.; Varshney, R.K.; Tar'an, B. Mapping quantitative trait loci for carotenoid concentration in three F_2 populations of chickpea. *Plant Genome* **2019**, *12*, 190067. [CrossRef]

189. Vahteristo, L.; Lehikoinen, K.; Ollilainen, V.; Varo, P. Application of an HPLC assay for the determination of folate derivatives in some vegetables, fruits and berries consumed in Finland. *Food Chem.* **1997**, *59*, 589–597. [CrossRef]

190. Han, J.Y.; Tyler, R.T. Determination of folate concentrations in pulses by a microbiological method employing trienzyme extraction. *J. Agric. Food Chem.* **2003**, *51*, 5315–5318. [CrossRef] [PubMed]

191. Rychlik, M.; Englert, K.; Kapfer, S.; Kirchhoff, E. Folate contents of legumes determined by optimized enzyme treatment and stable isotope dilution assays. *J. Food Compos. Anal.* **2007**, *20*, 411–419. [CrossRef]

192. Sen Gupta, D.; Thavarajah, D.; Thavarajah, P.; McGee, R.; Coyne, C.J.; Kumar, S. Lentils (*Lens culinaris* L.), a rich source of folates. *J. Agric. Food Chem.* **2013**, *61*, 7794–7799. [CrossRef]

193. Zhang, H.; Jha, A.B.; Warkentin, T.D.; Vandenberg, A.; Purves, R.W. Folate stability and method optimization for folate extraction from seeds of pulse crops using LC-SRM MS. *J. Food Compos. Anal.* **2018**, *71*, 44–55. [CrossRef]

194. Zhang, H.; Jha, A.B.; De Silva, D.; Purves, R.W.; Warkentin, T.D.; Vandenberg, A. Improved folate monoglutamate extraction and application to folate quantification from wild lentil seeds by ultra-performance liquid chromatography-selective reaction monitoring mass spectrometry. *J. Chromatogr. B Anal. Technol. Biomed. Life Sci.* **2019**, *1121*, 39–47. [CrossRef]

195. Combs, G.F.; Lü, J. Selenium as a cancer preventive agent. In *Selenium*; Hatfield, D.L., Ed.; Springer: Boston, MA, USA, 2001; pp. 205–217.

196. Thavarajah, D.; Vandenberg, A.; George, G.N.; Pickering, I.J. Chemical form of selenium in naturally selenium rich lentils (*Lens culinaris* L.) from Saskatchewan. *J. Agric. Food Chem.* **2007**, *55*, 7337–7341. [CrossRef]

197. Fuge, R.; Johnson, C. Iodine and human health, the role of environmental geochemistry and diet: A review. *Appl. Geochem.* **2015**, *63*, 282–302. [CrossRef]

198. Mackowiak, C.L.; Grossl, P.R. Iodate and iodide effects on iodine uptake and partitioning in rice (*Oryza sativa* L.) grown in solution culture. *Plant Soil* **1999**, *212*, 135–143. [CrossRef] [PubMed]

199. Smolen, S.; Sady, W.; Ledwozyw-Smolen, I.; Strzetelski, P.; Liszka-Skoczylas, M.; Rozek, S. Quality of fresh and stored carrots depending on iodine and nitrogen fertilization. *Food Chem.* **2014**, *159*, 316–322. [CrossRef] [PubMed]

200. Medrano-Macias, J.; Leija-Martinez, P.; Gonzales-Morales, S.; Juarez-Maldonado, A.; Benavides-Mendoza, A. Use of iodine to biofortify and promote growth and stress tolerance in crops. *Front. Plant Sci.* **2016**, *7*, 1146. [CrossRef] [PubMed]

201. McCallum, J.; Timmerman-Vaughan, G.; Frew, T.; Russel, A. Biochemical and genetic linkage analysis of green seed colour in field pea. *J. Am. Soc. Hortic. Sci.* **1997**, *122*, 218–225. [CrossRef]

202. Abbo, S.; Bonfil, D.J.; Berkovitch, Z.; Reifen, R. Towards enhancing lutein concentration in chickpea, cultivar and management effects. *Plant Breed.* **2010**, *129*, 407–411. [CrossRef]

203. Holasová, M.; Dostálová, R.; Fiedlerová, V.; Horáček, J. Variability of lutein content in peas (*Pisum sativum* L.) in relation to the variety, season and chlorophyll content. *Czech. J. Food Sci.* **2009**, *27*, S188–S191. [CrossRef]

204. Marles, M.A.S.; Warkentin, T.D.; Bett, K.E. Genetic abundance of carotenoids and polyphenolics in the hull of field pea (*Pisum sativum* L.). *J. Sci. Food Agric.* **2013**, *93*, 463–470. [CrossRef]

205. Lu, W.; Haynes, K.; Wiley, E.; Clevidence, B. Carotenoid content and colour in diploid potatoes. *J. Am. Soc. Hortic. Sci.* **2001**, *126*, 722–726. [CrossRef]

206. Beyer, P.; Al-Babili, S.; Ye, X.; Lucca, P.; Schaub, P.; Welsch, R.; Potrykus, I. Golden Rice: Introducing the beta-carotene biosynthesis pathway into rice endosperm by genetic engineering to defeat vitamin A deficiency. *J. Nutr.* **2002**, *132*, 506S–510S. [CrossRef]

207. Amorim, E.P.; Vilarinhos, A.D.; Cohen, K.O.; Amorim, V.B.O.; Santos-Serejo, J.A.D.; Silva, S.O.; Pestana, K.N.; Santos, V.J.D.; Paes, N.S.; Monte, D.C.; et al. Genetic diversity of carotenoid-rich bananas evaluated by Diversity Arrays Technology (DArT). *Genet. Mol. Biol.* **2009**, *32*, 96–103. [CrossRef]

208. Fernandez-Orozco, R.; Gallardo-Guerrero, L.; Hornero-Me'ndez, D. Carotenoid profiling in tubers of different potato (*Solanum* sp.) cultivars: Accumulation of carotenoids mediated by xanthophyll esterification. *Food Chem.* **2013**, *141*, 2864–2872. [CrossRef] [PubMed]

209. Shrestha, A.K.; Arcot, J.; Paterson, J. Folate assay of foods by traditional and trienzyme treatments using cryoprotected *Lactobacillus casei*. *Food Chem.* **2000**, *71*, 545–552. [CrossRef]

210. Chew, S.C.; Loh, S.P.; Khor, G.L. Determination of folate content in commonly consumed Malaysian foods. *Int. Food Res. J.* **2012**, *19*, 189–197.

211. Fajardo, V.; Alonso-Aperte, E.; Varela-Moreiras, G. Total folate content in ready-to eat vegetable meals from the Spanish market. *J. Food Compos. Anal.* **2017**, *64*, 223–231. [CrossRef]

212. De Brouwer, V.; Storozhenko, S.; Van De Steene, J.C.; Wille, S.M.; Stove, C.P.; Van Der Straeten, D.; Lambert, W.E. Optimisation and validation of a liquid chromatography-tandem mass spectrometry method for folates in rice. *J. Chromatogr. A* **2008**, *1215*, 125–132. [CrossRef]

213. De Brouwer, V.; Storozhenko, S.; Stove, C.P.; Van Daele, J.; Van der Straeten, D.; Lambert, W.E. Ultra-performance liquid chromatography–tandem mass spectrometry (UPLC-MS/MS) for the sensitive determination of folates in rice. *J. Chromatogr. B* **2010**, *878*, 509–513. [CrossRef]

214. Camara, J.E.; Lowenthal, M.S.; Phinney, K.W. Determination of fortified and endogenous folates in food-based standard reference materials by liquid chromatography-tandem mass spectrometry. *Anal. Bioanal. Chem.* **2013**, *405*, 4561–4568. [CrossRef]

215. Khanal, S.; Xue, J.; Khanal, R.; Xie, W.; Shi, J.; Pauls, K.P.; Navabi, A. Quantitative trait loci analysis of folate content in dry beans, *Phaseolus vulgaris* L. *Int. J. Agron.* **2013**, *2013*, 1–9. [CrossRef]

216. Konings, E.J.; Roomans, H.H.; Dorant, E.; Goldbohm, R.A.; Saris, W.H.; van den Brandt, P.A. Folate intake of the Dutch population according to newly established liquid chromatography data for foods. *Am. J. Clin. Nutr.* **2001**, *73*, 765–776. [CrossRef]

217. Henderson, G.I.; Perez, T.; Schenker, S.; Mackins, J.; Antony, A.C. Maternaltofetal transfer of 5-methyltetrahydrofolate by the perfused human placental cotyledon: Evidence for a concentrative role by placental folate receptors in fetal folate delivery. *J. Lab. Clin. Med.* **1995**, *126*, 184–203.

218. Scaglione, F.; Panzavolta, G. Folate, folic acid and 5-methyltetrahydrofolate are not the same thing. *Xenobiotica* **2014**, *44*, 480–488. [CrossRef] [PubMed]

219. Hotz, C.; Loechl, C.; de Brauw, A.; Eozenou, P.; Gilligan, D.; Moursi, M.; Munhaua, B.; van Jaarsveld, P.; Carriquiry, A.; Meenakshi, J.V. A large-scale intervention to introduce orange sweet potato in rural Mozambique increases vitamin A intakes among children and women. *Br. J. Nutr.* **2012**, *108*, 163–176. [CrossRef] [PubMed]

220. Hotz, C.; Loechl, C.; Lubowa, A.; Tumwine, J.K.; Ndeezi, G.; Nandutu Masawi, A.; Baingana, R.; Carriquiry, A.; de Brauw, A.; Meenakshi, J.V.; et al. Introduction of β-carotene-rich orange sweet potato in rural Uganda results in increased vitamin A intakes among children and women and improved vitamin A status among children. *J. Nutr.* **2012**, *142*, 1871–1880. [CrossRef] [PubMed]

221. Gannon, B.; Kaliwile, C.; Arscott, S.A.; Schmaelzle, S.; Chileshe, J.; Kalungwana, N.; Mosonda, M.; Pixley, K.; Masi, C.; Tanumihardjo, S.A. Biofortified orange maize is as efficacious as a vitamin A supplement in Zambian children even in the presence of high liver reserves of vitamin A: A community-based, randomized placebo-controlled trial. *Am. J. Clin. Nutr.* **2014**, *100*, 1541–1550. [CrossRef] [PubMed]

222. Finkelstein, J.L.; Mehta, S.; Udipi, S.A.; Ghugre, P.S.; Luna, S.V.; Wenger, M.J.; Murray-Kolb, L.E.; Przybyszewski, E.M.; Haas, J.D. A randomized trial of iron-biofortified pearl millet in school children in India. *J. Nutr.* **2015**, *145*, 1576–1581. [CrossRef]

223. Haas, J.; Luna, S.V.; Lung'aho, M.G.; Ngabo, F.; Wenger, M.; Murray-Kolb, L.; Beebe, S.; Gahutu, J.; Egli, I. Consuming iron biofortified beans significantly improved iron status in Rwandan women after 18 weeks. *J. Nutr.* **2017**, *146*, 1586–1592. [CrossRef]

224. Brinch-Pedersen, H.; Borg, S.; Tauris, B.; Holm, P.B. Molecular genetic approaches to increasing mineral availability and vitamin content of cereals. *J. Cereal Sci.* **2007**, *46*, 308–326. [CrossRef]

225. Warkentin, T.D.; Delgerjav, O.; Arganosa, G.; Rehman, A.U.; Bett, K.E.; Anbessa, Y.; Rossnagel, B.; Raboy, V. Development and characterization of low-phytate pea. *Crop Sci.* **2012**, *52*, 74–78. [CrossRef]

226. Liu, X.; Glahn, R.P.; Arganosa, G.C.; Warkentin, T.D. Iron bioavailability in low phytate pea. *Crop Sci.* **2015**, *55*, 320–330. [CrossRef]

227. Shewry, P.R.; Ward, J.L. Exploiting genetic variation to improve wheat composition for the prevention of chronic diseases. *Food Energy Secur.* **2012**, *1*, 47–60. [CrossRef]

228. Shi, J.; Wang, H.; Hazebroek, J.; Ertl, D.S.; Harp, T. The maize low-phytic acid 3 encodes a myo-inositol kinase that plays a role in phytic acid biosynthesis in developing seeds. *Plant J.* **2005**, *42*, 708–719. [CrossRef] [PubMed]

229. Brinch-Pedersen, H.; Sørensen, L.D.; Holm, P.B. Engineering crop plants: Getting a handle on phosphate. *Trends Plant Sci.* **2002**, *7*, 118–125. [CrossRef]

230. Campion, B.; Sparvoli, F.; Doria, E.; Tagliabue, G.; Galasso, I.; Fileppi, M.; Bollini, R.; Nielsen, E. Isolation and characterisation of an *lpa* (low phytic acid) mutant in common bean (*Phaseolus vulgaris* L.). *Theor. Appl. Genet.* **2009**, *118*, 1211–1221. [CrossRef]

231. Panzeri, D.; Cassani, E.; Doria, E.; Tagliabue, G.; Forti, L.; Campion, B.; Bollini, R.; Brearley, C.A.; Pilu, R.; Nielsen, E.; et al. A defective ABC transporter of the MRP family, responsible for the bean *lpa1* mutation, affects the regulation of the phytic acid pathway, reduces seed myo-inositol and alters ABA sensitivity. *New Phytol.* **2011**, *191*, 70–83. [CrossRef] [PubMed]

232. Shunmugam, A.S.K.; Bock, C.; Arganosa, G.C.; Georges, F.; Gray, G.R.; Warkentin, T.D. Accumulation of phosphorus-containing compounds in developing seeds of low-phytate pea (*Pisum sativum* L.) mutants. *Plants* **2015**, *4*, 1–26. [CrossRef]

233. Cominelli, E.; Confalonieri, M.; Carlessi, M.; Cortinovis, G.; Daminati, M.G.; Porch, T.G.; Losa, A.; Sparvoli, F. Phytic acid transport in *Phaseolus vulgaris*: A new low phytic acid mutant in the *PvMRP1* gene and study of the *PvMRPs* promoters in two different plant systems. *Plant Sci.* **2018**, *270*, 1–12. [CrossRef]

234. Petry, N.; Egli, I.; Campion, B.; Nielsen, E.; Hurrell, R. Genetic reduction of phytate in common bean (*Phaseolus vulgaris* L.) seeds increases iron absorption in young women. *J. Nutr.* **2013**, *143*, 1219–1224. [CrossRef]

235. Vermeris, W.; Nicholson, R. *Phenolic Compound Biochemistry*; Springer: Dordrecht, The Netherlands, 2006.

236. Lattanzio, V.; Lattanzio, V.M.T.; Cardinali, A. Role of phenolics in the resistance mechanisms of plants against fungal pathogens and insects. In *Phytochemistry: Advances in Research*; Imperato, F., Ed.; Research Signpost: Kerala, India, 2006; pp. 23–67.

237. Manach, C.; Scalbert, A.; Morand, C.; Rémésy, C.; Jimenez, L. Polyphenols: Food sources and bioavailability. *Am. J. Clin. Nutr.* **2004**, *79*, 727–747. [CrossRef]

238. Scalbert, A.; Manach, C.; Morand, C.; Remesy, C. Dietary polyphenols and the prevention of diseases. *Crit. Rev. Food Sci. Nutr.* **2005**, *45*, 287–306. [CrossRef]

239. Hart, J.J.; Tako, E.; Kochian, L.V.; Glahn, R.P. Identification of black bean (*Phaseolus vulgaris* L.) polyphenols that inhibit and promote iron uptake by Caco-2 cells. *J. Agric. Food Chem.* **2015**, *63*, 5950–5956. [CrossRef]

240. Jha, A.B.; Purves, R.W.; Elessawy, F.M.; Zhang, H.; Vandenberg, A.; Warkentin, T.D. Polyphenolic profile of seed components of white and purple flower pea lines. *Crop Sci.* **2019**, *59*, 2711–2719. [CrossRef]

241. Gregorio, G.B.; Senadhira, D.; Htut, H.; Graham, R.D. Breeding for trace mineral density in rice. *Food Nutr. Bull.* **2000**, *21*, 382–386. [CrossRef]

242. Haas, J.D.; Beard, J.L.; Murray-Kolb, L.E.; del Mundo, A.M.; Felix, A.; Gregorio, G.B. Iron-biofortified rice improves the iron stores of non-anemic Filipino women. *J. Nutr.* **2005**, *135*, 2823–2830. [CrossRef] [PubMed]

243. Tsakirpaloglou, N.; Mallikarjuna Swamy, B.P.; Acuin, C.; Slamet-Loedin, I.H. Biofortified Zn and Fe rice: Potential contribution for dietary mineral and human health. In *Nutritional Quality Improvement in Plants. Concepts and Strategies in Plant Sciences*; Jaiwal, P., Chhillar, A., Chaudhary, D., Jaiwal, R., Eds.; Springer: Cham, Switzerland, 2019; pp. 1–24.

244. Winger, R.; Konig, J.; House, D. Technological issues associated with iodine fortification of foods. *Trends Food Sci. Technol.* **2008**, *19*, 94–101. [CrossRef]

245. Nayyar, H.; Kaur, S.; Singh, S.; Upadhyaya, H.D. Differential sensitivity of Desi (small-seeded) and Kabuli (large-seeded) chickpea genotypes to water stress during seed filling: Effects on accumulation of seed reserves and yield. *J. Sci. Food Agric.* **2006**, *86*, 2076–2082. [CrossRef]

246. Tar'an, B.; Warkentin, T.D.; Tullu, A.; Vandenberg, A. Genetic mapping of ascochyta blight resistance in chickpea (*Cicer arietinum* L.) using a simple sequence repeat linkage map. *Genome* **2007**, *50*, 26–34. [CrossRef]

247. Bueckert, R.A.; Wagenhoffer, S.; Hnatowich, G.; Warkentin, T.D. Effect of heat and precipitation on pea yield and reproductive performance in the field. *Can. J. Plant Sci.* **2015**, *95*, 629–639. [CrossRef]

248. Atienza, S.G.; Palomino, C.; Gutiérrez, N.; Alfaro, C.M.; Rubiales, D.; Torres, A.M.; Ávila, C.M. QTLs for ascochyta blight resistance in faba bean (*Vicia faba* L.): Validation in field and controlled conditions. *Crop Pasture Sci.* **2016**, *67*, 216–224. [CrossRef]

249. Jha, A.B.; Tar'an, B.; Stonehouse, R.; Warkentin, T.D. Identification of QTLs associated with improved resistance to ascochyta blight in an interspecific pea recombinant inbred line population. *Crop Sci.* **2016**, *56*, 2926–2939. [CrossRef]

250. Jha, A.B.; Gali, K.K.; Tar'an, B.; Warkentin, T.D. Fine mapping of QTLs for ascochyta blight resistance in pea using heterogeneous inbred families. *Front. Plant Sci.* **2017**, *8*, 765. [CrossRef]

© 2020 by the authors. Licensee MDPI, Basel, Switzerland. This article is an open access article distributed under the terms and conditions of the Creative Commons Attribution (CC BY) license (http://creativecommons.org/licenses/by/4.0/).

Article

Gene Expression Pattern of Vacuolar-Iron Transporter-Like (VTL) Genes in Hexaploid Wheat during Metal Stress

Shivani Sharma [1,2], Gazaldeep Kaur [1], Anil Kumar [1], Varsha Meena [1], Hasthi Ram [1], Jaspreet Kaur [2] and Ajay Kumar Pandey [1,*]

[1] Department of Biotechnology, National Agri-Food Biotechnology Institute, Sector 81, Knowledge City, Mohali, Punjab 140306, India; shivani@nabi.res.in (S.S.); gazaldeep@nabi.res.in (G.K.); anilkumar@nabi.res.in (A.K.); meenavarsha8@nabi.res.in (V.M.); hasthi@nabi.res.in (H.R.)

[2] University Institute of Engineering and Technology, Sector 25, Panjab University, Chandigarh, Punjab 160015, India; jaspreet_uiet@pu.ac.in

* Correspondence: pandeyak@nabi.res.in or pandeyak1974@gmail.com; Tel.: +91-1725221124

Received: 29 November 2019; Accepted: 31 January 2020; Published: 11 February 2020

Abstract: Iron is one of the important micronutrients that is required for crop productivity and yield-related traits. To address the Fe homeostasis in crop plants, multiple transporters belonging to the category of major facilitator superfamily are being explored. In this direction, earlier vacuolar iron transporters (VITs) have been reported and characterized functionally to address biofortification in cereal crops. In the present study, the identification and characterization of new members of vacuolar iron transporter-like proteins (VTL) was performed in wheat. Phylogenetic distribution demonstrated distinct clustering of the identified *VTL* genes from the previously known *VIT* genes. Our analysis identifies multiple *VTL* genes from hexaploid wheat with the highest number genes localized on chromosome 2. Quantitative expression analysis suggests that most of the *VTL* genes are induced mostly during the Fe surplus condition, thereby reinforcing their role in metal homeostasis. Interestingly, most of the wheat *VTL* genes were also significantly up-regulated in a tissue-specific manner under Zn, Mn and Cu deficiency. Although, no significant changes in expression of wheat *VTL* genes were observed in roots under heavy metals, but *TaVTL2*, *TaVTL3* and *TaVTL5* were upregulated in the presence of cobalt stress. Overall, this work deals with the detailed characterization of wheat *VTL* genes that could provide an important genetic framework for addressing metal homeostasis in bread wheat.

Keywords: micronutrient uptake; *Triticum aestivum* L.; Zinc transport; biofortification; Iron deficiency

1. Introduction

Successful micronutrient biofortification of crops through biotechnology requires detailed knowledge of complex homeostatic mechanisms that tightly regulate the micronutrient concentrations in plants. Iron (Fe) is one of the important micronutrients that is involved in multiple important cellular and physiological processes in plants [1–3]. Some of the important functions include its importance in photosynthesis, nitrogen fixation and respiration [4,5]. Although Fe may be present in the soil, yet due to alkaline rhizospheric conditions or unfavorable circumstances, it is not being efficiently taken up by plants [6–9]. Moreover, the Fe is mobilized through a multistep process that overcomes transport bottlenecks and eventually is loaded in the developing grains [10–13]. Researchers worldwide are utilizing multiple approaches to either enrich Fe in grains or their storage with enhanced bioavailability [14–17]. To improve Fe content in cereal grains, multiple transporters and chelators have been targeted through multiple molecular approaches [14,15,18]. A number of additional micronutrient transporters have been identified, those are good candidates for micronutrient biofortification, including transporters belonging to the major facilitator superfamily (MFS) gene family [19,20]. Limited evidences are available that have performed molecular characterization of wheat genes or gene families those which are specifically involved in Fe and Zinc (Zn) homeostasis. Recent reports are emerging for the identification of few functional gene families belonging to, yellow stripe like transporters [21], nicotianamine synthase (NAS), deoxymugineic acid synthase (DMAS) [22], yet many genes families remained to be characterized in hexaploid wheat. Similarly, other wheat genes including Zinc–Induced Facilitator-Like Family (ZIFL) transporters have been characterized for their role in mobilizing the uptake of micronutrient such as Fe and Zn [23]. Most of these gene families are highly upregulated in roots subjected to Fe starvation conditions [24,25]. These works identify some of the important candidate genes as an important resource to strategize approaches for micronutrient biofortification in wheat [26].

Fe storage in seeds gets compartmentalized in major subcellular organelles including chloroplasts and vacuoles. For example, 95% of the iron is stored in vacuoles in the *Arabidopsis* seeds [27]. Vacuoles are an important site for Fe mobilization wherein, they are bound to various chelators like phytic acid, nicotianamine and other organic acids etc. Therefore, uptake of Fe into vacuoles could be an alternate strategy to enhance total micronutrient content with a minimized tradeoff for its toxicity in the tissue. To design such strategy, the role of vacuolar transporters needs to be addressed and exploited [14,28]. Previously, vacuolar iron transporters (VIT) were shown to be play an important role in maintaining Fe in the optimal physiological range and prevent cellular toxicity [14]. *VIT* genes from multiple plant species have been characterized and assessed for their ability to enhance Fe content in cereal crops [15]. These *VIT* genes show high homology with a small family of nodulin like protein containing a CCC-1 (Ca^{2+}-Sensitive Cross Complementer) like domain with yeast CCC1p1 [29]. CCC-1 like the domain was initially discovered in yeast encoded for the vacuolar iron transporter in yeast. Furthermore, mutant *ccc1* cells show increased sensitivity to external iron [27,30] *AtVIT1* is one of the early characterized genes showing the presence of CCC-1 like domain and transport of iron to vacuoles [27].

Utilizing the bioinformatics resources, subsequent studies led to the identification of many vacuolar iron transporters-like (VTL) proteins from different plant species. Model species, *Arabidopsis* genome encodes five VTL proteins and overexpression of the few genes have shown increased Fe content in seeds. AtVIT1 protein can transport iron into the vacuoles to counter the toxicity and support the seedling development under enhanced iron conditions [29,31].

Wheat is an important crop that is consumed in many developing countries, including India and is therefore being targeted for trait improvement for nutritional quality. Therefore, the characterization of vacuolar transporters in an important crop such as wheat becomes a prerequisite to address the global issue of biofortification. In the current work genome-wide identification of wheat *VTL* genes was performed. Further, expression studies during different regimes of Fe, Zn and multiple heavy metals was done to gain insight for the regulation of wheat *VTL* genes in a tissue-specific manner.

2. Results

2.1. Identification, Phylogenetic Analysis and Genomic Distribution of Wheat VTL Genes

Thirty-one wheat VIT family sequences were identified based on Ensembl Pfam search and bidirectional BLAST analysis (Table S1). Subsequently, to study the phylogenetic relationship among VIT and VTL family protein sequences from wheat, *Brachypodium*, maize, rice, *Arabidopsis* and *S. cerevisiae*, an unrooted neighbour-joining tree was constructed. This analysis separated the sequences into two distinct clades representing VTL and VIT proteins. This also led to the clustering of the wheat VIT family members into 8 VIT and 23 VTL sequences (Figure 1, Table S2). Due to the occurrence of homoeologs, the 23 VTL sequences were grouped into 4 *VTL* genes and named as *TaVTL1, TaVTL2, TaVTL4* and *TaVTL5* that corresponds to the rice orthologs followed by the chromosome number. None of the orthologs in wheat showed high confidence similarity with rice vacuolar iron transporter homolog 3. *TaVTL1* and *4* were found to have three homoeologs, while *TaVTL2* had four. In contrast, the phylogenetic analysis grouped 13 highly similar sequences together with rice vacuolar iron transporter homolog 5, these were named as *TaVTL5* (Figure S1).

TaVIT1 and *TaVIT2* have already been reported earlier [14]. Interestingly, another new wheat *VIT* with two homoeologs on chromosome 7 (sub-genomes A and D) was identified (referred as *TaVIT3*). *VIT* genes were located on chromosome groups 2, 5 and 7, while *VTL* genes were on chromosome groups 2, 4, and 6 with a maximum contribution from chromosome 2. Nine *VTL* genes were present in the B sub-genome, while seven each on A and D sub-genomes. The maximum number of VTL sequences were located on chromosome 2B (Figure 2A).

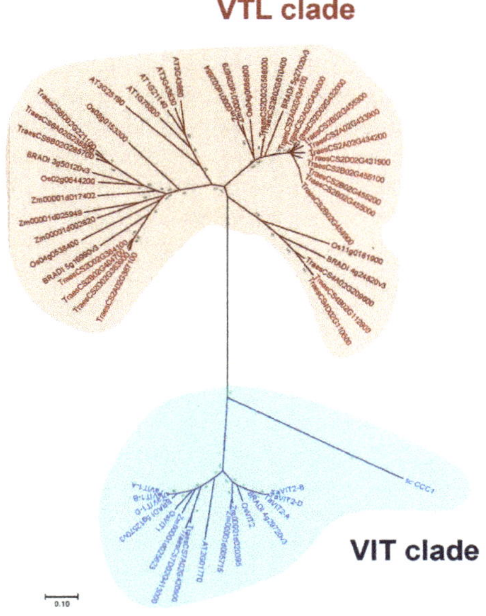

Figure 1. Phylogenetic analysis showing separation of vacuolar iron transporter (VIT) family in *Arabidopsis, Brachypodium, Oryza sativa, Zea mays* and *Triticum aestivum* into two distinct clades; vacuolar iron transporter-like (VTL) clade and VIT clade. The neighbour-joining phylogenetic tree was generated using MEGA. The numbers represent bootstrap values from 1000 replicates.

2.2. Gene, Protein Structure and Subcellular Localization

VIT genes in wheat have three and four intronic and exonic regions respectively, while *VTL* genes have a single exon each with the absence of any introns (Figure 2B), clearly dividing the *VIT* family into two sub-families based on gene structure also. CDS length was found to be varying from 657 to 747 nucleotides for wheat *VIT* genes. The CDS length for *VTL* genes was ranging from 549 to 810 nucleotides except for *TaVTL5-2A_3* that was 378 nucleotides long. The short length of one VTL gene is due to the missing sequence information at the stop site. The length of TaVIT peptides ranged from 218 to 256 while TaVTL protein length varied from 125 to 269 amino acids. The division of VIT and VTL proteins was also evident from the sub-cellular localization (Table S2); while TaVIT proteins were predicted to be predominantly localized on the plasma membrane and chloroplast thylakoid membrane, maximum TaVTL proteins were predicted to be present on the vacuolar membrane (87%). TaVTL4-4A was predicted to be localized on plasma membrane. VIT proteins had 3–4 predicted trans-membrane (TM) domains. TaVTL1, 2 and 4 had five TM domains majorly, except for TaVTL4-4B which was predicted to have 6 TM domains. Only TaVTL5-2D_3 had five TM domains; other paralogs/homoeologs of TaVTL5 had lesser number of TM domains probably due to gene duplication events or missing information. To summarize, TaVTLs have five TM domains predominantly, which are depicted in Table S2. VIT1 from *Eucalyptus grandis* (EgVIT1) crystal structure was deciphered recently [32] that was used to confirm the VIT family protein topology prediction using Phobius [33]. EgVIT1 was predicted to have only three TM domains while the crystal structure stated the presence of five TM domains. Therefore, VIT, as well as VTL protein sequences from wheat, were aligned to EgVIT1 to see the possible TM domains in addition to those predicted by Phobius (Figure S2).

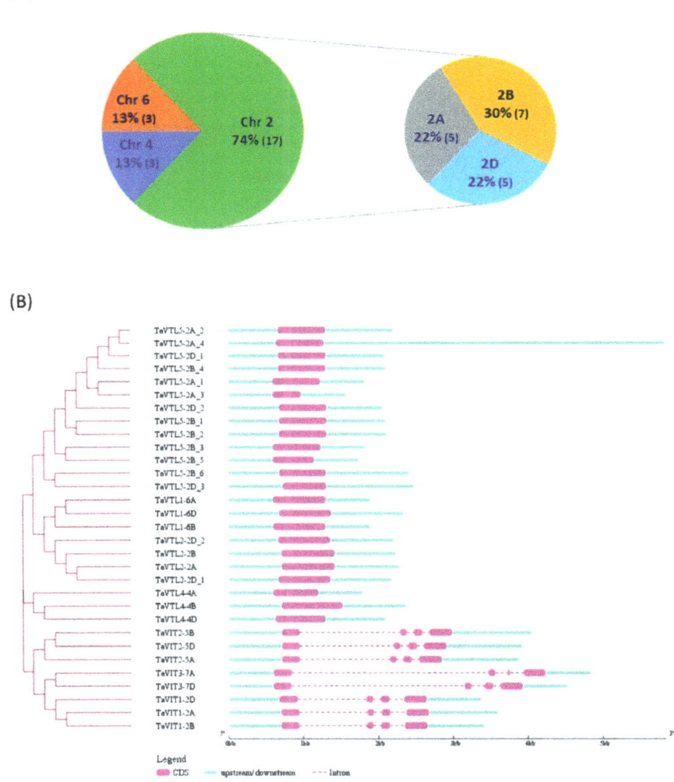

Figure 2. Genomic distribution and exon intron arrangements of *VTL* genes. (**A**) *VTL* genes genomic distribution. Wheat *VTL* genes were present on chromosome groups 2, 4 and 6 with maximum *VTL* genes on chromosome group 2, which was selected to show the *VTL* gene distribution on 2A, 2B and 2D chromosomes. (**B**) Genomic structure for wheat *VTL* and *VIT* genes. The intron-exon arrangement was identified using Gene Structure Display Server (GSDS). Exons and introns are represented using pink boxes and cyan lines, respectively. The scale determines the size of the genomic regions.

2.3. Conserved Domain and Motif Analysis

All the *VIT* and *VTL* genes were found to have the typical CCC1-like superfamily domains of yeast, which were demonstrated earlier for the iron and manganese transport from the cytosol to vacuole. Motif analysis using MEME webserver suggested that motifs 6, 9, and 10 are VIT specific with exceptions for motif 10 been absent in TaVIT3 and motif 6 absent in TaVIT3-7A. Similarly, motifs 5, 7, 8, 11 to 14 are VTL specific with the exception that motif 5 was absent in TaVTL5-2B_3 and TaVTL5-2B_5, where motif 7 was specific for TaVTL5 sequences except in TaVTL5-2A_3, TaVTL5-2B_6, TaVTL5-2D_3. Motif 8 was specific for TaVTL1, 2 and 4. Motif 11 for TaVTL1 and 2. Motifs 12 and 13 were unique for TaVTL2, whereas Motif 14 was present only in TaVTL1 (Figure 3, Table S3).

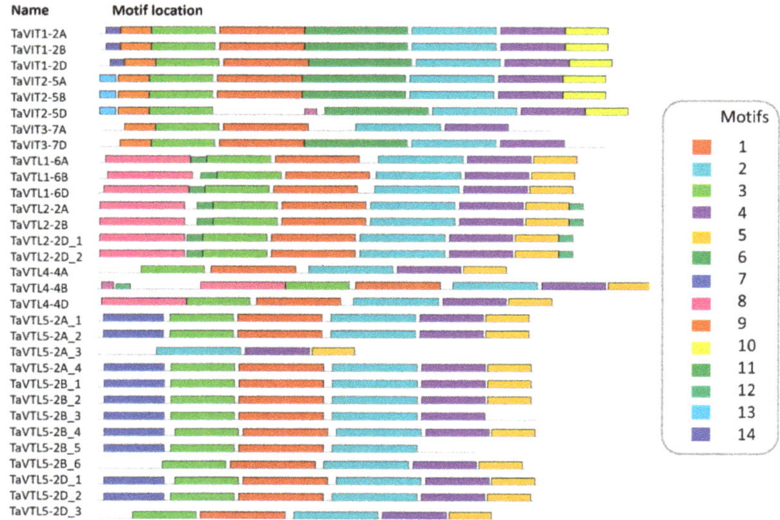

Figure 3. Conserved motifs identified for TaVIT and TaVTL proteins using MEME suite 5.1.0. The colored rectangles on each sequence represent specific conserved motifs numbered 1 through 14, as depicted by the color codes in the box.

2.4. Expression of Wheat VTL Genes under Fe Deficiency and Surplus Condition

To check the regulation of *VTL* genes at the transcriptional level, the promoters for the wheat *VTL* genes were scanned for the cis-elements responsive for Fe and heavy metals. The analysis revealed the presence of multiple such sequences, including iron-deficiency-responsive element 1 (IDE1), metal response element (MRE), heavy metal responsive element (HMRE) and iron-related bHLH transcription factor 2 (IRO2) binding site (Table S4). In the most abundant category, iron-deficiency-responsive element 1 (IDE1) was predominant. Interestingly, the IRO2 binding site was present only in the regulatory region of *TaVTL2B/D*. These observations suggest that *VTL* expression could be regulated by the presence-absence of specific metals including micronutrients such as Fe and Zn.

Previously, *VTL* genes were reported to have differential expression patterns under the changing regimes of Fe and Zn [31]. Therefore, we tested if wheat *VTL* genes could respond at the transcript level when subjected to changing Fe concentration. The expression in roots and shoots of wheat seedlings was measured after subjecting them for three and six days of starvation. Our expression analysis suggests that in roots all the *VTL* genes (*TaVTL1, TaVTL2, TaVTL4* and *TaVTL5*) were downregulated at both the days, whereas, only *TaVTL5* was upregulated at six days of starvation (Figure 4A).

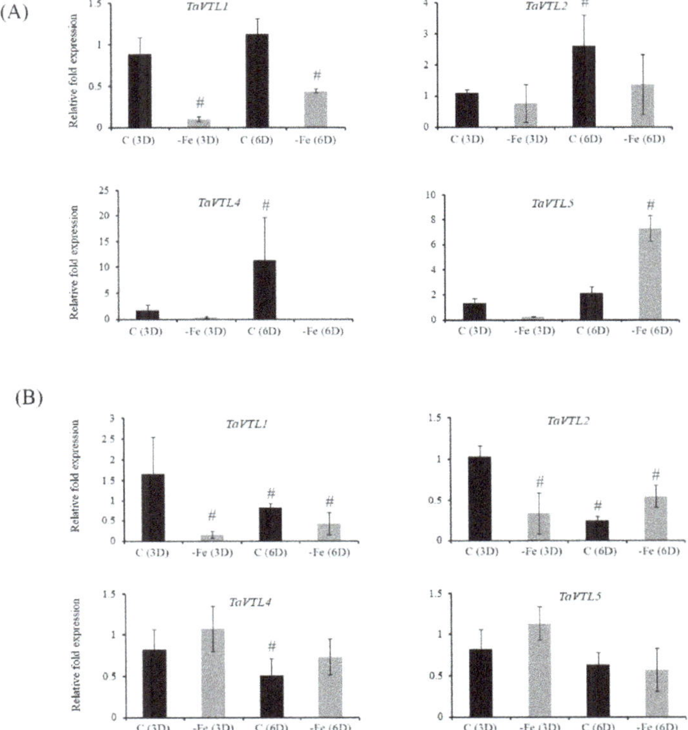

Figure 4. Tissue-specific qRT-PCR expression analysis of wheat VTL genes during Fe deficiency (-Fe) and in the control (C) conditions. Wheat seedlings were subjected to Fe deficiency for three and six days represented as -Fe(3D) and -Fe(6D). The controls for the respective time points are represented as C(3D) and C(6D). (**A**) Fold expression analysis was performed in roots and (**B**) in shoots. 2 µg of total RNA for the cDNA preparation. Relative fold expression levels were calculated relative to C(3D). C_t values were normalized using wheat *ARF1* as an internal control. Vertical bars represent the standard deviation. # represents the significant difference at $p < 0.05$ with respect to their respective control treatments.

Similarly, in shoots also all the expression of wheat *VTL* genes was suppressed except for *TaVTL2* that was upregulated only on six days post starvation (Figure 4B). These expression data demonstrate that under Fe deprivation *VTL* gene expression are negatively regulated in wheat seedling. Transcriptomic sequencing data from wheat seedlings after 20 days of Fe starvation (SRP189420) were also used to check expression response upon Fe starvation. Categorically, *TaVTL5* group genes were seen to be upregulated upto 12-fold, with *TaVTL5-2B_6* showing upregulation of ~60 fold, although the expression was not very high (Figure S3).

Next, we performed the gene expression analysis under the excess Fe regime. This was done to test if wheat *VTL* genes could be potentially involved in detoxification of excess Fe. Interestingly, we observed a significant up-regulation of all the *TaVTL* genes in roots at both the time points. Out of all, *TaVTL4* showed the highest fold gene expression (~100 fold) when compared to its control (Figure 5A).

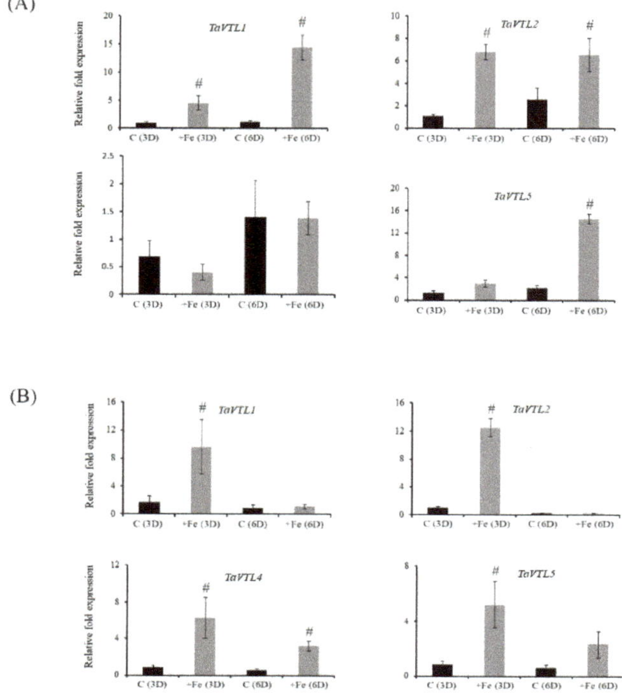

Figure 5. Tissue-specific qRT-PCR expression analysis of wheat VTL genes during Fe surplus (+Fe) and in the control (C) conditions. Wheat seedlings were subjected to Fe surplus for three and six days represented as +Fe(3D) and +Fe(6D). The controls for the respective time points are represented as C(3D) and C(6D). (**A**) Fold expression analysis was performed in roots and (**B**) in shoots. 2 μg of total RNA for the cDNA preparation and relative fold expression levels were calculated relative to C(3D). C_t values were normalized using wheat *ARF1* as an internal control. Vertical bars represent the standard deviation. # represents the significant difference at $p < 0.05$ with respect to their respective control treatments.

TaVTL2 show very early and high expression response, whereas both *TaVTL1* and *TaVTL5* were highly expressed at six days of treatment. At this time their gene expression level was more than ~14 fold compared to control. In contrast, in shoots most of the wheat *VTL* genes were expressed at the three days of treatment with *TaVTL1* and *TaVTL2* showing the transcript accumulation of 8–14 folds with respect to their control (Figure 5B).

2.5. Manganese, Zinc and Copper Deficiency Causes Differential Changes in VTL Expression

Wheat *VTL* genes showed high similarity to previously known *VIT* genes. In addition to Fe, *VIT* genes are known to be affected by the perturbed concentration of Mn [14]. Since many of these cation transporters are known for their reduced substrate specificity [34,35], therefore, expression of wheat *VTL* genes during Zn, Cu and Mn deprivation was also studied (Figure S4A). In general, during the changing regimes of Zn and Mn, wheat *VTL* genes showed specific expression in a tissue-specific manner (Figure 6). *TaVTL2* was the only gene showing enhanced accumulation of its transcript under Zn deficiency in both root and shoot tissue, whereas *TaVTL1* and *TaVTL5* showed high expression in roots only under Zn deficiency (Figure 6A and B). No significant changes in the expression of *TaVTL4* were observed for the studied time point under the changing Zn regime. In contrast, no induction of wheat *VTL* genes was observed in roots under Mn deficiency with respect to its control, whereas, in shoots, *TaVTL2* and *TaVTL4* showed high transcript accumulation (Figure 6A). Under Cu deficiency, all *VTL* genes showed an induced expression in shoots while only two of the genes, including *TaVTL1* and *TaVTL2* were upregulated in roots.

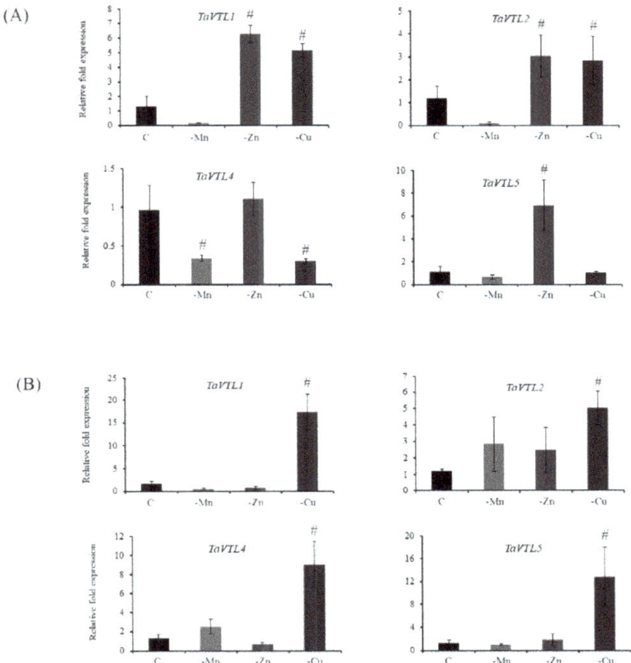

Figure 6. Tissue-specific qRT-PCR expression analysis of wheat *VTL* genes during Mn (-Mn), Zn (-Zn) and Cu (-Cu) deficiency with respect to the control (C) conditions. (**A**) Fold expression analysis was performed in roots and (**B**) in shoots. 2 µg of total RNA was used for the cDNA preparation and relative fold expression levels were calculated relative to control tissue (C). The C_t values were normalized using wheat *ARF1* as an internal control. Vertical bars represent the standard deviation. # represents the significant difference at $p < 0.05$ with respect to Control tissue.

2.6. Heavy Metal (Ni, Cd and Co) Mediated Expression of VTL Genes

To check the effect of the heavy metal stress on the gene expression pattern, wheat seedlings were subjected to treatment with Ni, Cd and Co and expression of *VTL* genes was performed. In the treated plants, decreased growth of the shoot and root length was observed, suggesting that heavy metals could affect the plant performance (Figure S4A). In general, the presence of heavy metals led to significant retardation in the growth of roots and shoots, thereby impacting the total plant growth (Figure S4B). Interestingly, none of the wheat *VTL* genes showed enhanced expression in roots after 15 days of heavy metal exposure, but downregulation was observed for *TaVTL1*, *TaVTL4* and *TaVTL5* (Figure 7A). In shoots, only Co stress could influence the gene expression when compared to the control. Only, *TaVTL2*, *TaVTL4* and *TaVTL5* genes were upregulated during the Co stress as compared to control shoot samples (Figure 7B). Altogether, this suggests the metal-specific expression of *VTL* genes in a tissue-specific manner. The previously reported wheat *VIT* genes showed grain specific expression data. Surprisingly, *VTL* genes showed very low or no expression in grains or their tissue parts, suggesting their probable roles in the specific organs of the plants (Figure S5).

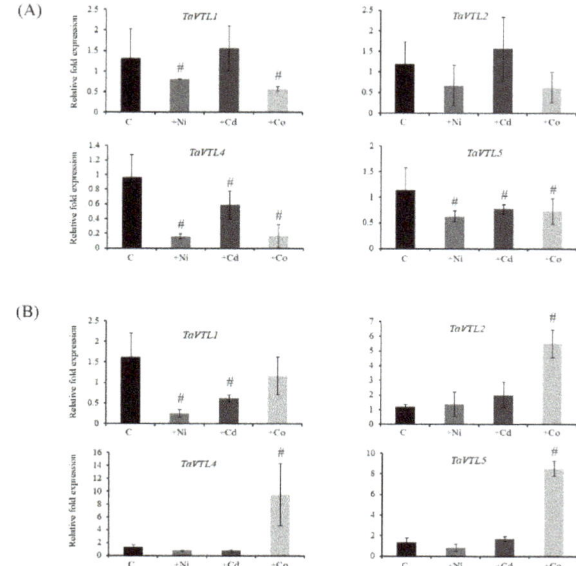

Figure 7. Tissue-specific qRT-PCR expression analysis of wheat *VTL* genes upon heavy metal treatments. Wheat seedlings were exposed to Ni (+Ni, 50 µm), Cd (+Cd, 50 µm) and Co (+Co, 50 µm). Control seedlings (C) without any exposure to heavy metals were compared with the treated ones. (**A**) Fold expression analysis was performed in roots and (**B**) in shoots. 2 µg of total RNA for the cDNA preparation and relative fold expression levels were calculated relative to control samples. C_t values were normalized using wheat *ARF1* as an internal control. Vertical bars represent the standard deviation. # represents the significant difference at $p < 0.05$ with respect to their respective control treatments.

3. Discussion

Fluctuation in the nutrient availability in the soil results in suitable adaptations by the plants. In general, plants rely on different physiological and molecular processes to minimize nutrient stress [36]. In this regard, MFS gene family plays an important role to provide the tolerance as well as mobilization of important minerals, including micronutrient translocation to the foliar parts including seeds [19]. In this study, the characterization of VTL was done in the hexaploid wheat. Our data reinforce the importance of *VTL* genes for their roles during metal homeostasis and substantiated them as a good candidate for micronutrient biofortification in cereal crops such as wheat and rice.

MFS family has been widely explored for its role as metal transporters and providing the necessary support for multiple functions in plants [37]. Previously, five *VTL* genes were reported in *Arabidopsis* and rice for this sub-class. Our study in wheat resulted in the identification of a maximum number of *VTL* genes from any crop plants. The high number of genes is due to the presence of multiple homoeologous and occurrence of the duplication of multiple wheat *VTL* genes. Interestingly, chromosome 2 having the highest number of wheat *VTL* genes has been linked with multiple quantitative trait loci (QTL) for the high grain Fe and Zn content [38]. Further dissection is required in this direction to identify if any of the wheat VTL could be linked with the loading of micronutrient in grains. Based on our expression analysis and the support from the previous studies, it could be suggested that *VTL* genes could also be involved in providing the tolerance to high levels of Fe and Zn in the soils [31]. In fact, the predicted localization data indicate that VTL could be localized at either the plasma membrane or the vacuolar membrane (Table S2). AtVTL1 was reported to be localized in the vacuolar membrane and others been associated with the plasma membrane [31]. Our PSORT analysis suggests that most of the wheat VTL proteins are localized in the vacuolar membrane, thus making them a suitable candidate for sequestering micronutrients such as Fe and Zn. AtVTL1 also rescued Δccc1 function in yeast by catalysing Fe uptake [31]. Vacuoles are the prime sites for the sequestration of micronutrients such as Fe. Whether, any of these predicted vacuolar TaVTL proteins could perform similar function as TaVIT2 for Fe biofortification needs to be studied [14].

The substrate specificity of the metal transporters is a major bottleneck to achieve high Fe and Zn in grains. Manipulating the specificity of these metal transporters to enrich the Fe and Zn remains the major challenge [23,34,39]. Therefore, studying the expression pattern of *VTL* genes in the presence of heavy metals could provide preliminary clues for employing such strategies. Consequently, the study was undertaken to see the influence of other metals like Ni, Cd and Co. The expression of wheat *VTL* genes in roots and shoots suggested an interesting phenomenon, where no significant changes in the expression of their transcript was observed when exposed to either Ni or Cd. In contrast, only Co was able to induce the expression of *TaVTL1*, *TaVTL2*, *TaVTL4* and *TaVTL5* in shoots only (Figure 7B). These data suggest the controlled expression of wheat *VTL* genes in a tissue specific manner. Additionally, besides Fe homeostasis, the vacuolar transporters are also linked with the impaired activity of Zn and Mn transport [39]. In our study only, *TaVTL2* was significantly induced by the Mn deficiency in shoots (Figure 6B). No such effects were observed in roots wherein, all the quantified wheat VTL showed downregulation under Mn deficiency (Figure 6A). Interestingly, *TaVTL1* and *TaVTL2* showed upregulation in roots under Zn deficiency (Figure 6A). The tissue dependent expression patterns of wheat *VTL* genes under the changing regimes of the metal exposure was observed. It has been observed that *VTL* genes from *Arabidopsis* showed transcriptional changes in response to Fe, Zn and Mn [31]. Based on the previous work and our results it could be suggested that regulation of the *VTL* genes at the transcript level could be conserved. Infact, wheat *VIT* genes can also transport Mn and Fe [14]. This suggest that *VTL/VIT* genes could be regulated only by Fe but also by other metals like Mn and Zn. Additionally, our gene expression data also corelate with the presence of multiple cis-elements in the promoter of wheat *VTL* genes. This indicated that primarily *VTL* genes could be involved during metal homeostasis related responses. Coupled with the localization information, it is possible that few wheat VTL proteins could sequester metals in an organelle specific manner.

Herein, a detailed inventory, structure and expression characterization of wheat *VTL* genes was performed. The expression analysis and analysis for the cis-elements in the promoters of wheat *VTL* genes implicated for their role in metals homeostasis including in Fe and Zn. Overall, the work presented here provide an important framework for identifying the molecular and physiological functions in bread wheat.

4. Materials and Methods

4.1. Plant Materials and Growth Conditions

For stress experiments, hexaploid wheat *Triticum aestivum* cv. C-306 (received from Punjab Agriculture University, Ludhiana) was used. Briefly, seeds were surface sterilized using 1.2% sodium hypochlorite prepared in 10% ethanol and then rinsed twice with autoclaved MQ. The seeds were kept on moist filter paper inside a Petri dish and stratified for 1 day at 4 °C in dark condition. Stratified seeds were further kept for germination for six days at room temperature. The remaining seed/endosperm were excised from seedlings at one leaf stage and was shifted to phytaboxes (10–12 seedling/phytabox) containing the Hoagland nutrient media for respective treatments. The standard composition of nutrient media for control includes 6 mM KNO3, 1 mM $MgSO_4$, 2 mM Ca(NO3), 2 mM $NH_4H_22PO_4$, 20 μM Fe-EDTA, 25 μM H3BO3, 2 μM MnSO4, 0.5 μM CuSO4, 2 μM ZnSO4, 50 μM KCl and 0.5 μM Na2MoO4. The variable concentrations used for treatments were excess Fe (+Fe; 200 μm), Fe starvation (-Fe; 2 μm), Zn deficiency (-ZnSO4; 0 μm), Mn deficiency (-MnSO4; 0 μm), Cu deficiency (-CuSO4; 0 μm), Cadmium stress (+Cd; 50 μm) [40], Cobalt stress (+Co; 50 μm) [41] and Nickel stress (+Ni; 50 μm) [42]. The aerobic condition was provided in hydroponics and the media was replaced every alternate day to avoid any contamination and drastic nutrient depletion. The respective roots and shoots samples belonging to iron deficient and sufficient plant groups were collected at three and six days after stress (D). For the rest of the treatments, root and shoot, samples were collected on the 15th Day of treatments. All the experiments were performed in a growth chamber under controlled environmental conditions at 22–24 °C temperature, 65%–70% humidity, at a photoperiod of 16 h day and 8 h night and 300 nm of light.

4.2. Identification of VIT Family and Classification of VTL Genes in Wheat

For the identification of wheat *VTL* genes, the Ensembl database was used to extract *VIT* family genes (Pfam ID: PF01988) for wheat. The identification was confirmed by bidirectional BLAST analysis. VIT family sequences from Arabidopsis, rice, maize, *Brachypodium* were also extracted using Pfam search. The identity of VIT family genes was further validated by confirming the presence of CCC1-like superfamily domain using NCBI-CDD domain search. CCC1 sequence for *S. cerevisiae* was also retrieved from its genome database. To separate out *VTL* genes from *VIT* genes and for further phylogenetic analysis, all the proteins were aligned through MUSCLE alignment and an unrooted neighbor-joining phylogenetic tree with 1000 bootstrap replicates was constructed with all the retrieved sequences. The tree was constructed through MEGA-7 [43]. Rice vacuolar iron transporter homolog 1–5 from UniProt were used for the nomenclature of the 23 *TaVTL* sequences based on the closest orthologs. The naming of the genes indicates the chromosome number and the sub-genome on which they are present.

4.3. Conserved Domains and Motif Detection, Analysis of Gene, Promoter and Protein Structure

Wheat VIT family genes were searched for conserved domains using NCBI-CDD database [44]. MEME suite v5.1.0 was used for further analysis to identify the common conserved motifs for both VIT and VTL proteins. The maximum number of motifs was set to 15 for MEME analysis. Gene structure for VITs and VTLs was studied using (GSDS) (http://gsds.cbi.pku.edu.cn/) [45] using genomic and CDS sequences. Sub-cellular localization and TM domains were predicted using web-based prediction programs Wolf PSORT and Phobius respectively [46]. For promoter analysis, ~2 Kb promoter elements of the corresponding wheat *VTL* genes were surveyed for the presence of the respective cis-elements. The promoter sequence was obtained for the respective genes using the IWGSC

4.4. Total RNA Isolation and cDNA Preparation

The collected root and shoot samples were ground separately in liquid nitrogen. Total RNA from respective samples was extracted by TRIZOL based method. The extraction was followed by the DNase treatment using Turbo DNAfree kit (Invitrogen, Carlsbad, CA, USA) to remove any genomic DNA contamination in the RNA samples. Subsequently, RNA purity was checked and quantified for the preparation of the cDNA. 2 µg of total RNA was used for cDNA synthesis using superScript III First-Strand Synthesis System (Invitrogen, Carlsbad, CA, USA). The cDNA quality was ascertained by using internal control and was further diluted 20X and used for gene expression studies.

4.5. Quantitative-Real Time PCR (qRT-PCR) Expression Analysis

To perform quantitative real time-PCR (qRT-PCR), forward and reverse primers of *TaVTL* genes were designed and used as listed in Table S5. The primers were designed from the conserved region of the all homoeolog of each gene. For *TaVTL5* the primers were designed from the conserved region of nine sequences, the significant conserved region was not found for remaining four homoeologs (TraesCS2B02G454900, TraesCS2B02G610400, TraesCS2D02G431900, TraesCS2D02G588000). qRT-PCR was performed in 7500 Real-Time PCR System (Applied Biosystems, Foster City, CA, USA) using 1/20 times dilution of the respective cDNAs. All qRT-PCR reactions were performed using SYBR Green I (QuantiFast® SYBR® Green PCR Kit, Qiagen, Hilden, Germany) chemistry and ARF (ADP-Ribosylation Factor: *TaARF1*—AB050957.1) as an internal control [40]. The efficiency of the qRT-PCR was checked and melt curve analysis was performed for each of the PCR reactions as per the guidelines. Gene expression analyses was carried out with three biological replicates and 2–3 technical replicates. Relative fold expression of genes was determined based on delta-delta CT-method ($2^{\Delta\Delta CT}$) [47].

4.6. RNA-Seq Expression Analysis for VIT Family Genes

To get the transcript expression levels for VIT family genes under Fe stress, RNAseq data from SRA project ID SRP189420 were utilized to extract transcript expression values (as FPKM) from control as well as Fe starved wheat root samples using the cufflinks pipeline. Subsequently, for expression analysis of *VTL* and *VIT* genes in wheat grain tissue developmental time course [48], expression values as Transcripts Per Kilobase Million (TPM) were retrieved from expVIP database [49]. Expression values from both studies were then used to plot heatmaps using MeV software (mev.tm4.org).

4.7. Statistical Analysis

Excel was used for data analysis. The mean values were calculated form the standard deviation including three technical replicates from at least three biological replicates. Student *t*-tests were used to observe the significant differences between the mean values of treatment and control plants. The significance threshold used was $^{\#}p < 0.05$.

5. Conclusions

The present work led to the identification of high number of *VTL* genes from hexaploid wheat. Because of polyploidization, a very high number of genes from this sub-family was identified. The presence of high number of VTL been restricted to only chromosome 2, 4 and 6 of the wheat genomes. The expression of these gene under metal stress including changes in the presence of Fe and Zn concentrations and exposure to heavy metals reinforce the importance of this gene-family during metal homeostasis. Our work will help in better understanding of the Fe transporters significance in metal homeostasis so as to biofortify wheat.

Supplementary Materials: The following are available online at http://www.mdpi.com/2223-7747/9/2/229/s1, **Table S1**: List of 31 VIT family genes extracted from ensembl biomart using Pfam ID: PF01988. **Table S2**: Subcellular localization (WolfPsort). **Table S3**: Conserved motifs identified in VTL and VIT proteins using MEME suite. The color code, consensus sequence logo, E-value and the number of proteins in which each motif was found are listed in the table. **Table S4**: Metal-responsive cis-elements found in *VTL* and *VIT* gene promoter regions. **Table S5**: List of gene specific primers used for qRT-PCR for *TaVTL* genes. **Figure S1**: Phylogenetic tree for VIT family genes from *Arabidopsis*, *Brachypodium*, *Oryza sativa*, *Zea mays* and *Triticum aestivum*. Sequences were extracted using Pfam ID followed by alignment by Muscle and construction of NJ tree using MEGA software. **Figure S2**: Transmembrane domains in wheat VIT family proteins. Figure shows potential TM domains in *TaVIT* and *TaVTL* proteins, based on the alignment with *EgVIT1* protein, using MUSCLE. **Figure S3**: Heatmap depicting the expression of VIT family genes (VIT and VTL genes) in Control (FPKM_Control) and Fe starved (FPKM_Fe) wheat roots. FPKM values were extracted using Cufflinks pipeline from SRA projectID SRP189420. Increasing intensity of blue colour shows increase in expression as shown by the colour bar above. **Figure S4**: Effect of different metals on the phenotype and growth of wheat seedlings. (A) Phenotype of wheat seedlings showing retarded growth of shoots and roots. (B) Impact of different metals on the growth (in cm) of roots and shoots. **Figure S5**: Heatmap depicting the expression of VIT family genes (VIT and VTL genes) in Control (FPKM_Control) and Fe starved (FPKM_Fe) wheat roots. FPKM values were extracted using Cufflinks pipeline from SRA project ID SRP189420. Increasing intensity of blue colour shows increase in expression as shown by the colour bar above.

Author Contributions: Conceptualization, S.S., A.K.P.; methodology, S.S., A.K., V.M. and A.K.P.; formal analysis, S.S., A.K.P., G.K. and H.R.; investigation, S.S., V.M.; writing—original draft preparation, A.K.P., A.K., J.K.; writing—review and editing, A.K.P., A.K., G.K., and H.R.; visualization, A.K.P.; funding acquisition, A.K.P. All authors have read and agreed to the published version of the manuscript.

Funding: This work was supported by the institutional NABI-CORE grant to AKP.

Acknowledgments: All the authors thank Executive Director, NABI for facilities and support. Support from International Wheat Genome Sequencing Consortium for providing the high-quality wheat genome resources is highly appreciated. DBT-eLibrary Consortium (DeLCON) is acknowledged for providing timely support and access to e-resources for this work.

Conflicts of Interest: The authors declare no conflict of interest.

References

1. Rout, G.R.; Sahoo, S. Role of Iron in Plant Growth and Metabolism. *Rev. Agric. Sci.* **2015**, *3*, 1–24. [CrossRef]
2. Briat, J.F.; Curie, C.; Gaymard, F. Iron utilization and metabolism in plants. *Curr. Opin. Plant Biol.* **2007**, *10*, 276–282. [CrossRef] [PubMed]
3. Morrissey, J.; Guerinot, M.L. Iron Uptake and Transport in Plants: The Good, the Bad, and the Ionome. *Chem. Rev.* **2009**, *109*, 4553–4567. [CrossRef] [PubMed]
4. Miller, G.W.; Huang, I.J.; Welkie, G.W.; Pushnik, J.C. Function of iron in plants with special emphasis on chloroplasts and photosynthetic activity. In *Iron Nutrition in Soils and Plants*; Springer: Dordrecht, The Netherlands, 1995; pp. 19–28.
5. Tang, C.; Robson, A.D.; Dilworth, M.J. A split-root experiment shows that iron is required for nodule initiation in *Lupinus angustifolius* L. *New Phytol.* **1990**, *115*, 61–67. [CrossRef]
6. Marschner, P. *Marschner's Mineral Nutrition of Higher Plants*, 3rd ed.; Elsevier Inc.: Amsterdam, the Netherlands, 2011; ISBN 9780123849052.
7. Krohling, C.A.; Eutrópio, F.J.; Bertolazi, A.A.; Dobbss, L.B.; Campostrini, E.; Dias, T.; Ramos, A.C. Ecophysiology of iron homeostasis in plants. *Soil Sci. Plant Nutr.* **2016**, *62*, 39–47. [CrossRef]
8. Jeong, J.; Guerinot, M.L. Homing in on iron homeostasis in plants. *Trends Plant Sci.* **2009**, *14*, 280–285. [CrossRef]
9. Colangelo, E.P.; Guerinot, M.L. Put the metal to the petal: Metal uptake and transport throughout plants. *Curr. Opin. Plant Biol.* **2006**, *9*, 322–330. [CrossRef]
10. Kim, S.A.; Guerinot, M.L. Mining iron: Iron uptake and transport in plants. *FEBS Lett.* **2007**, *581*, 2273–2280. [CrossRef]
11. DiDonato, R.J.; Roberts, L.A.; Sanderson, T.; Eisley, R.B.; Walker, E.L. Arabidopsis Yellow Stripe-Like2 (YSL2): A metal-regulated gene encoding a plasma membrane transporter of nicotianamine-metal complexes. *Plant J.* **2004**, *39*, 403–414. [CrossRef]
12. Bashir, K.; Inoue, H.; Nagasaka, S.; Takahashi, M.; Nakanishi, H.; Mori, S.; Nishizawa, N.K. Cloning and characterization of deoxymugineic acid synthase genes from graminaceous plants. *J. Biol. Chem.* **2006**, *281*, 32395–32402. [CrossRef]
13. Kobayashi, T.; Nishizawa, N.K. Iron Uptake, Translocation, and Regulation in Higher Plants. *Annu. Rev. Plant Biol.* **2012**, *63*, 131–152. [CrossRef] [PubMed]
14. Connorton, J.M.; Jones, E.R.; Rodríguez-Ramiro, I.; Fairweather-Tait, S.; Uauy, C.; Balk, J. Wheat vacuolar iron transporter TaVIT2 transports Fe and Mn and is effective for biofortification. *Plant Physiol.* **2017**, *174*, 2434–2444. [CrossRef] [PubMed]
15. Aggarwal, S.; Kumar, A.; Bhati, K.K.; Kaur, G.; Shukla, V.; Tiwari, S.; Pandey, A.K. RNAi-Mediated Downregulation of Inositol Pentakisphosphate Kinase (IPK1) in Wheat Grains Decreases Phytic Acid Levels and Increases Fe and Zn Accumulation. *Front. Plant Sci.* **2018**, *9*, 259. [CrossRef] [PubMed]
16. Singh, S.P.; Keller, B.; Gruissem, W.; Bhullar, N.K. Rice NICOTIANAMINE SYNTHASE 2 expression improves dietary iron and zinc levels in wheat. *Theor. Appl. Genet.* **2017**, *130*, 283–292. [CrossRef] [PubMed]
17. Boonyaves, K.; Wu, T.Y.; Gruissem, W.; Bhullar, N.K. Enhanced grain iron levels in iron-regulated metal transporter, nicotianamine synthase, and ferritin gene cassette. *Front. Plant Sci.* **2017**, *8*, 130. [CrossRef] [PubMed]
18. Masuda, H.; Ishimaru, Y.; Aung, M.S.; Kobayashi, T.; Kakei, Y.; Takahashi, M.; Higuchi, K.; Nakanishi, H.; Nishizawa, N.K. Iron biofortification in rice by the introduction of multiple genes involved in iron nutrition. *Sci. Rep.* **2012**, *2*, 543. [CrossRef]
19. Ricachenevsky, F.K.; Sperotto, R.A.; Menguer, P.K.; Sperb, E.R.; Lopes, K.L.; Fett, J.P. ZINC-INDUCED FACILITATOR-LIKE family in plants: Lineage-specific expansion in monocotyledons and conserved genomic and expression features among rice (Oryza sativa) paralogs. *BMC Plant Biol.* **2011**, *11*, 20. [CrossRef]
20. Peng, H.; Han, S.; Luo, M.; Gao, J.; Liu, X.; Zhao, M. Roles of Multidrug Transporters of MFS in Plant Stress Responses. *Int. J. Biosci. Biochem. Bioinform.* **2011**, *1*, 109. [CrossRef]
21. Kumar, A.; Kaur, G.; Goel, P.; Bhati, K.K.; Kaur, M.; Shukla, V.; Pandey, A.K. Genome-wide analysis of oligopeptide transporters and detailed characterization of yellow stripe transporter genes in hexaploid wheat. *Funct. Integr. Genom.* **2019**, *19*, 75–90. [CrossRef]

22. Beasley, J.T.; Bonneau, J.P.; Johnson, A.A.T. Characterisation of the nicotianamine aminotransferase and deoxymugineic acid synthase genes essential to Strategy II iron uptake in bread wheat (*Triticum aestivum* L.). *PLoS ONE* **2017**, *12*, e0177061. [CrossRef]

23. Sharma, S.; Kaur, G.; Kumar, A.; Meena, V.; Kaur, J.; Pandey, A.K. Overlapping transcriptional expression response of wheat zinc-induced facilitator-like transporters emphasize important role during Fe and Zn stress. *BMC Mol. Biol.* **2019**, *20*, 22. [CrossRef] [PubMed]

24. Kaur, G.; Shukla, V.; Kumar, A.; Kaur, M.; Goel, P.; Singh, P.; Shukla, A.; Meena, V.; Kaur, J.; Singh, J.; et al. Integrative analysis of hexaploid wheat roots identifies signature components during iron starvation. *J. Exp. Bot.* **2019**, *70*, 6141–6161. [CrossRef] [PubMed]

25. Wang, M.; Kawakami, Y.; Bhullar, N.K. Molecular Analysis of Iron Deficiency Response in Hexaploid Wheat. *Front. Sustain. Food Syst.* **2019**, *3*, 67. [CrossRef]

26. Connorton, J.M.; Balk, J. Iron Biofortification of Staple Crops: Lessons and Challenges in Plant Genetics. *Plant Cell Physiol.* **2019**, *60*, 1447–1456. [CrossRef]

27. Kim, S.A.; Punshon, T.; Lanzirotti, A.; Li, A.; Alonso, J.M.; Ecker, J.R.; Kaplan, J.; Guerinot, M.L. Localization of iron in Arabidopsis seed requires the vacuolar membrane transporter VIT1. *Science* **2006**, *314*, 1295–1298. [CrossRef]

28. Martinoia, E. Vacuolar transporters -Companions on a longtime journey. *Plant Physiol.* **2018**, *176*, 1384–1407. [CrossRef]

29. Gollhofer, J.; Schläwicke, C.; Jungnick, N.; Schmidt, W.; Buckhout, T.J. Members of a small family of nodulin-like genes are regulated under iron deficiency in roots of Arabidopsis thaliana. *Plant Physiol. Biochem.* **2011**, *49*, 557–564. [CrossRef]

30. Li, L.; Chen, O.S.; Ward, D.M.V.; Kaplan, J. CCC1 Is a Transporter That Mediates Vacuolar Iron Storage in Yeast. *J. Biol. Chem.* **2001**, *276*, 29515–29519. [CrossRef]

31. Gollhofer, J.; Timofeev, R.; Lan, P.; Schmidt, W.; Buckhout, T.J. Vacuolar-iron-transporter1-like proteins mediate iron homeostasis in arabidopsis. *PLoS ONE* **2014**, *9*, e110468. [CrossRef]

32. Kato, T.; Kumazaki, K.; Wada, M.; Taniguchi, R.; Nakane, T.; Yamashita, K.; Hirata, K.; Ishitani, R.; Ito, K.; Nishizawa, T.; et al. Crystal structure of plant vacuolar iron transporter VIT1. *Nat. Plants* **2019**, *5*, 308–315. [CrossRef]

33. Käll, L.; Krogh, A.; Sonnhammer, E.L.L. A combined transmembrane topology and signal peptide prediction method. *J. Mol. Biol.* **2004**, *338*, 1027–1036. [CrossRef] [PubMed]

34. Khan, M.A.; Castro-Guerrero, N.; Mendoza-Cozatl, D.G. Moving toward a precise nutrition: Preferential loading of seeds with essential nutrients over non-essential toxic elements. *Front. Plant Sci.* **2014**, *5*, 51. [CrossRef] [PubMed]

35. Conte, S.S.; Chu, H.H.; Chan-Rodriguez, D.; Punshon, T.; Vasques, K.A.; Salt, D.E.; Walker, E.L. Arabidopsis thaliana yellow stripe1-like4 and yellow stripe1-like6 localize to internal cellular membranes and are involved in metal ion homeostasis. *Front. Plant Sci.* **2013**, *4*, 283. [CrossRef] [PubMed]

36. Morgan, J.B.; Connolly, E.L. Plant-Soil Interactions: Nutrient Uptake. *Nat. Educ. Knowl.* **2013**, *4*, 2.

37. Haydon, M.J.; Cobbett, C.S. A novel major facilitator superfamily protein at the tonoplast influences zinc tolerance and accumulation in Arabidopsis. *Plant Physiol.* **2007**, *143*, 1705–1719. [CrossRef] [PubMed]

38. Krishnappa, G.; Singh, A.M.; Chaudhary, S.; Ahlawat, A.K.; Singh, S.K.; Shukla, R.B.; Jaiswal, J.P.; Singh, G.P.; Solanki, I.S. Molecular mapping of the grain iron and zinc concentration, protein content and thousand kernel weight in wheat (*Triticum aestivum* L.). *PLoS ONE* **2017**, *12*, e0174972. [CrossRef] [PubMed]

39. Sinclair, S.A.; Kraemer, U. The Zinc homeostasis network of land plants. *Biochim. Biophys. Acta* **2012**, *1823*, 1553–1567. [CrossRef]

40. Bhati, K.K.; Alok, A.; Kumar, A.; Kaur, J.; Tiwari, S.; Pandey, A.K. Silencing of ABCC13 transporter in wheat reveals its involvement in grain development, phytic acid accumulation and lateral root formation. *J. Exp. Bot.* **2016**, *67*, 4379–4389. [CrossRef]

41. Lwalaba, J.L.W.; Louis, L.T.; Zvobgo, G.; Richmond, M.E.A.; Fu, L.; Naz, S.; Mwamba, M.; Mundende, R.P.M.; Zhang, G. Physiological and molecular mechanisms of cobalt and copper interaction in causing phyto-toxicity to two barley genotypes differing in Co tolerance. *Ecotoxicol. Environ. Saf.* **2020**, *187*, 109866. [CrossRef]

42. Uruç Parlak, K. Effect of nickel on growth and biochemical characteristics of wheat (*Triticum aestivum* L.) seedlings. *NJAS-Wagening. J. Life Sci.* **2016**, *76*, 1–5. [CrossRef]

43. Kumar, S.; Nei, M.; Dudley, J.; Tamura, K. MEGA: A biologist-centric software for evolutionary analysis of DNA and protein sequences. *Br. Bioinform.* **2008**, *9*, 299–306. [CrossRef] [PubMed]

44. Marchler-bauer, A.; Lu, S.; Anderson, J.B.; Chitsaz, F.; Derbyshire, M.K.; Deweese-scott, C.; Fong, J.H.; Geer, L.Y.; Geer, R.C.; Gonzales, N.R.; et al. CDD: A Conserved Domain Database for the functional annotation of proteins. *Nucleic Acids Res.* **2011**, *39*, 225–229. [CrossRef] [PubMed]

45. Hu, B.; Jin, J.; Guo, A.; Zhang, H.; Luo, J. Genome analysis GSDS 2.0: An upgraded gene feature visualization server. *Bioinformatics* **2015**, *31*, 1296–1297. [CrossRef] [PubMed]

46. Horton, P.; Park, K.; Obayashi, T.; Fujita, N.; Harada, H.; Nakai, K. WoLF PSORT: Protein localization predictor. *Nucleic Acids Res.* **2007**, *35*, 585–587. [CrossRef] [PubMed]

47. Livak, K.J.; Schmittgen, T.D. Analysis of relative gene expression data using real-time quantitative PCR and the $2^{-\Delta\Delta}$CT method. *Methods* **2001**, *25*, 402–408. [CrossRef] [PubMed]

48. Pfeifer, M.; Kugler, K.G.; Sandve, S.R.; Zhan, B.; Rudi, H.; Hvidsten, T.R.; Mayer, K.F.X.; Olsen, O.A.; Rogers, J.; Doležel, J.; et al. Genome interplay in the grain transcriptome of hexaploid bread wheat. *Science* **2014**, *345*, 1250091. [CrossRef] [PubMed]

49. Borrill, P.; Ramirez-Gonzalez, R.; Uauy, C. expVIP: A customizable RNA-seq data analysis and visualization platform. *Plant Physiol.* **2016**, *170*, 2172–2186. [CrossRef] [PubMed]

© 2020 by the authors. Licensee MDPI, Basel, Switzerland. This article is an open access article distributed under the terms and conditions of the Creative Commons Attribution (CC BY) license (http://creativecommons.org/licenses/by/4.0/).

 *plants* 

Article

Mineral Element Composition in Grain of Awned and Awnletted Wheat (*Triticum aestivum* L.) Cultivars: Tissue-Specific Iron Speciation and Phytate and Non-Phytate Ligand Ratio

Paula Pongrac [1,*], Iztok Arčon [1,2], Hiram Castillo-Michel [3] and Katarina Vogel-Mikuš [1,4]

[1] Jožef Stefan Institute, Jamova 39, SI-1000 Ljubljana, Slovenia; iztok.arcon@ung.si (I.A.);
     katarina.vogelmikus@bf.uni-lj.si (K.V.-M.)
[2] Laboratory for quantum optics, University of Nova Gorica, Vipavska 13, SI-5000 Nova Gorica, Slovenia
[3] European Synchrotron Radiation Facility, 38043 Grenoble, France; hiram.castillo_michel@esrf.fr
[4] Biotechnical Faculty, University of Ljubljana, Jamnikarjeva 101, SI-1000 Ljubljana, Slovenia
* Correspondence: paula.pongrac@ijs.si; Tel.: +386-51-222-963; Fax: +386-477-31-51

Received: 10 December 2019; Accepted: 6 January 2020; Published: 8 January 2020

Abstract: In wheat (*Triticum aestivum* L.), the awns—the bristle-like structures extending from lemmas—are photosynthetically active. Compared to awned cultivars, awnletted cultivars produce more grains per unit area and per spike, resulting in significant reduction in grain size, but their mineral element composition remains unstudied. Nine awned and 11 awnletted cultivars were grown simultaneously in the field. With no difference in 1000-grain weight, a larger calcium and manganese—but smaller iron (Fe) concentrations—were found in whole grain of awned than in awnletted cultivars. Micro X-ray absorption near edge structure analysis of different tissues of frozen-hydrated grain cross-sections revealed that differences in total Fe concentration were not accompanied by differences in Fe speciation (64% of Fe existed as ferric and 36% as ferrous species) or Fe ligands (53% were phytate and 47% were non-phytate ligands). In contrast, there was a distinct tissue-specificity with pericarp containing the largest proportion (86%) of ferric species and nucellar projection (49%) the smallest. Phytate ligand was predominant in aleurone, scutellum and embryo (72%, 70%, and 56%, respectively), while nucellar projection and pericarp contained only non-phytate ligands. Assuming Fe bioavailability depends on Fe ligands, we conclude that Fe bioavailability from wheat grain is tissue specific.

Keywords: biofortification; phytate; iron; awn; X-ray fluorescence; X-ray absorption spectrometry; phosphorus; sulphur; nicotianamine

1. Introduction

Mineral micronutrient sufficiency—a prerequisite for human well-being—can be ensured by diet diversification or consumption of mineral-dense produce [1]. In human diets, mineral micronutrients are predominantly acquired from plant-based sources, in particular staple grain [2,3]. However, most mineral micronutrients (manganese (Mn), iron (Fe), copper (Cu) and zinc (Zn)) in grain are tightly bound in phytate (myo-inositol hexakisphosphate), a phosphorus (P)-rich salt, which cannot be digested by mammals. This makes phytate-bound mineral elements poorly bioavailable and ineffectively exploited for normal body functions [4]. Furthermore, mineral density of the cereal grain has been for a long time regarded as of minor importance compared to the crop yield [5] resulting in prevalent micronutrient deficiencies in humans [6].

Efforts to increase bio-available concentrations of mineral elements in staple crops to remedy mineral micronutrient deficiencies—particularly in marginal populations—have been invested recently, and are referred to as biofortification [7,8]. Of the seven mineral elements often lacking in our diets, Fe deficiency is most widespread, affecting up to 60% of the global population [7]. However, increasing bio-available Fe concentration through the agronomic and genetic approaches in crops is challenging [9] for several reasons: (i) poor Fe availability in the soils limits uptake into plant roots [10], (ii) strict metabolic control over Fe accumulation and sequestration in plants tissues (sufficiency ranging between 50 and 150 mg Fe·kg^{-1} dry weight in leaves of crop plants [11]), (iii) removal of Fe-rich layers during the processing of staple grain [12], and (iv) poor Fe bioavailability from phytate-rich produce such as cereal grain [13].

A large degree of variation in the accumulation of Fe in grain and seeds has been found in different crops, which is not a result from just the environmental factors. For example, in bread wheat (*Triticum aestivum* L.) grain, the variation in total Fe concentration, i.e., a ratio between the minimum and maximum total Fe concentration in grain, was up to 1.76 [14], in barley (*Hordeum vulgare* L.) the variation was up to 4.5 [15], in rice (*Oryza sativa* L.) up to 10.7 in flooded conditions and up to 288 in unflooded conditions [16], in pearl millet (*Pennisetum glaucum* (L.) R. Br.) up to 4.4 [17], in chickpea (*Cicer arietinum* L.) up to 3.2, and in pea (*Pisum sativum* L.) up to 3.5 [18]. Following predominantly classic breeding strategies, the existing natural variation in Fe density has been exploited for the development of biofortified varieties within the HarvestPlus programme [8], which demonstrated, for different crops and in different populations, that consumption of Fe-biofortified crops provides significantly more bioavailable Fe.

Despite its obvious importance for human nutrition, the filling of the staple grain with Fe, and understanding tissue-specific partitioning of Fe and Fe ligands remains a poorly understood subject [19]. Most grain filling processes take place through phloem tissues, which deliver Fe remobilised from the leaves. The presence of awns (bristle such as structures extending from lemmas), exhibiting photosynthetic activity accompanied by transpiration activity in wheat [20], may therefore play a role in the grain filling. This connection has not been investigated so far. It is, however, well-accepted that the level of phloem-mobility of a mineral element significantly affects its concentration and location in the grain, with elements such as calcium (Ca), exhibiting poor phloem mobility, not easily reaching the filial tissues of the grain and mostly remaining in the pericarp (maternal) tissues of the grain [12,21,22]. Iron has intermediate phloem mobility [23,24], so relatively large concentrations (exceeding those in leaves) of Fe have been found in some filial tissues of different staple grain, particularly the aleurone and embryo with values in the range from 200 to 400 mg Fe·kg^{-1} and from 100 to 200 mg Fe·kg^{-1}, respectively [12,15,25–28]. In these grain tissues, Fe was found to strongly co-localise with P [12,28–30]. Since approximately 80% of total P in the grain is in the form of phytate stored mainly in the aleurone cells [31], it has been inferred that the majority of Fe is bound to phytate in these tissues. Co-localisation analyses can, however, only predict potential ligands, not unambiguously determine the Fe binding environment, so conclusions must be drawn carefully when P is being used as a proxy for phytate. Using X-ray absorption near edge structure (XANES), which enables simultaneous analysis of Fe chemical form (speciation) and the type of complexing agents, it has been shown that around 80% of Fe in the whole grain of different wheat cultivars is bound to phytate, 15% to 24% as Fe^{2+} (ferrous) and 57% to 85% as Fe^{3+} (ferric) species [25]. Obtaining Fe K-edge XANES spectra of sufficiently high signal-to-noise ratio is a challenging task, with Fe concentrations typically found in grains and particularly in endosperm (<20 mg Fe·kg^{-1} in wheat [12,32], barley [15], and Tartary buckwheat (*Fagopyrum tataricum* Gaertn.) [21]), thus such studies remain scarce. One way to circumvent these technical challenges is to combine reliable Fe distribution mapping, which identifies tissues or cell-types with the largest Fe concentrations, with micro-XANES analysis. One such study, conducted on cotyledons (containing on average 187 mg Fe·kg^{-1} [21]) of Tartary buckwheat grain showed that 47% of Fe was bound to phytate, 22%, of that as Fe^{2+} and 25% as Fe^{3+}, while the remaining Fe^{3+} was bound to citrate [33]. Furthermore, in wheat aleurone, modified aleurone (surrounding the crease) and in nucellar projection the micro-XANES analysis indicated that Fe was bound to phytate/citrate, phytate, and Fe-nicotianamine/Fe oxide-hydroxide, respectively [28]. However, tissue-specific Fe speciation was not resolved [28] and the XANES analysis in pericarp and embryo have not been acquired so far.

Therefore, the aim of the study was to compare mineral element composition of the awned and awnletted (those that have short or no awns) cultivars and to determine tissue-specific Fe speciation and Fe ligands in the contrasting cultivars to test the following hypotheses: (i) the presence of awns affects the mineral element composition of the wheat grain, (ii) majority of Fe is bound to phytate in different tissues of wheat grain, (iii) Fe speciation and Fe ligands across different wheat cultivars are stable, and (iv) Fe ligand profile depends on local Fe concentration.

2. Results

2.1. Total Concentrations of Mineral Elements in Whole Wheat Grain

A significantly larger total concentration of Ca and Mn, but significantly smaller total concentration of Fe was found in whole grain of awned wheat cultivars than in awnletted cultivars (Figure 1).

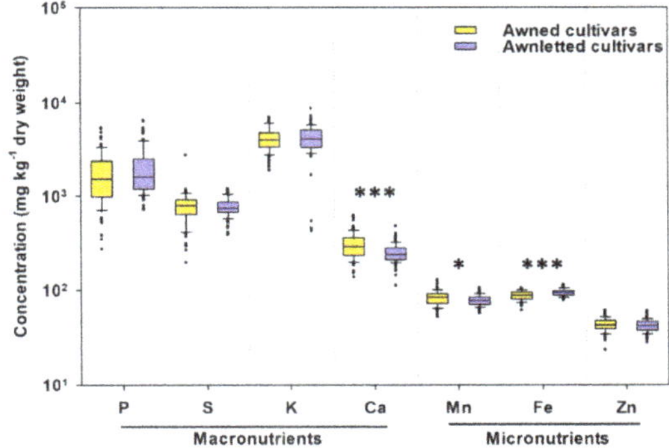

Figure 1. Variability in the total concentration of phosphorus (P), sulphur (S), potassium (K), calcium (Ca), manganese (Mn), iron (Fe), and zinc (Zn) in whole grain of wheat (*Triticum aestivum* L.) cultivars differing in the awn type (awned cultivars have long awns and awnletted cultivars have short or no awns) grown in the same field. Shown are boxplots representing 25th and 75th percentile of the data, with the middle line representing the median, whiskers representing the 5th and 95th percentile and the black dots representing outliers ($n = 87$ and $n = 100$ data points of nine awned and 11 awnletted wheat cultivars, respectively). Asterisks indicate significant differences between the awned and the awnletted cultivars (Student *t*-test; *** $p < 0.001$ and * $p < 0.05$).

The total concentration of Fe in the awned wheat cultivars ranged from 50.8 to 94 mg·kg⁻¹ dry weight (1.85-fold variability) and in the awnletted wheat genotypes from 67.3 to 103 mg·kg⁻¹ dry weight (1.53-fold variability). Considering all wheat cultivars studied, total Fe concentration in whole grain varied 2.03-fold.

There was no difference in 1000-grain weight (Table S1) and in concentrations of P, sulphur (S), potassium (K) and Zn in the whole grain between the awned and the awnletted wheat cultivar group (Figure 1), nor was there any apparent separation of awned and awnletted cultivars when the whole elemental profile was considered (Figure S1). The hierarchical clustering indicated that Fe and Zn were grouped apart from the rest of the mineral elements, among which P, K, and S grouped apart from Ca and Mn (Figure S1). A significant positive correlation was observed between grain concentrations of P and those of K, S, and Fe, but no significant correlation between grain concentrations of P and Ca, Mn, and Zn (Figure S2a). Positive correlation was found between grain concentration of Fe and Mn, and of Fe and Zn (Figure S2b).

Four wheat cultivars with contrasting Fe concentrations were selected for further in-depth analyses: Two awned cultivars (cv. Vulkan and cv. Soissons) of low-Fe accumulation (the average total concentrations in the grain was 73.4 and 77.0 mg Fe·kg^{-1} dry weight, respectively), and two awnletted cultivars (cv. Katarina and cv. Super Zitarka) of high-Fe accumulation (the average total Fe concentrations in the grain was 83.7 and 91.3 mg Fe·kg^{-1} dry weight, respectively; Figure 2). In agreement with observations for all wheat cultivars studied, there was a positive correlation between grain P concentrations and grain S, K, and Fe concentration (Figure S3a) and between grain Fe and grain Zn (but not Mn) concentrations (Figure S3b) in these four wheat cultivars.

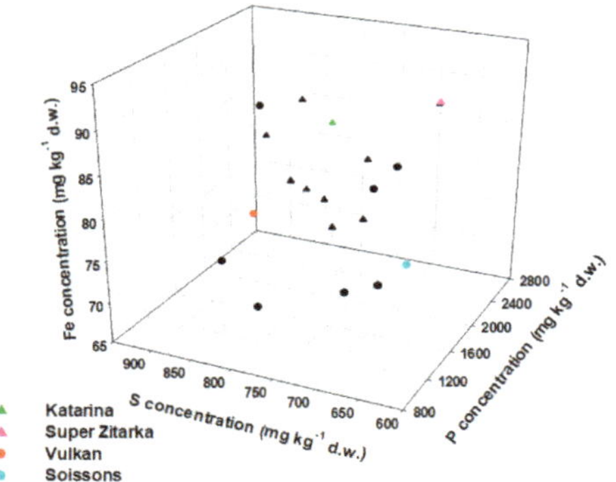

Figure 2. Total iron (Fe), sulphur (S) and phosphorus (P) concentrations in the grain of wheat (*Triticum aestivum* L.) cultivars differing in the awn type (awned cultivars have long awns and awnletted cultivars have short or no awns) grown in the same field. Circles represent awned cultivars and triangles represent awnletted cultivars. The four cultivars (Katarina, Super Zitarka, Vulkan and Soissons) selected for further in-depth analyses are highlighted in colour. Shown are averages (n = 6–12). d.w.—dry weight.

2.2. Tissue-Specific Iron, Phosphorus and Sulphur Concentrations, Iron Speciation and Iron Ligands

Iron species and Fe ligands were studied in two different regions of interests (Figure 3) of the frozen-hydrated grain cross-sections of the four wheat cultivars. The first region of interest comprised nucellar projection, modified aleurone, endosperm, transfer cells, and scutellum. The second region of interest comprised aleurone, scutellum, embryo, endosperm, and pericarp.

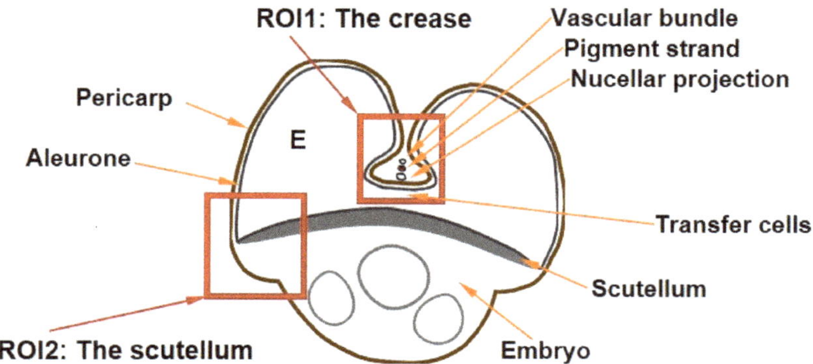

Figure 3. A representative wheat (*Triticum aestivum* L.) grain cross section with the two regions of interest (ROI) highlighted with red squares, namely ROI1 (the crease) and ROI2 (the scutellum). E—endosperm.

To identify higher Fe signal pixels—selected for subsequent micro-XANES analysis—the regions of interest were first subjected to a fast (to avoid photoreduction of Fe by the focused X-ray beam) micro-XRF mapping at the ID21 beamline at ESRF to localise Fe, P, and S (the quantitative maps are shown in Figures S4 and S5). By identifying Fe hotspots, the best signal-to-noise ratio in Fe K-edge micro-XANES spectra was ensured. In endosperm, the concentrations of Fe were too small (on average 3.5 mg·kg^{-1} fresh weight in imbibed grains of awned cultivars and 11.4 mg·kg^{-1} fresh weight in imbibed grains of awnletted cultivars) to yield micro-XANES spectra of sufficient quality. Similarly, larger Fe concentrations were found in aleurone and in pericarp of the awnletted cultivars (82.3 and 24.6 mg·kg^{-1} Fe fresh weight, respectively) than in aleurone of the awned cultivars (49.2 and 12.2 mg·kg^{-1} Fe fresh weight, respectively). By contrast, in embryo and nucellar projection the average Fe concentration of awned cultivars (33.8 and 160 mg·kg^{-1} Fe fresh weight, respectively) was larger than in awnletted cultivars (12.5 and 92 mg·kg^{-1} Fe fresh weight, respectively). In scutellum, both cultivars contained similar Fe concentration (60.5 mg·kg^{-1} Fe fresh weight in awned cultivars and 64 mg·kg^{-1} Fe fresh weight in awnletted cultivars).

The micro-XANES spectra from selected Fe hotspots (2 to 4 per section as indicated on the Fe, P and S co-localisation maps in Figures 4 and 5) were compared to the spectra of the Fe reference compounds and complexes (Figure S6; reported previously [25,33]).

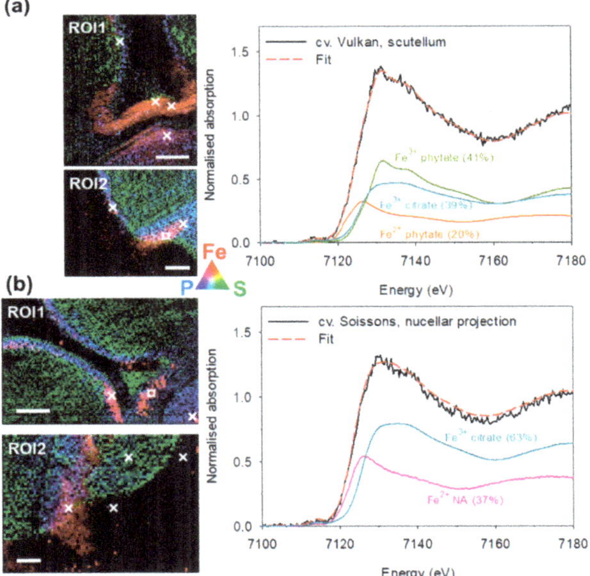

Figure 4. Co-localisation images of iron (Fe) in red, sulphur (S) in green and phosphorus (P) in blue in the two regions of interest (ROI): the crease (ROI1) and the scutellum (ROI2) in the frozen-hydrated grain of wheat (*Triticum aestivum* L.) cultivar Vulkan (**a**) and Soissons (**b**), the awned wheat cultivars. × indicates pixels where Fe K-edge micro-XANES spectra were recorded and ¤ indicates where the selected Fe K-edge micro-XANES spectra (solid line) was recorded and is displayed on the right-hand side. The best linear combination fit (red dashed line) was obtained by the spectra of the reference Fe compounds. The relative amount of each component is given in parentheses. eV—electron volts; NA—nicotianamine. Scale bars = 200 μm. Quantitative distribution maps of Fe, P and S can be found in Figure S4.

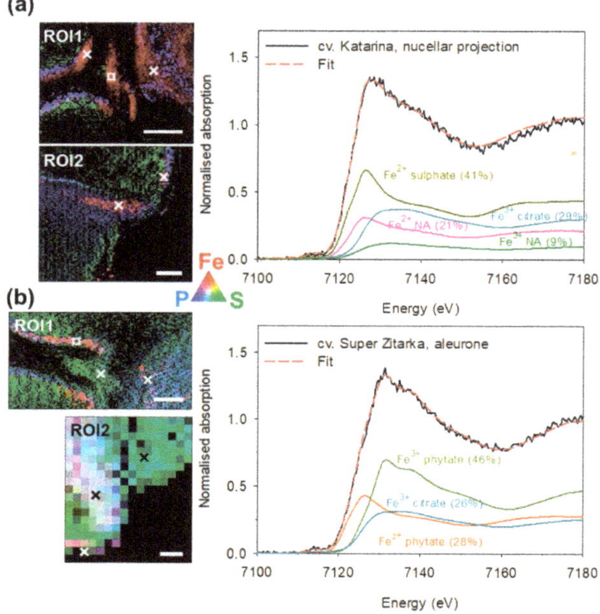

Figure 5. Co-localisation images of iron (Fe) in red, sulphur (S) in green and phosphorus (P) in blue in the two regions of interest (ROI): the crease (ROI1) and the scutellum (ROI2) in the frozen-hydrated grain of wheat (*Triticum aestivum* L.) cultivar Katarina (**a**) and Super Zitarka (**b**), the awnletted wheat cultivars. × indicates pixels where Fe K-edge micro-XANES spectra were recorded and ¤ indicates where the selected Fe K-edge micro-XANES spectra (solid line) was recorded and is displayed on the right-hand side. The best linear combination fit (red dashed line) was obtained by the spectra of the reference Fe compounds. The relative amount of each component is given in parentheses. eV—electron volts; NA—nicotianamine. Scale bars = 200 μm. Quantitative distribution maps of Fe, P and S can be found in Figure S5.

The Fe K-edge micro-XANES spectra could be described as linear combinations of the Fe K-edge XANES spectra of the following Fe complexes: Fe^{2+} phytate, Fe^{2+} sulphate, Fe^{2+} nicotianamine, Fe^{3+} phytate, Fe^{3+} nicotianamine, Fe^{3+} citrate, and α-$Fe^{3+}OOH$ (Fe oxide-hydroxide; goethite). Relative amount of each Fe complex in the combination (Table S2) was obtained from the best fit with a ± 1% error for the Fe^{2+}/Fe^{3+} complex ratio and a ± 5% error for the Fe^{3+} phytate/Fe^{3+} non-phytate ratio.

On average, the four cultivars did not differ in the Fe species and Fe ligand composition (Figure 6a,c). Ferric species was predominant in all four cultivars, with 64% of the total Fe found in this form (Figure 6a), which was a cumulation of 26% being phytate ligand and 38% non-phytate ligands. The remaining Fe was present as ferrous species (36%) in all four cultivars, which was a cumulation of 27% bound to phytate and 9% to non-phytate ligands. In total, 53% of Fe was found bound to phytate and the remaining 47% to non-phytate ligands (Figure 6b and Table S2).

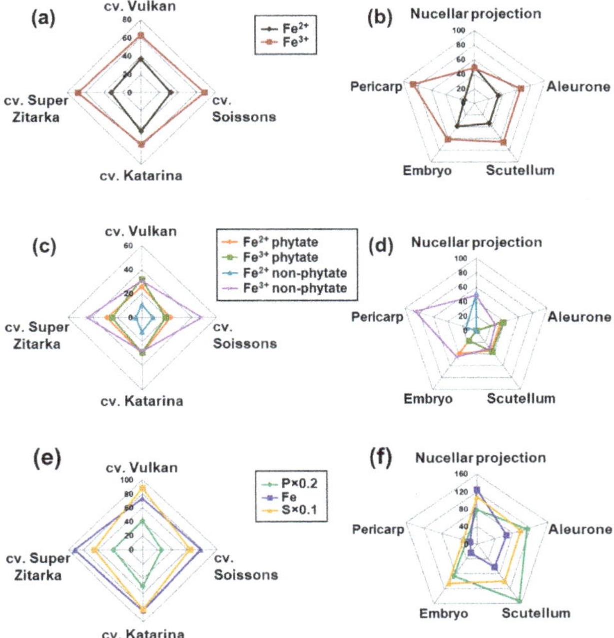

Figure 6. Average relative amounts (%) of iron (Fe) species, Fe ligands and Fe, phosphorus (P) and sulphur (S) concentration in the grain of wheat (*Triticum aestivum* L.) cultivars Vulkan and Soissons (awned wheat cultivars) and Katarina and Super Zitarka (awnletted wheat cultivars) (**a,c,e**) and in grain tissues (**b,d,f**). Phosphorus, Fe and S concentrations are in mg·kg^{-1} dry weight (**e**) or fresh weight (**f**). In (**b,d,f**) results indicate average across all four cultivars.

A significant tissue-specificity for Fe speciation and Fe ligand composition was observed (Figure 6b,d). In all tissues studied, the majority of Fe species were ferric—except in nucellar projection, with equal contribution of ferric and ferrous species (Figure 6b). In the order of decreasing content of ferric species, the list of tissues is: pericarp < aleurone = scutellum < embryo < nucellar projection (Figure 6b). By the proportion of phytate ligands, the tissues were ordered as: aleurone < scutellum << embryo << nucellar projection = pericarp, with the latter two tissues having no phytate ligands, but only non-phytate ligands (Figure 6d and Table S2). Of non-phytate ligands in nucellar projection Fe^{3+} citrate was most prominent (34%), followed by Fe^{2+} nicotianamine (29%), Fe^{2+} sulphate (22%) and Fe^{3+} nicotianamine. By contrast, the pericarp contained mostly Fe^{3+} oxide-hydroxide (52%), followed by Fe^{3+} nicotianamine (23%), Fe^{2+} sulphate (14%) and Fe^{3+} citrate (12%). Of non-phytate ligands in other tissues, only Fe^{3+} citrate was found, with the largest proportions in embryo (45%), followed by scutellum (30%) and aleurone (27%). No clear correlation between the total and local Fe, P and S concentration and the Fe ligand profile in tissues could be discerned (Figure 6e,f). The nucellar projection contained the largest concentration of Fe, while pericarp contained the smallest concentration of Fe, P and S (Figure 6f).

3. Discussion

The frequent lack in intake of essential mineral elements in human diets can be significantly improved by shaping agronomic practice and/or designing staple crops to generate mineral-dense produce [7]. Agronomic approaches have been efficient when (i) the soil contains insufficient amounts of certain element(s), which can be added to the agricultural system as fertilizers or (ii) when changes to phytoavailability of elements in the rhizosphere are required and pH-related intervention can offer solutions. On the other hand, variability in the elemental composition of the edible produce can be exploited to (i) introduce cultivars with superior mineral-use efficiency, provided there is no penalty to agronomically important traits or (ii) to identify candidate genes for future genetic optimisation [9]. After these interventions are implemented and a produce with the largest possible inherent concentration is available, the Fe status of the individual and other food components (e.g., dietary fibre, organic acids) will still play crucial roles in the availability of a certain element, with Fe being particularly problematic [34], further complicating the efforts to ensure optimal nutrition in humans.

3.1. Awned Cultivars Accumulate More Ca and Mn But Less Fe in Grain than Awnletted Wheat Cultivars

We investigated the diversity in grain mineral element accumulation in 20 wheat cultivars and found that there is a link between the awn length and the Ca, Mn and Fe concentrations (Figure 1). In wheat, awns have been shown to have transpiration and photosynthetic activity [20], thus their presence could contribute to the translocation of elements taken up by roots on the one side and/or to the phloem-driven (re)allocation of assimilates on the other side, thereby affecting mineral element density in the grain. This connection has, however, not been investigated so far. Awnletted wheats have been shown to produce significantly more grains per unit area and per spike, resulting in a significant reductions in grain size and an increased frequency of small, shrivelled grains [35]. Our observations did not fully support this report, since there was no significant difference in 1000-grain weight between the awned and awnletted cultivars, both for the established agronomic values and for those from our experiment (Table S1). The observed differences in elemental concentrations are conceivably not a consequence of dilution by grain weight, but rather arise from genetic differences in uptake, allocation and/or mobilisation or Ca, Mn and Fe in these cultivars. It may however be, that by accumulating larger Fe concentration in the embryo and particularly in the nucellar projection in awned cultivars compared to awnletted cultivars as observed in our study, less Fe is being translocated to other grain tissues, resulting in larger total Fe concentration in the whole grain.

On average, total Fe concentrations in grain from our experiment showed 2.03-fold variability (Figure 2), which is in a similar range as the value 1.76 reported for 150 bread wheat cultivars [14]. The whole grain Fe concentrations in all studied cultivars exceed the reported maximum value (50.8 mg Fe·kg^{-1}) for the bread wheat cultivars [14], but were within the range of observations in spelt (*Triticum spelta* L.) grain, for which up to 99 mg Fe·kg^{-1} was found, but with significant variation due to the year and the location [36]. Iron concentration in barley grain exceeding 100 mg·kg^{-1} has also been reported [37].

The positive correlation between P and Fe concentrations was observed (Figures S2 and S3) in agreement with findings in bread wheat [14] and in spelt [36], suggesting that the increased Fe concentration may be accompanied by a decreased bioavailability (i.e., due to phytate), presenting a further challenge for biofortification. However, the positive relationship has not been observed in all instances [38,39] and a strong genotype, environment and/or genotype × environment interaction has been shown to affect the relationship [40,41]. The positive correlation between grain Fe and Zn observed also in our experiment (Figures S2 and S3), seems to be quite stable as it has been consistently reported, over different seasons and locations, for example in wheat [14,38], durum wheat [39], spelt [36], and barley [15,37]. It could be attributed to the limited specificity of transporters and metal ligands for either Zn or Fe [42], suggesting that the increased density could be achieved simultaneously for a larger number of trace elements. Still, the issues with the bioavailability of these trace elements will have to be addressed before any such observations are implemented into breeding strategies.

3.2. Iron Speciation and Iron Ligands in Wheat Grain Are Stable across Cultivars Differing in Total Iron Concentration

To complement the current knowledge on Fe speciation and Fe ligands in whole wheat grain [25] and some of its tissues [28] we studied these traits in five grain tissues of four wheat cultivars. Initial X-ray fluorescence mapping in frozen hydrated cross-sections of the grain revealed Fe hotspots and provided information on P and S distribution as well. Based on the co-localisation of Fe, P, and S (Figures 4 and 5), selected regions of interest (Figure 3) were easily distinguished and high-Fe pixels were investigated by micro-XANES. There was no apparent difference in the Fe speciation or Fe ligands in the grain of awned and awnletted cultivars, regardless of the differences in the Fe concentrations (Figure 6a,c) indicating that the total Fe concentration in the grain does not influence Fe species or Fe ligands. Similar results were found by Singh et al. [25], who included in the analysis a wild relative of common wheat, *Aegilops kotschyi* Boiss., which contains up to three times larger Fe concentrations than grain of wheat landraces. The relative amounts of ferrous (36%) species in wheat grain—up to one third of total Fe as assessed by micro-XANES in our experiment (Figures 4–6)—was somewhat larger than in previous findings in whole wheat with values between 14% and 24% [25]. Furthermore, on average 53% of Fe was bound to phytate. The proportion of non-phytate ligands is in agreement with another study [28] on wheat and Tartary buckwheat grain, where 22% of total Fe was bound to non-phytate ligands [33]. Perplexingly, no direct association of the Fe ligand profile could be found with Fe bioavailability (assessed using Caco-2 cell system) in Tartary buckwheat sprouts containing a much larger proportion (55%) of Fe^{3+} citrate [33]. In legume seeds, which store large amounts of Fe in ferritin, progressive accumulation of phytate with seed maturity limits Fe bioavailability, as demonstrated by comparing immature and mature pea (*Pisum sativum* L.) seeds [43]. Apparently, more studies are required to reach consensus on the connection between Fe ligands and Fe bioavailability from different plant-based sources.

3.3. Distinct Tissue Specificity in Iron Speciation, Iron Ligands and Iron Concentration in Wheat Grain

At the tissue level a large variability in the Fe speciation and Fe ligands was found (Figure 6) in line with a previous report [28], suggesting differences in bioavailability of Fe from different grain tissues. All tissues contained >60% of ferric species in line with analyses in whole grain [25]. The only exception was nucellar projection, where equal amounts of Fe^{3+} and Fe^{2+} were found (Figure 6b). Regarding the Fe ligands, the presence of non-phytate ligands in the pericarp and nucellar projection was particularly striking (Figure 6d). The pericarp is a maternal tissue and is typically accessed by xylem, which may be a reason for the Fe-citrate pool. There is some evidence that Fe from the Tartary buckwheat pericarp [33] and from the wheat bran [44] is relatively bioavailable. While Tartary buckwheat pericarp is not edible, the inclusion of wheat bran in the meal would, despite markedly increased phytate concentration, outweigh the negative effect of phytate-induced losses in bioavailability.

The major part of cereal grain loading with micro and macronutrients is presumed to take place through vasculature, which in mature grain is shrunk and borders the pigment strand, which in turn borders the nucellar projection (Figure 3). During grain maturity the tissues within the crease undergo a series of transformations, and nutrients passing into the endosperm must cross the pigment strand, the nucellar projection, and the endosperm transfer cells [45]. The nucellar projection is a part of the nucellar tissue that faces the vascular tissue, has a morphology characteristic of transfer cells [46] and contains large concentration of Fe, while the pigment strand was rich in Mn, making these two tissues clearly distinguishable [12,25,32]. As already follows from the apparent lack of co-localisation of Fe and P in the nucellar projection (Figures 4 and 5; [12,25]) only non-phytate ligands were found there. Our analysis also confirms the presence of nicotianamine in the nucellar projection reported previously in [28]. For the first time, ferric and ferrous nicotianamine compounds are demonstrated to exist in a grain tissue (together 45% of Fe ligands). In addition, some Fe was bound to sulphate (22%), in agreement with previous reports [28] and predicted from S localisation in this tissue (Figures 4–6, Figures S4 and S5) and exclusion of P [25].

In embryo, scutellum and aleurone, the only non-phytate ligand found was citrate, which is in agreement with results for cotyledons in Tartary buckwheat grain [33]. Most of the Fe present in the liquid endosperm in pea was found as a mixture of Fe^{3+} citrate and malate [47]. The mixture has been shown to undergo an ascorbate-driven reduction [47] which makes Fe more mobile within the developing seedling. Ferric citrate-malate complexes have been demonstrated as the main form of Fe circulating in pea (*Pisum sativum* L.) plants [19]. In pea, the mother plant transports Fe^{3+} malate/citrate complexes via the seed coat to the embryo, which in turn secretes ascorbate to reduce Fe^{3+} to Fe^{2+} for uptake during germination [47]. These observations suggest active participation of all grain tissues and not only crease tissues in Fe loading of the grain and seed.

4. Conclusions

Because awns presumably have photosynthetic activity, awned and awnletted wheat cultivars were compared for their 1000-grain weight, mineral composition and Fe speciation and Fe local chemical environment. While there was no difference in 1000-grain weight, a larger Ca and Mn, but smaller Fe concentrations, were found in whole grain of awned than in awnletted cultivars. Genetic and/ or metabolic reasons behind the observed differences in mineral composition will need to be studied in future experiments. The evaluation of Fe speciation and Fe ligands revealed that differences in total Fe concentration were not accompanied by differences in Fe speciation (on average 64% of Fe existed as ferric and 36% as ferrous species) or Fe ligands (on average 53% were phytate and 47% were non-phytate ligands) in the two awned and two awnletted cultivars studied using micro-XANES. Contrastingly, there was a distinct tissue-specificity with pericarp containing the largest proportion (86%) of ferric species and nucellar projection (49%) the smallest. Iron was predominantly bound to phytate in aleurone, scutellum and embryo (72%, 70%, and 56%, respectively), while in nucellar projection and pericarp Fe was bound only to non-phytate ligands. Assuming Fe bioavailability depends on Fe ligands, we conclude that Fe bioavailability from wheat grain is tissue specific.

5. Materials and Methods

5.1. Plant Material and Total Element Concentration in Grain

The grain of 20 different wheat (*Triticum aestivum* L.) cultivars were obtained from the Agricultural Institute of Slovenia, with their agronomic characteristics reported in Table S1. Among these wheat cultivars, nine were awned (cv. Euclide, cv. Lukullus, cv. Vulkan, cv. Isengrain, cv. Renan, cv. Soissons, cv. Ingenio, cv. Bologna, and cv. Element) and 11 were awnletted (cv. BC Nina, cv. BC Renata, cv. Rosario, cv. Felix, cv. Renata, cv. Katarina, cv. BC Zdenka, cv. Anđelka, cv. Gracia, cv. Bastide, and cv. Super Zitarka). The plants were grown in the field of the Infrastructure centre Jablje (central Slovenia: 46°8′59″ N, 14°33′31″ E, 307 m above sea level) in 2014/2015 on pseudogley-gley soil type, which has a silty clay texture. The previous crop on the field site was grain maize (*Zea mays* L.). The field was fertilised with 205 kg·ha^{-1} N (in 5 rations), 90 kg·ha^{-1} P_2O_5 and 120 kg·ha^{-1} K_2O; 400–750 germinative seeds m^{-2} were sowed respecting interrow spacing of 12.5 cm. Trial layout was a randomized block design with four repetitions, with each plot having 7.5 m^2. At maturity, the plants were harvested, the 1000-grain weight was determined (g), the grain was homogenised and milled or stored for the localisation analyses. The ground material was pressed into pellets (6–12 for each cultivar) using a pellet die and hydraulic press. Total concentrations of P, S, K, Ca, Mn, Fe, and Zn were measured in whole grain samples and in standard reference material NIST SRM 1573a (tomato leaves) for quality assurance using X-ray fluorescence, as described previously [48]. Based on the total Fe concentration in the whole grain of the 20 wheat cultivars, four contrasting cultivars were selected. Cultivars Katarina and Super Zitarka, the two awnletted wheat cultivars, had a larger total Fe concentration than the two awned cultivars Vulkan and Soissons.

5.2. Sample Preparation and Micro-XRF Mapping and Micro-XANES Analysis

The grain of the four wheat cultivars was soaked in MiliQ water for 2 h at 4 °C. Whole grain was hand-cut transversely at the embryo location under a stereomicroscope into approximately 200 µm thick sections using new stainless-steel platinum-coated razor blades and frozen immediately in liquid isopentane (Sigma-Aldrich, St. Louis, MO, USA) cooled by liquid nitrogen [49]. Frozen-hydrated sections were sandwiched between two Ultralene® foils (each 4 µm thick), deposited on custom made Cu holders and analysed at the ID21 beamline (ESRF, Grenoble, France) as described previously [33]. In short, measurements of Fe K-edge micro-XANES (recorded in the energy region from 7040 to 7250 eV) were performed on high Fe pixels ($n = 2$–9) identified by fast mapping of grain sections using synchrotron radiation micro-X-ray fluorescence (micro-XRF) at the same beamline. This approach enabled the best quality micro-XANES spectra of the two regions of interest in the wheat grain: the crease and the scutellum. Specific attention was paid to avoid radiation damage and photoreduction of Fe^{3+} [50], as described in detail in [25,33].

The qualitative distribution maps of Fe, P, and S were quantitatively analysed, as described previously [51,52] and the quantitative distribution maps and co-localisation maps were generated with PyMCA software [53]. The Fe K-edge micro-XANES spectra were analysed with the IFEFFIT program package Athena [54], where linear combination fitting was performed using the following reference Fe^{2+} and Fe^{3+} complexes: Fe^{2+} phytate, Fe^{2+} nicotianamine, Fe^{2+} sulphate, Fe^{3+} phytate, Fe^{3+} nicotianamine, Fe^{3+} citrate and α-Fe^{3+}OOH - oxide/hydroxide (goethite). The preparation and analysis of these reference Fe complexes has been described elsewhere [25,33], except for Fe nicotianamine standards, which were synthesized by mixing water solution of nicotianamine (CAS 34441-14-0, Santa Cruz Biotechnology, Inc., USA) with $FeCl_2 \times 4H_2O$ (Sigma-Aldrich) or $FeCl_3 \times 6H_2O$ (Sigma-Aldrich) water solution at the molar ratio of 10:1 in water. Final pH was adjusted to 6.5–7.0. XANES spectra of the reference compounds Fe^{2+} nicotianamine and Fe^{3+} nicotianamine were measured at XAFS beamline of synchrotron Elettra (Trieste, Italy) in transmission detection mode, on homogeneous pellets with optical thickness of about 2 above Fe absorption K-edge. Iron species and Fe ligands for each wheat cultivar were obtained by averaging across all tissue-specific XANES results in each cultivar.

5.3. Statistical Analysis

Pairwise comparisons were tested by Student *t*-test at $p < 0.05$ and linear regression analysis was performed in SigmaPlot version 13.0 (Systat Software, San Jose, CA). Clustering and heatmap was created using z-transformed average total concentrations in whole grain using function heatmap.2 in gplots package within R i386 3.0.2 software.

Supplementary Materials: The following are available online at http://www.mdpi.com/2223-7747/9/1/79/s1. Figure S1: Hierarchical clustering and heatmap of grain elemental composition of all wheat cultivars studied. Figure S2: Correlations between element concentrations in whole grain for all wheat cultivars studied. Figure S3: Correlations between element concentrations in whole grain for the selected four wheat cultivars. Figure S4: Quantitative distribution maps in the grain of awned wheat cultivars. Figure S5: Quantitative distribution maps in the grain of awnletted wheat cultivars. Figure S6: Iron K-edge XANES of reference compounds used in the linear combination fitting. Table S1: Agronomic characteristics of the wheat cultivars studied. Table S2: Relative amounts of iron ligands in the grain wheat cultivars and in different grain tissues.

Author Contributions: Formal analysis, P.P., I.A. and K.V.-M.; investigation, P.P., I.A., H.C.-M. and K.V.-M.; project administration, K.V.-M.; visualization, P.P.; writing—original draft, P.P., H.C.-M. and K.V.-M. All authors have read and agreed to the published version of the manuscript.

Funding: The study was financed by the Slovenian Research Agency (ARRS) through programmes (P1-0212, P1-0112 and I0-0005), and projects (N7-077, J7-9418, J7-9398, N1-0105 and N1-0090).

Acknowledgments: The European Synchrotron Radiation Facility (ESRF), Grenoble, France, is acknowledged for the provision of synchrotron radiation facilities at beamline ID21 (experiment number LS-2225) and Elettra, Trieste, Italy, for access to the synchrotron radiation facilities at beamline XAFS (experiment number 20185165). The authors are grateful to Giuliana Aquilanti and Simone Pollastri from XAFS beamline of ELETTRA for expert advice on beamline operation and to Darinka Koron from the Agricultural Institute of Slovenia for kindly providing the grain of the wheat cultivars studied.

Conflicts of Interest: The authors declare no conflict of interest.

References

1. Shenkin, A. Micronutrients in health and disease. *Postgrad. Med. J.* **2006**, *82*, 559–567. [CrossRef]
2. DellaPenna, D. Nutritional genomics: Manipulating plant micronutrients to improve human health. *Science* **1999**, *285*, 375–379. [CrossRef]
3. White, P.J.; George, T.S.; Gregory, P.J.; Bengough, A.G.; Hallett, P.D.; McKenzie, B.M. Matching roots to their environment. *Ann. Bot.* **2013**, *112*, 207–222. [CrossRef]
4. Clemens, S. Zn and Fe biofortification: The right chemical environment for human bioavailability. *Plant Sci.* **2014**, *225*, 52–57. [CrossRef]
5. Fan, M.-S.; Zhao, F.-J.; Fairweather-Tait, S.J.; Poulton, P.R.; Dunham, S.J.; McGrath, S.P. Evidence of decreasing mineral density in wheat grain over the last 160 years. *J. Trace Elem. Med. Biol.* **2008**, *22*, 315–324. [CrossRef]
6. Sands, D.C.; Morris, C.E.; Dratz, E.A.; Pilgeram, A.L. Elevating optimal human nutrition to a central goal of plant breeding and production of plant-based foods. *Plant Sci.* **2009**, *177*, 377–389. [CrossRef]
7. White, P.J.; Broadley, M.R. Biofortification of crops with seven mineral elements often lacking in human diets—Iron, zinc, copper, calcium, magnesium, selenium and iodine. *New Phytol.* **2009**, *182*, 49–84. [CrossRef]
8. Bouis, H.E.; Saltzman, A. Improving nutrition through biofortification: A review of evidence from HarvestPlus, 2003 through 2016. *Glob. Food Secur.* **2017**, *12*, 49–58. [CrossRef]
9. Connorton, J.M.; Balk, J. Iron biofortification of staple crops: Lessons and challenges in plant genetics. *Plant Cell Physiol.* **2019**, *60*, 1447–1456. [CrossRef]
10. Colombo, C.; Palumbo, G.; He, J.-Z.; Pinton, R.; Cesco, S. Review on iron availability in soil: Interaction of Fe minerals, plants, and microbes. *J. Soils Sediments* **2014**, *14*, 538–548. [CrossRef]
11. White, P.J.; Brown, P.H. Plant nutrition for sustainable development and global health. *Ann. Bot.* **2010**, *105*, 1073–1080. [CrossRef]
12. Pongrac, P.; Kreft, I.; Vogel-Mikuš, K.; Regvar, M.; Germ, M.; Vavpetič, P.; Grlj, N.; Jeromel, L.; Eichert, D.; Budič, B.; et al. Relevance for food sciences of quantitative spatially resolved element profile investigations in wheat (*Triticum aestivum*) grain. *J. R. Soc. Interface* **2013**, *10*. [CrossRef]
13. Collings, R.; Harvey, L.J.; Hooper, L.; Hurst, R.; Brown, T.J.; Ansett, J.; King, M.; Fairweather-Tait, S.J. The absorption of iron from whole diets: A systematic review. *Am. J. Clin. Nutr.* **2013**, *98*, 65–81. [CrossRef]
14. Zhao, F.J.; Su, Y.H.; Dunham, S.J.; Rakszegi, M.; Bedo, Z.; McGrath, S.P.; Shewry, P.R. Variation in mineral micronutrient concentrations in grain of wheat lines of diverse origin. *J. Cereal Sci.* **2009**, *49*, 290–295. [CrossRef]
15. Detterbeck, A.; Pongrac, P.; Rensch, S.; Reuscher, S.; Pečovnik, M.; Vavpetič, P.; Pelicon, P.; Holzheu, S.; Krämer, U.; Clemens, S. Spatially resolved analysis of variation in barley (*Hordeum vulgare*) grain micronutrient accumulation. *New Phytol.* **2016**, *211*, 1241–1254. [CrossRef]
16. Pinson, S.R.M.; Tarpley, L.; Yan, W.; Yeater, K.; Lahner, B.; Yakubova, E.; Huang, X.-Y.; Zhang, M.; Guerinot, M.L.; Salt, D.E. Worldwide genetic diversity for mineral element concentrations in rice grain. *Crop Sci.* **2015**, *55*, 294. [CrossRef]
17. Bashir, E.M.A.; Ali, A.M.; Ali, A.M.; Melchinger, A.E.; Parzies, H.K.; Haussmann, B.I.G. Characterization of Sudanese pearl millet germplasm for agro-morphological traits and grain nutritional values. *Plant Genet. Resour. Character. Util.* **2014**, *12*, 35–47. [CrossRef]
18. Available online: https://npgsweb.ars-grin.gov/gringlobal/descriptors.aspx? (accessed on 8 January 2020).
19. Grillet, L.; Mari, S.; Schmidt, W. Iron in seeds—Loading pathways and subcellular localization. *Front. Plant Sci.* **2014**, *4*, 1–8. [CrossRef]
20. Li, X.-F.; Bin, D.; Hong-Gang, W. Awn anatomy of common wheat (*Triticum aestivum* L.) and its relatives. *Caryologia* **2010**, *63*, 391–397. [CrossRef]
21. Pongrac, P.; Vogel-Mikuš, K.; Jeromel, L.; Vavpetič, P.; Pelicon, P.; Kaulich, B.; Gianoncelli, A.; Eichert, D.; Regvar, M.; Kreft, I. Spatially resolved distributions of the mineral elements in the grain of tartary buckwheat (*Fagopyrum tataricum*). *Food Res. Int.* **2013**, *54*, 125–131. [CrossRef]
22. Vogel-Mikuš, K.; Pelicon, P.; Vavpetič, P.; Kreft, I.; Regvar, M. Elemental analysis of edible grains by micro-PIXE: Common buckwheat case study. *Nucl. Instrum. Methods Phys. Res. Sect. B Beam Interact. Mater. Atoms* **2009**, *267*, 2884–2889. [CrossRef]

23. Garnett, T.P.; Graham, R.D. Distribution and remobilization of iron and copper in wheat. *Ann. Bot.* **2005**, *95*, 817–826. [CrossRef] [PubMed]

24. White, P.J. Long-distance Transport in the Xylem and Phloem. In *Marschner's Mineral Nutrition of Higher Plants*, 3rd ed.; Marschner, P., Ed.; Academic Press: San Diego, CA, USA, 2012; Chapter 3; pp. 49–70. ISBN 978-0-12-384905-2.

25. Singh, S.P.; Vogel-Mikuš, K.; Arčon, I.; Vavpetič, P.; Jeromel, L.; Pelicon, P.; Kumar, J.; Tuli, R. Pattern of iron distribution in maternal and filial tissues in wheat grains with contrasting levels of iron. *J. Exp. Bot.* **2013**, *64*, 3249–3260. [CrossRef] [PubMed]

26. Eroglu, S.; Karaca, N.; Vogel-Mikuš, K.; Kavčič, A.; Filiz, E.; Tanyolac, B. The conservation of VIT1-dependent iron distribution in seeds. *Front. Plant Sci.* **2019**, *10*, 907. [CrossRef] [PubMed]

27. Ibeas, M.A.; Grant-Grant, S.; Coronas, M.F.; Vargas-Pérez, J.I.; Navarro, N.; Abreu, I.; Castillo-Michel, H.; Avalos-Cembrano, N.; Paez Valencia, J.; Perez, F.; et al. The diverse iron distribution in eudicotyledoneae seeds: From arabidopsis to quinoa. *Front. Plant Sci.* **2019**, *9*, 1–10. [CrossRef] [PubMed]

28. De Brier, N.; Gomand, S.V.; Donner, E.; Paterson, D.; Smolders, E.; Delcour, J.A.; Lombi, E. Element distribution and iron speciation in mature wheat grains (*Triticum aestivum* L.) using synchrotron X-ray fluorescence microscopy mapping and X-ray absorption near-edge structure (XANES) imaging. *Plant. Cell Environ.* **2016**, *39*, 1835–1847. [CrossRef]

29. Regvar, M.; Eichert, D.; Kaulich, B.; Gianoncelli, A.; Pongrac, P.; Vogel-Mikuš, K.; Kreft, I. New insights into globoids of protein storage vacuoles in wheat aleurone using synchrotron soft X-ray microscopy. *J. Exp. Bot.* **2011**, *62*, 3929–3939. [CrossRef]

30. Kyriacou, B.; Moore, K.L.; Paterson, D.; de Jonge, M.D.; Howard, D.L.; Stangoulis, J.; Tester, M.; Lombi, E.; Johnson, A.A.T. Localization of iron in rice grain using synchrotron X-ray fluorescence microscopy and high resolution secondary ion mass spectrometry. *J. Cereal Sci.* **2014**, *59*, 173–180. [CrossRef]

31. O'Dell, B.L.; De Boland, A.R.; Koirtyohann, S.R. Distribution of phytate and nutritionally important elements among the morphological components of cereal grains. *J. Agric. Food Chem.* **1972**, *20*, 718–723. [CrossRef]

32. Singh, S.P.; Vogel-Mikuš, K.; Vavpetič, P.; Jeromel, L.; Pelicon, P.; Kumar, J.; Tuli, R. Spatial X-ray fluorescence micro-imaging of minerals in grain tissues of wheat and related genotypes. *Planta* **2014**, *240*, 277–289. [CrossRef]

33. Pongrac, P.; Scheers, N.; Sandberg, A.S.; Potisek, M.; Arčon, I.; Kreft, I.; Kump, P.; Vogel-Mikuš, K. The effects of hydrothermal processing and germination on Fe speciation and Fe bioaccessibility to human intestinal Caco-2 cells in Tartary buckwheat. *Food Chem.* **2016**, *199*, 782–790. [CrossRef] [PubMed]

34. Hurrell, R.; Egli, I. Iron bioavailability and dietary reference values. *Am. J. Clin. Nutr.* **2010**, *91*, 1461S–1467S. [CrossRef] [PubMed]

35. Rebetzke, G.J.; Bonnett, D.G.; Reynolds, M.P. Awns reduce grain number to increase grain size and harvestable yield in irrigated and rainfed spring wheat. *J. Exp. Bot.* **2016**, *67*, 2573–2586. [CrossRef] [PubMed]

36. Gomez-Becerra, H.F.; Erdem, H.; Yazici, A.; Tutus, Y.; Torun, B.; Ozturk, L.; Cakmak, I. Grain concentrations of protein and mineral nutrients in a large collection of spelt wheat grown under different environments. *J. Cereal Sci.* **2010**, *52*, 342–349. [CrossRef]

37. Detterbeck, A.; Nagel, M.; Rensch, S.; Weber, M.; Börner, A.; Persson, D.P.; Schjoerring, J.K.; Christov, V.; Clemens, S. The search for candidate genes associated with natural variation of grain Zn accumulation in barley. *Biochem. J.* **2019**, *476*, 1889–1909. [CrossRef]

38. Morgounov, A.; Gómez-Becerra, H.F.; Abugalieva, A.; Dzhunusova, M.; Yessimbekova, M.; Muminjanov, H.; Zelenskiy, Y.; Ozturk, L.; Cakmak, I. Iron and zinc grain density in common wheat grown in Central Asia. *Euphytica* **2007**, *155*, 193–203. [CrossRef]

39. Ficco, D.B.M.; Riefolo, C.; Nicastro, G.; De Simone, V.; Di Gesù, A.M.; Beleggia, R.; Platani, C.; Cattivelli, L.; De Vita, P. Phytate and mineral elements concentration in a collection of Italian durum wheat cultivars. *Field Crop. Res.* **2009**, *111*, 235–242. [CrossRef]

40. Murphy, K.M.; Hoagland, L.A.; Yan, L.; Colley, M.; Jones, S.S. Genotype × environment interactions for mineral concentration in grain of organically grown spring wheat. *Agron. J.* **2011**, *103*, 1734. [CrossRef]

41. Joshi, A.K.; Crossa, J.; Arun, B.; Chand, R.; Trethowan, R.; Vargas, M.; Ortiz-Monasterio, I. Genotype × environment interaction for zinc and iron concentration of wheat grain in eastern Gangetic plains of India. *Field. Crop. Res.* **2010**, *116*, 268–277. [CrossRef]

42. Li, S.; Zhou, X.; Huang, Y.; Zhu, L.; Zhang, S.; Zhao, Y.; Guo, J.; Chen, J.; Chen, R. Identification and characterization of the zinc-regulated transporters, iron-regulated transporter-like protein (ZIP) gene family in maize. *BMC Plant Biol.* **2013**, *13*, 114. [CrossRef]

43. Moore, K.L.; Rodríguez-Ramiro, I.; Jones, E.R.; Jones, E.J.; Rodríguez-Celma, J.; Halsey, K.; Domoney, C.; Shewry, P.R.; Fairweather-Tait, S.; Balk, J. The stage of seed development influences iron bioavailability in pea (*Pisum sativum* L.). *Sci. Rep.* **2018**, *8*, 1–11. [CrossRef]

44. Morris, E.R.; Ellis, R. Isolation of Monoferric Phytate from Wheat Bran and its Biological Value as an Iron Source to the Rat. *J. Nutr.* **1976**, *106*, 753–760. [CrossRef] [PubMed]

45. Thiel, J. Development of endosperm transfer cells in barley. *Front. Plant Sci.* **2014**, *5*, 1–12. [CrossRef] [PubMed]

46. Sreenivasulu, N.; Borisjuk, L.; Junker, B.H.; Mock, H.-P.; Rolletschek, H.; Seiffert, U.; Weschke, W.; Wobus, U. Barley Grain Development: Toward an Integrative View. In *International Review of Cell and Molecular Biology*; Jeon, K.W., Ed.; Academic Press: Amsterdam, The Netherlands, 2010; Volume 281, Chapter 2; pp. 49–89.

47. Grillet, L.; Ouerdane, L.; Flis, P.; Hoang, M.T.T.; Isaure, M.-P.; Lobinski, R.; Curie, C.; Mari, S. Ascorbate efflux as a new strategy for iron reduction and transport in plants. *J. Biol. Chem.* **2014**, *289*, 2515–2525. [CrossRef] [PubMed]

48. Nečemer, M.; Kump, P.; Ščančar, J.; Jaćimović, R.; Simčič, J.; Pelicon, P.; Budnar, M.; Jeran, Z.; Pongrac, P.; Regvar, M.; et al. Application of X-ray fluorescence analytical techniques in phytoremediation and plant biology studies. *Spectrochim. Acta Part B At. Spectrosc.* **2008**, *63*, 1240–1247. [CrossRef]

49. Vogel-Mikuš, K.; Pongrac, P.; Pelicon, P. Micro-PIXE elemental mapping for ionome studies of crop plants. *Int. J. PIXE* **2014**, *24*, 217–233. [CrossRef]

50. Wang, P.; McKenna, B.A.; Menzies, N.W.; Li, C.; Glover, C.J.; Zhao, F.J.; Kopittke, P.M. Minimizing experimental artefacts in synchrotron-based X-ray analyses of Fe speciation in tissues of rice plants. *J. Synchrotron Radiat.* **2019**, *26*, 1272–1279. [CrossRef]

51. Koren, Š.; Arčon, I.; Kump, P.; Nečemer, M.; Vogel-Mikuš, K. Influence of $CdCl_2$ and $CdSO_4$ supplementation on Cd distribution and ligand environment in leaves of the Cd hyperaccumulator *Noccaea (Thlaspi) praecox*. *Plant Soil* **2013**, *370*, 125–148. [CrossRef]

52. Kump, P.; Vogel-Mikuš, K. Quantification of 2D elemental distribution maps of intermediate-thick biological sections by low energy synchrotron μ-X-ray fluorescence spectrometry. *J. Instrum.* **2018**, *13*. [CrossRef]

53. Solé, V.A.; Papillon, E.; Cotte, M.; Walter, P.; Susini, J. A multiplatform code for the analysis of energy-dispersive X-ray fluorescence spectra. *Spectrochim. Acta Part B At. Spectrosc.* **2007**, *62*, 63–68. [CrossRef]

54. Ravel, B.; Newville, M.A. ATHENA, ARTEMIS, HEPHAESTUS: Data analysis for X-ray absorption spectroscopy using IFEFFIT. *J. Synchrotron Radiat.* **2005**, *12*, 537–541. [CrossRef] [PubMed]

© 2020 by the authors. Licensee MDPI, Basel, Switzerland. This article is an open access article distributed under the terms and conditions of the Creative Commons Attribution (CC BY) license (http://creativecommons.org/licenses/by/4.0/).

Commentary

Low phytic acid Crops: Observations Based On Four Decades of Research

Victor Raboy

USDA-ARS Small Grains and Potato Research Unit, 1691 South 2700 West, Aberdeen, ID 83210, USA; vraboy@gmail.com

Received: 29 November 2019; Accepted: 17 January 2020; Published: 22 January 2020

Abstract: The *low phytic acid (lpa)*, or "low-phytate" seed trait can provide numerous potential benefits to the nutritional quality of foods and feeds and to the sustainability of agricultural production. Major benefits include enhanced phosphorus (P) management contributing to enhanced sustainability in non-ruminant (poultry, swine, and fish) production; reduced environmental impact due to reduced waste P in non-ruminant production; enhanced "global" bioavailability of minerals (iron, zinc, calcium, magnesium) for both humans and non-ruminant animals; enhancement of animal health, productivity and the quality of animal products; development of "low seed total P" crops which also can enhance management of P in agricultural production and contribute to its sustainability. Evaluations of this trait by industry and by advocates of biofortification via breeding for enhanced mineral density have been too short term and too narrowly focused. Arguments against breeding for the low-phytate trait overstate the negatives such as potentially reduced yields and field performance or possible reductions in phytic acid's health benefits. Progress in breeding or genetically-engineering high-yielding stress-tolerant low-phytate crops continues. Perhaps due to the potential benefits of the low-phytate trait, the challenge of developing high-yielding, stress-tolerant low-phytate crops has become something of a holy grail for crop genetic engineering. While there are widely available and efficacious alternative approaches to deal with the problems posed by seed-derived dietary phytic acid, such as use of the enzyme phytase as a feed additive, or biofortification breeding, if there were an interest in developing low-phytate crops with good field performance and good seed quality, it could be accomplished given adequate time and support. Even with a moderate reduction in yield, in light of the numerous benefits of low-phytate types as human foods or animal feeds, should one not grow a nutritionally-enhanced crop variant that perhaps has 5% to 10% less yield than a standard variant but one that is substantially more nutritious? Such crops would be a benefit to human nutrition especially in populations at risk for iron and zinc deficiency, and a benefit to the sustainability of agricultural production.

Keywords: phytic acid; phosphorus; phytate; low phytic acid; lpa; seed; genetics; animal nutrition; human nutrition; sustainability

---

1. Introduction: The *low phytic acid* Trait

Phytic acid (myo-inositol-1,2,3,4,5,6-hexakisphospate; Figure 1A) is the storage form of phosphorus (P) in seeds, typically representing from 75 ± 10% of seed total P [1]. Following synthesis during seed development, it accumulates and is deposited as mixed "phytate" or "phytin" salts primarily of potassium (K) and magnesium (Mg). These salts may also contain iron (Fe) and zinc (Zn) and other mineral cations. Prior to 1980 (the year I began my PhD studies of genetic and environmental factors impacting soybean phytic acid), there was no "Mendelian" genetics of seed phytic acid; there were no known mutations or variant alleles that perturbed seed phytic acid synthesis or accumulation. The first *low phytic acid (lpa)* mutants of any crop species, maize (*Zea mays* L.) *low phytic acid* 1-1 (*Zmlpa*1-1) and

Zmlpa2-1, were isolated around 1990 in my then fairly new USDA-ARS (Agricultural Research Servic) lab [2]. Homozygosity for these first two maize mutations reduced seed phytic acid (myo-inositol hexakisphosphate) by 50% to 70% while having little discernible effect on seed total phosphorus (P) (illustrated in Figure 1B for maize *lpa*1-1). Isolation of maize *lpa*1-1 and *lpa*2-1 was followed in my lab and many others with the isolation of mutations in a third maize locus [3], *lpa* mutants in barley (*Hordeum vulgare* L.), rice (*Oryza sativa* L.), wheat (*Triticum aestivum* L.), soybean [*Glycine max* (L.) Merr.], common bean (*Phaseolus vulgaris* L.), in several additional crops and model systems such as *Arabidopsis thaliana* (L. (Heynh)), and increasingly via genetic engineering in a number of crops [4–6]. In seed homozygous for most but not all *lpa* mutations or genotypes, reductions in phytic acid P are largely matched by increases in inorganic P, with little or no change in seed total P. In some cases (such as maize *lpa*2) reductions in phytic acid P result in increased inorganic P and increases in other inositol phosphates such as myoinositol tris-, tetra*kis*, or penta*kis*phosphate, but the seed total P remains similar to wild-type (data not shown).

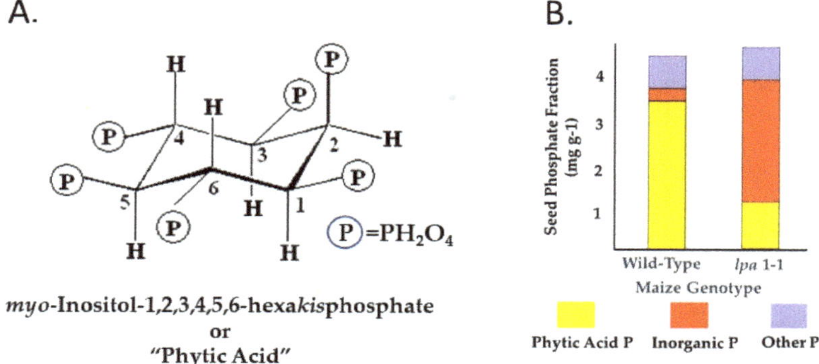

Figure 1. (**A**) Phytic acid, myo-inositol-1,2,3,4,5,6-hexakisphosphate. (**B**) The basic seed phosphorus (P) phenotype of "normal phytic acid" or wild-type and *low phytic acid* (*lpa*) genotypes, as illustrated by the initial analyses of wild-type and *low phytic acid* 1-1 maize isolines [2]. While seed total P may vary in any given genotype depending on various environmental, non-genetic factors, the relative contributions of phytic acid P, inorganic P and "other P" to seed total P typically remain fairly constant. "Other P" refers to all P other than phytic acid P and inorganic P, such as P found in other inositol phosphates, DNA, RNA, protein, carbohydrate, lipids, etc.

Here, I will present what I think are a few interesting observations and views about the practical or "applied" side of seed phytic acid and the potential role of *lpa* crops, first in reference to phytic acid's role in animal and human nutrition and second in reference to questions about breeding or engineering the low-phytate trait. The two main problems that the *lpa* trait can be used to address are (1) P management in non-ruminant animal production and; (2) mineral deficiency in humans. For many years there have been attractive alternative approaches and technologies to address these problems. For the P management issue, there is the use of phytase (phytic acid-specific phosphohydrolases; myo-inositol hexakisphosphate 3- and 6-phosphohydrolase; EC 3.1.3.8 and EC 3.1.3.26) as a feed additive [7,8] and for the human mineral nutrition problem, there is "biofortification" breeding for enhanced mineral density [9]. Both of these alternatives are less challenging to implement and more readily accessible than breeding the *lpa* trait and perhaps in specific cases have advantages over it, resulting in much wider acceptance, utilization and recognition. Does the *lpa* approach have a useful place in addressing these two problems? Does the *lpa* trait possibly have any advantages over these alternative approaches?

I will address these questions and present views developed through the course of nearly four decades of work. In summary, I have two main observations. First, in regards to the "low-phytate" trait and the alternative approaches to the problems it presents in human health and animal agriculture, people tend to "miss the forest for the trees". People tend to look at or address isolated effects or problems, but do not consider the bigger picture which is the sum of the positive effects described below. For example, studies might address the effect of reducing dietary phytate on iron retention but not on zinc, calcium or magnesium retention. Or biofortification breeding programs might target iron-dense crops, or zinc-dense crops, but usually not both in the same germplasm. Few if any have addressed what I define as "global" mineral nutritional quality; the nutritional quality of food in terms of all the major mineral nutrients. Programs or studies might address the problem of P management in animal production or the problem of mineral deficiency in humans, but not both. Second, I believe people's efforts or views are often too short-term, such that one might miss the long-term goals and benefits. For example, the development of a high-yielding low-phytate crop might require long-term, sustained effort in breeding or genetic engineering. This sort of long-term effort is not typical of most institutions.

First, an estimate of the potential value of the low-phytate trait will provide a basis for subsequent discussion. This will include a brief review of the role seed phytic acid plays as a major bottleneck in the flow of P through the world's agricultural ecosystem. Next, the fundamentally important question of the role of phytic acid in animal and human nutrition will be addressed. Last, a few points regarding the breeding the low-phytate crops will be addressed.

2. An Estimate of the Potential Value of the *low phytic acid* Trait

Studies conducted over the years with *lpa* mutants and genotypes of maize and other species indicate that the positive impact of this single-gene or simply-inherited trait on seed nutritional quality compares very favorably with any other single- or multigenic trait described to-date. As a preliminary example, in the case of maize *lpa*1-1 anecdotal evidence (unpublished) from researchers in the early 1990s indicated that pigs whose diet consisted solely of *lpa*1-1 maize, with no additional supplementation, did fairly well. They did not display the severe P and mineral deficiency and poor bone density observed if they were solely fed wild-type maize, which has low bioavailable P. The single *lpa*1-1 mutation erased these deficiencies and bone disease, converting maize into a nearly complete food. Can one imagine a situation where it might be beneficial to have a staple food and feed grain crop that is a nearly complete food, requiring little additional supplementation?

Seed phytic acid P was recognized as early as 1939 as a "non-available" form of P for monogastric animals (poultry, swine, fish) [10–12]. In the 20th century, there was growing concern over the contribution of phytic acid-derived P in animal waste to water pollution and eutrophication [13]. So, as the first major benefit of the *lpa* trait, studies utilizing *lpa* lines (reviewed in [6]) revealed that use of these lines in non-ruminant animal feeds enhanced seed-derived P bioavailability proportional to the decrease in seed phytic acid P and increase in inorganic P, and depending on diet formulation, could substantially reduce animal waste P.

Since the mid-20th century, seed-derived dietary phytic acid was well known to play an important role in the nutritional quality of human foods, via its negative impact on mineral (iron, zinc, magnesium, calcium, phosphorus) retention and utilization [14,15]. Mineral deficiency, especially iron and zinc deficiency, was widely recognized as a major international public health problem [16,17]. As the second major benefit of the *lpa* trait, studies indicated that human and non-ruminant mineral nutrition including iron, zinc, and calcium nutrition, is enhanced following consumption of foods or feeds prepared with *lpa* types as compared with normal-phytate types (reviewed in [6]).

Perturbation of phytic acid synthesis or accumulation may also favorably alter the distribution of minerals across the tissues of the cereal grain, in some cases resulting in higher mineral levels in the central starchy endosperm, in turn resulting in higher mineral density in milled products like white rice (*Oryza sativa* L.; reviewed in [6]). It is also likely that protein utilization will be enhanced by reductions in seed-derived dietary phytic acid [7]. The biosynthetic pathways to both phytic acid and the raffinosaccharide series of sugars share a common precursor, myo-inositol [18]. Knock-out of the soybean (*Glycine max* L. (Merr.)) genome's seed-specific myo-Inositol 1-phosphate synthase ("MIPS") gene, encoding the enzyme which is the sole biosynthetic source of the myo-inositol ring, resulted in seeds with both reduced phytic acid and reduced raffinosaccharides, the later an undigestible and undesirable component of foods [18].

A study published in 2000 found that pigs consuming *lpa*1-1 diets as compared with normal-phytate diets were leaner, had enhanced muscle density and had less "backfat" [19]. A patent was then awarded to a reputable company claiming a method to produce eggs with reduced overall cholesterol and enhanced cholesterol quality (reduced relative levels of low-density lipoproteins), this method being consumption of maize *lpa1-1* as compared with wild-type [20]. Since crop lines with varying, genetically-determined levels of endogenous seed phytic acid were not available prior to the isolation of these first *lpa* mutants, one might predict that their subsequent use in nutrition studies would yield unexpected findings or outcomes not anticipated based on prior science. The finding of the potential benefit of the low-phytate trait in cholesterol management represents an early example of this. Mechanisms that might contribute to enhanced "leanness" and muscle density, desirable traits in humans as well as pigs, include the possibility that consumption of *lpa* feeds leads to enhanced overall mineral nutritional health, enhanced P/Ca status, and enhanced protein digestion resulting in enhanced amino-acid nutritional status.

Any discussion of real-world problems relating to seed phytic acid should include the fact that it represents a major bottleneck in P flux through the world-wide agricultural ecosystem (Figure 2). The total amount of seed phytic acid P annually produced by major crops represents a sum equivalent to nearly 65% of fertilizer P manufactured annually worldwide [21]. This bottleneck represents a potentially valuable target in efforts to reduce the negative environmental impact of agricultural production. Agricultural P runoff contributes to surface water pollution and eutrophication, which in turn leads to oxygen depletion, die-off and "dead zones" [22]. This is a major, ongoing, world-wide and newsworthy (https://www.nytimes.com/interactive/2019/12/25/world/europe/farms-environment.html) problem. For example, a recent survey [23] found that in terms of eutrophication, greater than 95% of the Baltic sea is considered a "problem area". Low-phytate genetics can help with the global eutrophication problem in two ways: (1) first and foremost via its beneficial change in seed chemistry where total P remains unchanged but substantially more of that P is bioavailable for non-ruminants, resulting in more P in "product" and less P in "waste" (Figure 2); (2) via *lpa* alleles that both alter seed chemistry and condition reduced seed total P amount. For example, if one could reduce seed P by 20% but not impact yield, that would be equivalent to increasing the "fuel efficiency" of that crop, at least in terms of the macronutrient P; one would obtain the same amount of grain per unit of production but during harvest remove or "mine" 20% less P from the field, leaving it in the field for subsequent years of production [24].

That is exactly the case with barley *lpa*1-1 [25–27] (reviewed in [24]). In addition to a ~50% reduction in grain phytic acid P, it also conditions a 15% to 20% reduction in grain total P, while having little or no effect on crop yield (see below). Thus barley *lpa*1-1 represents the first low-seed total P crop variant necessary for this novel approach to enhancing P management in crop production. In fact, it conditions a seed P phenotype that represents the ideal for all end uses: reduced seed total P accompanied by reduced phytic acid P.

It has estimated that for annual global rice production alone, a 20% reduction in soil P mining during crop production, achieved via a genetic reduction of a similar extent in seed total P, could save producers "several hundred million US dollars annually in fertilizer inputs" [28]. It would also help address a potentially even more important long-term problem, "Peak Phosphorus" (reviewed in [22]). Phosphorus used for fertilizer is obtained from a potentially limiting resource, rock phosphate. Future reserves may for both political and technical reasons become limiting to agricultural production. A

20% reduction in phosphate use that has little impact on crop productivity could significantly enhance the long-term sustainability of agricultural production.

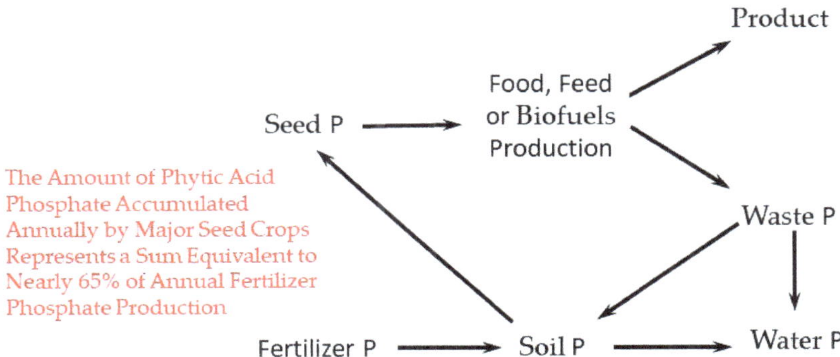

Figure 2. Phosphorus (P) flow through the world-wide agricultural ecosystem.

Barley *lpa*1-1's good field performance and yield has led to the breeding and subsequent release of two barley cultivars with this unique seed P phenotype; "Herald" [29] and "Clearwater" [30]. The release of such cultivars represents a very solid proof-of-principle and validation of the feasibility of both (1) the "low-phytate" approach to enhancing the nutritional value of crops and (2) the "low-seed total P" approach to enhancing P management in agriculture.

The gene perturbed in barley *lpa*1-1 was subsequently identified as encoding a member of the sulfate family of transporters and thus termed *HvST (H. vulgare Sulfate Transporter*; [31]). Some members of this large gene family have functions other than sulfate transport. Thus the ortholog perturbed in barley *lpa*1-1 probably functions in P transport specific to phytic acid synthesis. This possibility was confirmed via analysis of a rice ortholog of this gene, termed SULT-like Phosphorus Distribution Transporter (SPDT; [32]). A knock-out of this rice ortholog resulted in a seed P phenotype almost identical to that of barley *lpa*1-1, reduced phytate and a 20% to 30% reduction in seed total P, and like barley *lpa*1-1, appeared to have little impact on plant performance in an initial field trial.

The work with the barley *lpa*1-1 and the rice the SPDT knockouts also illustrate an important value of the classical "forward genetics" approach represented by the isolation of *lpa* mutants. Forward genetics is defined as the isolation of a mutant phenotype followed by identification of the mutation and gene conditioning that phenotype. That value is that it provides the information needed to then conduct "reverse genetics", where perturbations of that gene are first obtained and the phenotypes they condition subsequently described. In the case of the low-phytate trait, the first example of this process was the isolation of maize *lpa1-1* [2], which then led to the identification of the gene perturbed in maize *lpa*1-1, the sequence of which was then subsequently used to genetic engineer the trait into soybean. [33].

As a summary of this estimate of the potential value of the *low phytic acid* trait: taken together, these various nutritional and P-management benefits, of value to human nutrition and animal production, and to efforts to reduce the environmental impact and enhance the sustainability of agricultural production, are pretty amazing for a single-gene or simply-inherited seed chemistry trait.

In 1987, when starting my career in the USDA-ARS, it was helpful to have the positive rationale that isolating *lpa* mutants would be useful in breeding of *lpa* crops, which in turn could be used to reduce water pollution from animal agriculture, then a popular concern. However, I viewed breeding low-phytate crops as a long-term and even speculative goal. I had two other initial goals that I thought were more straightforward in the short-term: to use such genetics resources to study the biology of phytic acid in plants, such as figuring out phytic acid's biosynthetic pathway; to use the near-isogenic lines we could develop to better analyze the role of seed phytic acid in human and animal nutrition. In the 1980's it was not yet widely understood that inositol hexaphosphate is the most abundant inositol phosphate in nature, and a central metabolic pool in the eukaryotic cell [34,35]. The "inositol phosphate pathways" are central to cellular sensing and signaling including P sensing and signaling [4], and several other critically important cellular processes in all eukaryotes, such as RNA processing. In this context, the first maize *lpa* mutants were essentially the first inositol phosphate pathways mutants identified in any eukaryotic species. The isolation of these and other *lpa* mutants has led to the identification of several novel genes and functions and thus has contributed to our understanding of the biosynthesis of phytic aid during seed development and has contributed to basic cell biology [3,4,33,36,37]. For example, the maize *lpa*1 gene was the first inositol hexaphosphate transporter identified in any species, and such transporters are key to the roles inositol hexaphosphate plays in cellular metabolism and signaling [33].

But here I will mostly focus on the applied side of things, the role of seed phytic acid in animal and human nutrition and the breeding/genetic engineering of *lpa* crops.

3. Lessons Learned From Animal and Human Nutrition Studies

Working for four decades in the field of genetics, breeding, and engineering food and feed crops for enhanced nutritional value gives one some perspective. Below I will attempt to explain why: (1) I think that in the case of the use of the *lpa* trait in animal feeds, decisions were made early on that focused too much on short-term profitability and too much solely on P management and; (2) in breeding or genetics to enhance nutritional value of crops for human foods, too great a focus was placed on biofortification via enhanced mineral density to the detriment of support for breeding aimed at antinutrients like phytic acid. In both the feed and food cases, these decisions then impacted long-term strategies and long-term funding support possibly to the detriment of agricultural production's efficiency and sustainability, and environmental and public health. This is only an opinion and I, of course, could be biased or wrong, or perhaps simply just "venting" about what I perceive as the lack of support for the low-phytate approach.

3.1. The Low Phytic Acid Trait and "Available Phosphorus" in Non-Ruminant Animal Agriculture

First let us take the example of how the *lpa* trait was initially viewed and evaluated by an Ag biotech company when it was first available for use in crop breeding in the early 1990s. The main interest in seed phytic acid in the agricultural industry in the U.S. and Europe was and largely remains its role as the major P fraction in grains and legumes destined for use in poultry and swine feeds. Since phytic acid P is "non-available P" whereas essentially all other forms of P are "available P" (Figure 3A; reviewed in [6]), the low-phytate trait in maize was initially referred to as "High Available P" or "HAP" corn [38]. The approach initially used by this Ag biotech company to determine the value of the *lpa* trait was in terms of the dollar value of the increase in available P, using then-current (early 1990s) P prices. For example, using the market cost for rock phosphate (https://www.indexmundi.com/commodities/?commodity=rock-phosphate&months=300); if an *lpa* allele converts half of the phytic acid P in seeds from "non-available P" to "available P", in 1990 the available P in a low-phytate line would represent ~$5 billion (US) (Figure 3B, right), as compared with $1.66 billion (US) for the available P in the "normal phytate" line (Figure 3B, left).

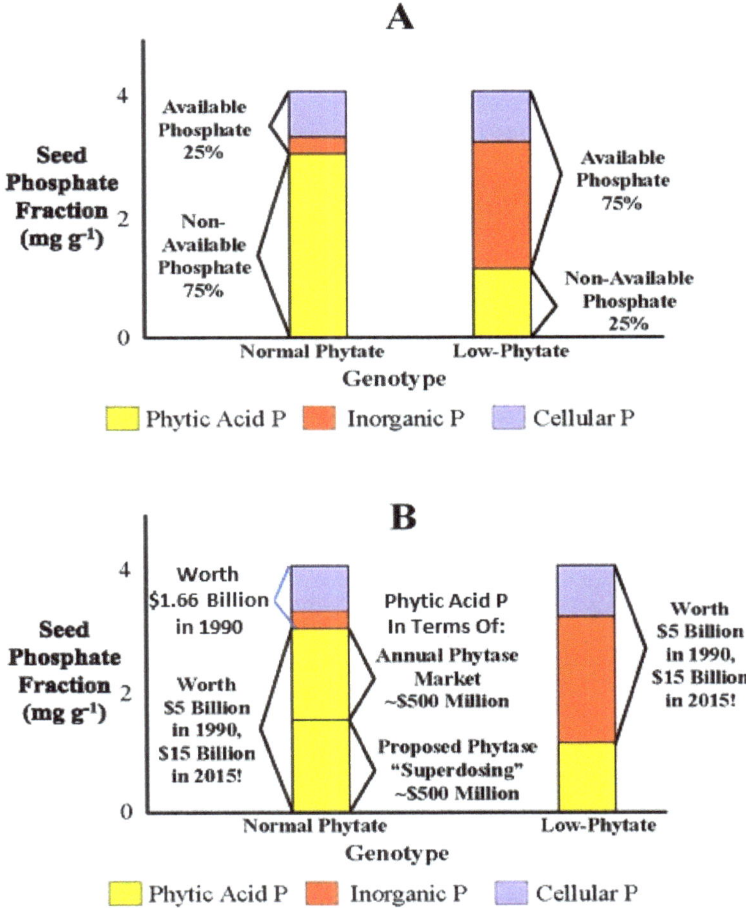

Figure 3. (**A**) Typical seed phosphorus (P) fractions and their bioavailability to non-ruminant animals, in a wild-type or "normal phytate" crop type versus a low-phytate line. "Available phosphate" and "non-available phosphate" refers to that fraction of seed total P that would be absorbed and utilized following consumption by non-ruminants. This is only an illustration for discussion purposes and does not represent actual data for any given crop or low-phytate line. Cereal crops usually have 3.0 to 4.0 mg Total P g⁻¹, such as in the illustration. Legume and oil-seed crops can often have 5.0 or more mg Total P g⁻¹, but the relative proportions of phytic acid P, inorganic P and "cellular P" remain similar to that observed in cereal grains. "Cellular P" refers to all P other than phytic acid P and inorganic P, such as P found in DNA, RNA, protein, carbohydrate, lipids, etc. (**B**) The same bar graph of seed P fractions but with: (1) the relative value in U.S. dollars of the respective seed P fractions, expressed in terms of the market price for rock phosphate (adjusted for inflation), in 1990 versus 2015 and (2) the relative value of the current "phytase market" and the proposed "superdosing" of phytase.

Since P was relatively inexpensive in 1990, valuating the low-phytate trait in terms of "feed P equivalents" created several hurdles to development and production. Rather than using the well-proven "penetration pricing" strategy, where new products are initially sold at relatively low prices to capture market share [39,40], industry participants at that time wanted to attach a "technology surcharge" to

HAP hybrid seed, in my opinion, a short-sighted business strategy. The problem was the "feed P value" of the first *lpa* types, based on market prices for P in the early 1990s, was not sufficient to accommodate this "technology surcharge". But what if the price of P for use in feeds and fertilizers dramatically increased over the next 10 to 20 years? What if the price of P will continue to increase or experience price spikes? It takes a minimum of ten but often many more years to go from the initial development of a new biotech trait to its widespread marketing [41]. Should not that early decision been based on future projections of P prices? Today the increasing cost and potential future scarcity of rock phosphate for use in fertilizer P and feed P production is changing views of the importance of seed phytic acid. By 2015, the cost of rock phosphate had roughly tripled since 1990, so that net available P in a low-phytate line where half the phytate is converted to available P would represent $15 billion (US) (Figure 3B, right). Also, using this same estimate of the dollar value of seed P, if the 20% to 30% reduction in seed total P trait of barley *lpa*1-1 or the rice SPDT knock-outs were engineered into all major grain and legume seed crops, that could potentially represent a savings of up to$ 4.5 billion (US), annually.

A second way to look at the value of the seed P represented by phytic acid P (non-available P) is to consider the total market value for the feed additive phytase needed to break down a given amount of feed phytic acid (Figure 3B, middle). In this discussion, it is useful to remember that the world market for phytase as a feed additive is the largest market for any industrial enzyme. In 2015, the total market value for the standard rate of the feed additive phytase (500 units enzyme/kilo feed, an application rates designed to breakdown ~50% of feed-derived phytic acid), was >$500 million (US). The objective of this application rate is not to improve animal health or productivity, but rather to meet regulatory standards designed to ameliorate water pollution by dictating acceptable levels of animal waste P. Subsequently, participants in the industry started advocating phytase "superdosing"; application rates twice, three times or more than 500 units per kilo feed (http://phytate.info/superdosing-phytase). The primary rationale for phytase superdosing was that the resulting application rates achieved a near-complete breakdown of dietary phytic acid and that this had added benefits beyond the 50% improvement in dietary P utilization achieved via the "500 enzyme unit" application rate. These additional benefits were enhanced overall animal health, resulting largely from optimized mineral nutritional health, leading to enhanced productivity. A secondary benefit was more efficient use of increasingly expensive P and optimized reduction in waste P. If a proposed "superdosing" use of phytase becomes industry standard, this would then add at a minimum another >$500 mMillion (US) (Figure 3B, middle).

So if annual phytase costs were used to value a genetically-determined 50% reduction in seed phytic acid, it would at a minimum be worth $0.5 Billion for maize alone. But who gets that value? Is it the grain grower, the phytase producer, or the livestock producer? If the dollar value of enhancing P management and reducing P waste is captured in the grain, breeding low-phytate crops should ultimately bring profits to the grain grower.

The rationale for phytase superdosing was based in part on the findings of the first generation of animal nutrition studies using the initial maize and barley *lpa* near-isogenic lines my lab USDA-ARS lab-produced: genetically-determined reductions in crop seed phytic acid translated into enhanced mineral nutritional health in a global sense (enhanced P, Ca, Zn, Fe, Mg etc.), and in other possible benefits such as enhanced protein utilization, that in sum resulted in healthier, more productive animals. This validated one of my major initial objectives in studying seed phytic acid genetics: that the resulting development of sets of near-isogenic lines that largely differ only in seed phytic acid P level will be valuable in studying the role of phytic acid in animal and human nutrition. Thus the resulting research was not only of value in studying the role of dietary phytic acid but also led to rationalizing increased phytase use!

That an opportunity was missed in the 1990s by the maize breeding industry is highlighted by the fact that little known subsequent studies have shown that both corn farmers who might grow HAP corn and consumers of foods produced using the non-GMO HAP corn favorably viewed its environmental benefits [42,43]. A survey of the Delmarva Peninsula (on the eastern side of the Chesapeake Bay, US) corn farmers found a willingness to grow non-GMO HAP corn due to the environmental benefits but increasing resistance if doing so was accompanied by higher production costs or reduced yield [42]. A separate study found consumers looked favorably on chicken produced using HAP corn [43], again due to the environmental benefits. This also serves as a good example of how focusing just on P management limits a full appreciation of the value of the *lpa* trait. What if those farmers or consumers were instead asked if they would look favorably on use of a new corn type in feeds that are "high available P", with the resulting environmental benefits, but also results in healthier animals due to a global enhancement of mineral nutritional health and possibly healthier animal products, due to reduced cholesterol and increased "leanness"?

One critical hurdle to the adoption of HAP corn was that initial yield studies of lines into which an *lpa* allele had been crossed indicated that it was associated with a yield loss of perhaps 5% to 10%. Since *lpa* corn is competing with commodity "yellow-dent" corn for the end-use strictly as a feed ingredient, in the horizontally-organized agricultural system in the U.S., where grain growers in most cases derive their income entirely from the amount of grain they can produce, that initial *lpa*-associated yield loss represented a financial hurdle. There was little consideration at the time of how sustained breeding and research, meaning over five to 10 years or longer, might result in elite performance of a non-GMO *lpa* type. That is simply too great a time span for any private-sector company's balance sheet. Further, in these initial deliberations in the early 1990's, consideration of environmental issues or sustainable management of P issues were secondary to the simple question of the cost of feed P. No predictions of possible changes in feed P costs over the long-term, meaning over decades, were included.

These ways of estimating the value of the *lpa* trait are meant only to provide a historical perspective on views and attitudes that impact decision making in the agricultural industry. Looking at this trait simply in terms of the value of the P in seeds as determined by market rates for P fertilizer or feed P undervalues the trait over the longer term, and does not take into account sustainability, the value of reduced environmental impact and water quality, nor the potential full benefits for animal health and productivity independent of P nutrition.

3.2. Low Phytic Acid Near-Isogenic Lines: A Powerful Model to Study the Impact of Seed-Derived Dietary Phytic Acid in Human and Animal Nutrition

The power of *lpa* near-isogenic lines as an experimental tool in human and animal research is that one can produce test foods or feeds with very accurately known levels of seed-derived endogenous phytic acid that are stable and vary from wild-type levels in a step-wise fashion, through moderate reductions to lines which produce seed with a near-absence of phytic acid. No additional experimental ("artificial") manipulations are needed. As a result, when using these lines, the impact of dietary phytic acid on some aspect of nutrition can be assayed more quantitatively, and with a higher degree of accuracy, than is possible using other experimental approaches. Furthermore, any difference in nutritional outcomes observed between wild-type versus *lpa* near-isogenic lines can be attributed to the single-gene allelic difference conditioning the trait. As an illustration, consider two nutrition studies, one with pigs (*Sus scrofa domesticus* (Erxleben)) and one with trout (*Oncorhynchus mykiss* (Walbaum)), that were conducted with a set of four barley near-isogneic lines (Figure 4) [44,45]. These four lines produce seed with either wild-type levels of seed phytic acid (cv. Harrington), or seed with step-wise reductions in phytic acid: a ~40% reduction in *Hvlpa*1-1, a ~70% reduction in *Hvlpa*3-1, and a >95% reduction in *Hvlpa*M955. This provides a nicely linear set of feed phytic acid treatments that in both studies revealed a highly linear negative correlation between grain phytic acid level and calcium bioavailability ("% apparent digestibility coefficient" or (ADC) in the trout study and "% retention/intake" in the pig study).

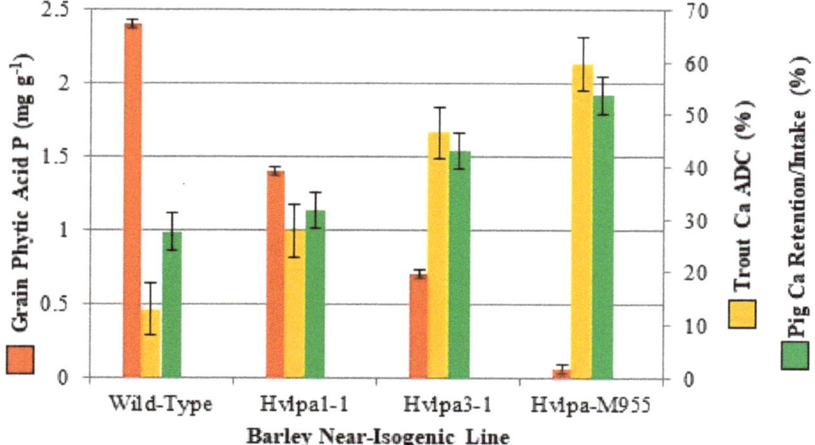

Figure 4. The highly linear negative relationship between dietary calcium bioavailability and grain phytic acid level, as observed in two separate studies, one with pigs [44] and one with trout [45]. Calcium bioavailability was measured in trout as "apparent digestibility coefficient" (ADC) and in pigs as "% retention/intake". Animals were fed diets prepared with four barley near-isogenic lines produced in the same location: (1) wild-type (for grain phytic acid, the cv. Harrington); *Hvlpa*1-1 with a ~40% reduction in grain phytic acid; *Hvlpa*3-1 with a ~70% reduction in grain phytic acid; and *Hvlpa*-M955 with a >95% reduction in grain phytic acid. Error bars are the standard deviation of the mean or "standard error" (SE) for each variable: grain phytic acid P, SE = 0.03, n = 3; Trout Ca ADC %, SE = 5.01, n = 2; Pig Ca retention/intake %, SE = 3.50, n = 5.

Calcium nutritional health is both important to animal health and productivity and is a significant public health issue worldwide, impacting at least 200 million people, many of whom consume diets rich in phytic acid. A study with human subjects confirmed the findings in studies with pigs or trout (Figure 4): that genetically-determined reductions in seed phytic acid (in this case maize) translated into enhanced calcium bioavailability [46]. Tortillas were prepared using seed produced by a pair of maize near-isogenic "Dent Corn" hybrids: a "wild-type" hybrid that produced seed with wild-type levels of phytic acid, or *Zmlpa*1-1 hybrid that produced seed with a ~66% reduction in phytic acid. "Fractional calcium absorption" was 43% greater following consumption of the *Zmlpa*1-1 tortillas compared with that observed following consumption of "wild-type" tortillas.

One important conclusion arrived at via observation of linearity of response has to do with the historical use of thresholds of dietary phytic acid in studying its impact on iron and zinc bioavailability. While defining such thresholds is important if not essential to progress in understanding the impact of dietary phytic acid, a large number of studies utilizing wild-type and *lpa* isolines in various crop species, and conducted with both human subjects and animal models, clearly illustrates that genetically-determined, step-wise reductions in grain- or seed-derived dietary phytic acid typically results in what appears to be fairly linear increases in iron and zinc bioavailability. For example, an oft-quoted paper established a widely-accepted threshold that a reduction in dietary phytic acid of at least 90% was essential to see benefits for iron bioavailability [47]. But studies with maize *lpa* isolines revealed that iron absorption was 49% greater in human subjects following consumption of foods prepared with maize grain with reductions in phytic acid of only 66%, as compared with normal-phytate grain [48].

In the case of zinc, reductions in the phytic acid:zinc molar ratio to <10 to 20 (depending on levels of phytic acid, zinc, calcium, and other dietary constituents) have been viewed as necessary to observe positive impacts in zinc bioavailability [49,50]. An initial study with human subjects that evaluated a pair of "near-isogenic" maize hybrids, one a "normal-phyate" hybrid and the second an *lpa* hybrids, found that the reduction of the phytic acid:zinc molar ratio from 36 in the normal-phytate grain to 17 in the *lpa* grain nearly doubled fractional zinc absorption (from 0.17 to 0.30) [51]. A second study using two sets of near-isogenic hybrids reported a linear negative relationship between phytic acid:zinc molar ratios (ranging from 7.5 to 35), and fractional zinc absorption [52]. While the phytic acid:zinc molar ratios in both low-phytate types was <20, what is important is that there appeared to be a linearity of response [52]. In addition to the importance these observations have for public policy and the development of strategies to address mineral nutritional health, they also should inform the selection of targets for breeding *lpa* crops: even moderate reductions as compared with wild-type would result in enhanced iron and zinc bioavailability.

3.3. "Biofortification" via Breeding for Elevated Zinc or Iron vs. the Low Phytic Acid Approach? Has the Harvest Plus Project Been Too One-Sided? Has it Missed an Opportunity to Support a Type of Crop Breeding that would Benefit Human Health?

Biofortification via breeding crops for elevated micronutrient density is an effective approach to addressing micronutrient deficiency in at-risk populations, both in terms of crop genetics but also as an "appropriate technology", the latter defined in part as a technology that is practical for and suitable to the social and economic conditions of specific locale, culture or society [53,54]. In recognition of the importance of this strategy to enhancing human health in at-risk populations through crop breeding, one of its main advocates, Howarth Bouis of the International Food Policy Research Institute (IFPRI), was recognized in 2016 with the World Food Prize (https://www.ifpri.org/news-release/howarth-bouis-wins-world-food-prize). As of 2017, it is estimated that 20 million people in farm households in the developing world were growing biofortified crops [54]. That is impressive. This was in part the result of the vision and the ultimately successful acquisition of support and funding that did not come without struggle and substantial persistence (https://www.harvestplus.org/about/our-history).

I've often wondered though that this success may have resulted in "paradigm block" and pushed other approaches to the sidelines. It may have resulted in the "canalization" of funding. But in the case of mineral deficiency, if one considers not just iron and zinc but other important mineral nutrients such as calcium and magnesium, is breeding for elevated levels of micronutrients going to be efficacious if phytic acid levels remain high? In at least one case, the biofortification breeding of common bean (*Phaseolus vulgaris* L.) for elevated iron levels, the short answer is no [55]. I will return to these particular studies and the recommendations that resulted from them in more detail below.

If the goal is to enhance mineral nutritional health globally (to enhance mineral status for all nutritionally-important minerals), can one simultaneously breed or engineer high levels of many of the most nutritionally-important minerals? The argument has been made that in some cases selection or breeding for elevated levels of one element may also result in elevated levels of other elements because there are positive correlations between the seed concentrations of some elements [56]. This may occur in some cases, but I find arguments for this approach to understate its complexities. For example, seed calcium levels are not well correlated with zinc or iron levels, and the distribution of calcium, iron, and zinc in the grain differ greatly, impacting levels in food products made from differing grain fractions. In contrast, the results of animal and human trials of low-phytate types described here indicate that reducing dietary phytic acid via the low-phytate approach would result in enhanced global mineral nutrition.

Furthermore, there are other substantial considerations that argue for combining the low-phytate approach with the "high mineral" approach. Consider the cases of zinc and iron. In the case of zinc, the negative impact of dietary phytic acid is both on the zinc consumed in a meal, but also on endogenous zinc encountered in the intestinal tract, and this latter effect may contribute substantially to net zinc loss [57]. Thus in the case of populations that consume substantial amounts of phytic acid in the grain- and legume-based diets, the full benefits of elevated dietary zinc may be reduced by high phytate levels.

Consider the results of a study that evaluated zinc nutrition with chicks consuming feeds prepared with a normal-phytate barley and barley *lpa*-M955, in which seed phytic acid is reduced 95% (Figure 5) [58]. Chicks were fed six different diets. These were prepared such that the sole source of phytic acid was either a wild-type (for grain phytic acid) barley near-isogenic line or a sibling near-isogenic line ("M955" for *Hvlpa*-M955) in which grain phytic acid is reduced by >90%. Further, diets prepared with each barley were supplemented with either 0.0, 10 mg/kg or 20 mg/kg zinc. The two grain types had similar endogenous zinc levels (about 23 to 24 mg Zn/kg). The results (Figure 5) clearly illustrate that zinc nutritional health, as measured by tibia zinc, was enhanced by zinc supplementation only in those animals consuming wild-type barley. Zinc nutritional health as assayed by tibia zinc was optimal in animals consuming the M955 barley regardless of zinc supplementation level. Thus, in the case of this low-phytate barley, increasing levels of zinc supplementation resulted in no added benefit. This clearly illustrates that the endogenous zinc levels of barley grain are adequate for optimal zinc health, if not for the endogenous phytic acid. From this perspective, zinc supplementation functions only to overcome reduced bioavailability due to dietary phytic acid. Should not this be a consideration in policy-making for approaches to deal with world-wide zinc deficiency?

Still, these results (Figure 5) do illustrate the value of zinc supplementation or biofortification for those crops/foods where seed-derived dietary phytic acid is at the high levels typical of many cereal- and legume-based foods. A good example of this is a recent evaluation of fractional and total zinc absorption in young rural Zambian children from meals prepared with zinc-biofortified maize [59]. Despite relatively high phytate levels, consumption of meals prepared with zinc-biofortified maize (34 µg Zn/g grain) resulted in a near-doubling of total absorbed zinc (1.1 mg Zn absorbed/day) when compared with that resulting from consumption of meals prepared with standard maize (21 µg Zn/g grain; 0.6 mg Zn absorbed/d). The authors note that this increase in total zinc absorption from the biofortified grain "occurred despite the high phytate concentration and high dietary phytate:zinc molar ratio of the biofortified grain" [59]. Of course one might wonder how combining "high grain zinc" with low-phytate might have further enhanced zinc bioavailability?

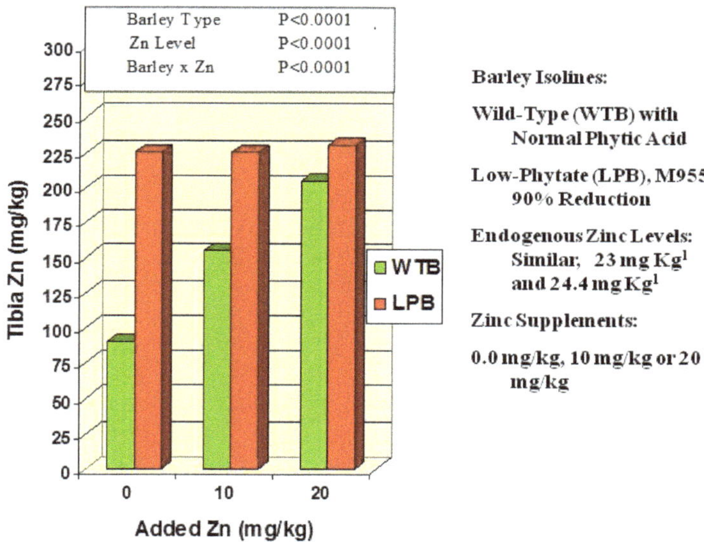

Figure 5. Effect of barley grain phytic acid level and supplemental zinc on chick tibia zinc at 21 days of age. The standard deviation of the mean or "standard error" ("SE") was 6.08, n = 4.

In the case of iron, consider that as part of the Harvest Plus program (https://www.harvestplus.org/), common bean was chosen as one of the main targets for iron biofortification efforts. But in fact, the negative impact of phytic acid in foods prepared from a high-iron common bean produced via biofortification breeding led leading experts in the field to the conclusion that the common bean has "limited potential as a vehicle for iron biofortification" [60]. These stable isotope studies with Rwandese women found that any benefit in the consumption of meals prepared with iron-biofortified beans versus "normal iron" beans was negated by the phytic acid content of the bean. In a subsequent study, this same group found that fractional iron absorption was increased 60% to 130% (depending on polyphenol content) and total iron absorption was increased 60% to 163% (again depending on the polyphenol content) when foods were prepared using an *lpa* bean with a 90% phytic acid reduction as compared with a normal-phytate bean [60]. This group subsequently concluded [61] that "due to the low bioavailability of bean iron . . . exclusively breeding for high iron concentration may not provide enough additional absorbable iron to impact iron status" and that "the focus should now be on PA reduction". Of relevance to questions of agronomic performance, the low-phytate bean line used in this study had previously been found to have little effect on plant performance or yield [62]. I will return to this last point below.

This group's results and recommendations are informative when addressing what I view as the "paradigm-lock" in the biofortification field. The paradigm in question, as repeatedly stated by proponents of the biofortification approach [9,53,63], can be summarized as follows: (1) breeding for enhanced mineral density is feasible for many different staple food crops and is an "appropriate technology" for addressing micronutrient deficiency in the most at-risk populations in the developing world; (2) it will have little or no negative effects on crop yields but rather probably will enhance crop yields; (3) breeding for low-phytate may be problematic since dietary phytic acid may also have positive nutritional benefits that might be lost; (4) those who attempt breeding for low-phytate should proceed with caution since phytic acid is so important to a plant's basic biology that breeding for low-phytate will inevitably harm plant growth and productivity. One can get the impression that in the case of breeding for enhanced mineral nutritional quality, biofortification breeding for enhanced mineral density is all good and low-phytate is all bad!

I have a few problems with this paradigm. Take the case of zinc. Leading experts in the international nutrition field have identified phytate "as the most important inhibitor of zinc absorption in adult human diets" [64]. The model and equation currently used for determining the bioavailable zinc in food [65] requires knowledge of the phytic acid content of that food. The European Food Safety Authority (EFSA) used this model to generate a new set of dietary zinc recommendations for adults based on four levels of dietary phytate [66]. Seed-derived dietary phytic acid plays a similar negative role in iron bioavailability and deficiency [15,16,47]. Therefore it seems obvious that addressing the phytic acid content of foods is critical to the question of iron and zinc deficiency. In this regard, it is telling that in two recent reviews of progress in iron and zinc biofortification [54,67] there is not one mention of phytic acid or phytic acid:zinc molar ratios. In these reviews, curiously, there is also no mention of negative or inconclusive results, such as those reported by Petry et al. [55].

But in cases such as foods prepared with legumes or whole grains, breeding for enhanced mineral density may not be needed at all, if one breeds for reduced phytic acid. As indicated in the studies with barley (zinc) and beans (iron) discussed above [58,60], simply reducing phytic acid may provide sufficient enhancement of both zinc and iron bioavailability in many foods. Furthermore, consider the results of analysis of iron bio-availability from grain of maize wild-type and *lpa*1-1 lines, the later producing grain with a ~66% reduction in phytic acid, when assayed using the human Caco-2 (colon adenocarcinoma) cell in vitro assay (Figure 6; unpublished results kindly provided by Ray Glahn, USDA-ARS). The Caco-2 assay was designed to study the intestinal absorption of nutrients [68,69]. Use of this assay indicated that the addition of ascorbic acid, a known enhancer of iron absorption, to wild-type maize flour more than doubled iron uptake observed in assays of wild-type maize with no ascorbic acid (Figure 6). Importantly, iron uptake from *lpa*1-1 maize flour was essentially identical to that observed from wild-type flour plus ascorbic acid. Furthermore, the addition of ascorbic acid to *lpa*1-1 maize flour further increased iron uptake by 40%.

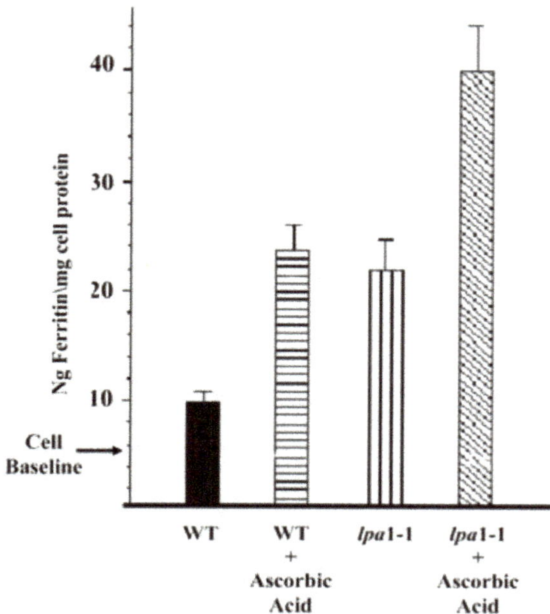

Figure 6. Caco-2 Assay for iron bioavailability from flour produced from maize wild-type (WT) and *lpa*1-1 isohybrid grain. The *lpa*1-1 grain contained 66% less phytic acid than WT. Bars indicated standard deviations, n = 3 (data provided by Ray Glahn, USDA-ARS Plant, Soil and Nutrition Lab, Cornell University, Ithaca, NY, USA).

These Caco-2 results provide further evidence that the inherent nutritional value of *lpa* grain can provide benefits equal to those achieved via supplementation or via the use of additives like ascorbic acid. This view is supported by the results of a recent analysis of the iron content and bioavailability, as determined via the Caco-2 assay, of the maize germ versus endosperm [70]. In maize, the germ fraction contains ~80% of the grain's total phytic acid, and 27–54% of its iron. Yet this iron is poorly bioavailable. Removal of the germ enhanced the iron bioavailability of the remaining grain fraction (endosperm plus aleurone), indeed to the extent that there was more bioavailable iron, as determined by the Caco-2 assay, in de-germed maize flour than in common bean flour, despite the fact that the common bean flower contained five times the absolute level of iron. Phytic acid is clearly the culprit here.

Most biofortification breeding efforts and nutritional studies of biofortified foods only address one micronutrient, iron or zinc or calcium for example. That simply reflects the standard approach to these types of programs and studies. But in viewing the real world of "global" mineral nutrition, what about the fact that greatly reducing dietary phytic acid will probably enhance bioavailability of all of these? Of course, there are many cases such as foods obtained following milling of grain crops where following the removal of bran the resulting food is a relatively poor source of iron or zinc yet is also low in phytic acid. In those cases, such as white rice, breeding for enhanced mineral density has many straightforward beneficial applications where the low-phytate trait might have little benefit.

The *lpa* beans used in the studies of Petry et al. [60,61] were not isolated with support from Harvest Plus [62]. In fact to the best of my knowledge, to date Harvest Plus and international agricultural centers participating in Gates Foundation-supported biofortification efforts have not provided support for the development of any low-phytate crops. Has this lack of support for "the *lpa* approach" by the most prominent programs in crop nutritional quality breeding for the international community been to the benefit or detriment of those populations at greatest risk for mineral deficiency? Consider the truly global improvement in mineral nutritional health obtained via the consumption of *lpa* types documented above. When field-dominating programs develop broad strategies that target specific approaches, a "canalization effect' can occur: support for other approaches might be negatively impacted. Consider the progress that might have been made in the breeding of low-phytate crops if it had the support of such well-funded programs, not post 2015 following the results of Petry et al. [60], but about 1995 about when low-phytate genetic variants first became available?

Biofortification breeding for mineral density may have another problem: it ignores the fact that in many societies, grain and legume crops or products made from them are used both in human foods and animal feeds. Thus an *lpa* crop might be more broadly useful and simply more practical. Consider the cases of rice. The vast majority of rice is consumed as white rice consisting almost entirely of central endosperm, the product of milling/polishing process that removes the germ and aleurone layer as the "bran fraction". Since in the cereal grains nearly all the phytic acid and most mineral stores are found in the germ and aleurone layer, they end up in the bran fraction. Rice bran is an important and valuable side-product in areas of major rice production and is used in animal feeds [71]. As a result, while a low-phytate mutation that primarily impacts the chemistry of the bran fraction might be relatively less valuable for white rice improvement, low-phytate bran would be far more nutritious than a "normal-phytate" bran when used in animal feeds. When chickens were fed a diet where rice bran, wheat bran, corn bran, soy bran, and oat hulls were used as fiber, only the ones fed rice bran had reduced body growth [71]. This was attributed to a higher phytic acid level in the rice bran diet (1.3%) as compared with that in other diets (0–0.4%). Furthermore, where rice is consumed as a whole-grain "brown rice", or for that matter any cereal grain crop when consumed as a whole-grain, the low-phytate trait would be of value for both human and animal consumption. Therefore, policy and research direction should more often take into account the value of *lpa* types both in human and animal feeds.

The quality of judgment about questions pertaining to the agronomics of breeding for enhanced mineral density versus low-phytate may have also been less than optimal. For example, the claim that breeding for enhanced mineral density will have little or no effect on yield but rather will probably enhance yield is not uniformly supported by published results. A thorough and well-conducted study of the relationship between grain yield and grain zinc concentration of zinc-biofortified rice cultivars, conducted over four locations and five years, found "high grain Zn concentration was at the cost of grain yield" [72]. Of course, a negative relationship between high seed mineral density and yield will not always be the case, but statements that imply that such a negative relationship will not likely be the case are misleading for several reasons [56]. For example, it is well known that there is often an inverse relationship between grain yield and grain mineral density [56]. There is a straightforward biological explanation for this: as "harvest index" (the amount of grain produced by a unit of crop biomass) increases, the concentration of minerals per unit of grain decreases. Thus the inverse of this suggests that one of the easiest ways to select for enhanced seed mineral density would be to select for reduced yield! The well-known negative relationship between grain yield and grain mineral density may explain the observation that enhanced iron and zinc density resulting from biofortification breeding is sometimes accompanied by increased phytic acid [55,59,73]. Of course, this could be due to other factors such as selection for transport functions that enhance the concentration of all three minerals simultaneously. But given the published observations, should not one ask if breeding for enhanced gain iron or zinc density typically results in elevated phytate levels?

With all of that said, an important recent study found that the adoption of improved "iron biofortified beans" by smallholder farmers in Rwanda led to a 23% increase in yields and a "potential" 24% increase in farmer income [74]. While non-breeding or non-biological factors not related to enhanced grain mineral density might have contributed to these benefits, such as the simple introduction of advanced cultivars or other cultural or economic factors, this represents a notable outcome and will factor into future policy supporting the adoption of biofortified crops. However, the assumption in the above "biofortification paradigm" that *lpa* types will inevitably have poor field performance, is also not supported by all agronomic studies or by crop breeding theory (see below).

What was the rationale behind this lack of support for the "low-phytate approach", given its potential? Perhaps it simply reflects a justifiably conservative approach to crop improvement strategies for those crops providing staple foods for at-risk populations in the developing world. After all, there may be some fundamental flaws for the low-phytate trait. For example, perhaps the change in seed P chemistry that results in reduced phytic acid P accompanied by elevated inorganic P might increase the incidence of fungal infection, which could increase the incidence of mycotoxin contamination [75]. Alternatively, the rationale might reflect a concern that dietary phytic acid also may have health-beneficial roles, as an anti-oxidant and anti-cancer agent or as an inhibitor of renal stone formation (reviewed in [76]). While highly significant, these are largely problems of aging in the developed world, but the vast majority of persons in mineral deficiency at-risk populations are primarily infants, children and women of child-bearing age of lower economic status in the developing world. For these individuals, malnutrition leading to "stunting" in adolescents, or iron-deficiency anemia in women, is of more pressing concern than is cancer or kidney stones.

But there is a second issue with the claim of health benefits for dietary phytic acid; some of the claims may be problematic. While the anti-oxidant and mineral chelation role of dietary phytic acid in the digestive tract is based on good science, claims for health benefits that require uptake and transport in bodily fluids such as blood or urine, for example, the claimed role in preventing renal stone formation, probably are unfounded, since a recent study demonstrated that there is no phytate in human plasma or urine [77]. These authors, in fact, conclude that administering phytate as a dietary supplement "in the hope that it might impact directly or indirectly on cancer or other pathologies seems highly inadvisable" [77]. The problem of widespread transmission of this false claim is not helped by a research article with the deceptive title "[3H] Phytic acid (inositol hexaphosphate) is absorbed and distributed to various tissues in rats" [78]. If one reads that paper, one finds that it was not phytic acid but primarily it's inositol backbone that was absorbed and distributed. myo-Inositol can have health beneficial effects, but myo-inositol is not phytic acid. Thus dietary phytate no doubt has some beneficial roles, but they probably are much fewer then is often claimed, including claims cited in papers advocating for biofortification breeding of high mineral density.

4. The Low-Phytate Trait, Yield and Seed/Grain Quality: Issues and Attitudes

4.1. Does the Low-Phytate Trait Mean Low Yield or Poor Seed Quality? Yes, If You Only Have Six Months to Breed or Engineer It!

One would be hard-pressed to find research reports or reviews of the low-phytate trait that do not begin with the statement to the effect that while it has many benefits, it is obligately associated with negative effects on plant and seed performance and yield. But as time goes there have been some reports of low-phytate genotypes of various crops, whether developed via conventional breeding or via genetic engineering, which have good field performance and yield [27,32,62,79]. Perhaps this attitude might slowly change as these developments become more widely known [80].

To be clear, studies indicate that alleles of genes that condition the *lpa* trait, including variant alleles of the same gene, can have greatly differing impacts on plant or seed performance. An excellent example of this is the maize *lpa*1 gene, alleles of which range from having a modest impact on plant and seed performance and yield [81] to lethality [82]. Maize *lpa*1 encodes one of the maize genome's multiple multidrug resistance-associated protein (MRP) genes, which encodes an ABC transporter specific to phytic acid transport [33]. Of the first 20 or so alleles of maize *lpa*1 that we isolated, two were lethal as homozygotes [82]. Furthermore, studies of another allele of maize *lpa*1, *lpa*1-241, found it negatively impacted seed viability via reduced protection against oxidative stress during seed maturation and storage, indicating a protective role for phytic acid as an anti-oxidant [83]. In contrast, recent studies of a rice *lpa* mutant (perturbed gene unknown), 9311-*lpa*, indicated that it's delayed or impaired germination, as compared with its non-mutant parental line, may be due to decreased ROS (reactive oxygen species) in seed tissues during germination [84]. This, in turn, is possibly due to a higher level of InsP$_3$ rather than to detoxification of the ROS burst during germination.

Alleles of an MRP gene provides another instructive example via studies of the low-phytate soybean and common bean lines conditioned by recessive alleles of their genomes' MRP genes. Each of these two close relatives' genomes contains multiple copies of this gene, two in the common bean and three in the soybean. The soybean low-phytate line CX1834 line [85], homozygous for the first soybean *lpa* mutations, was found to be conditioned by mutations in two copies of this particular MRP gene [86]. Initial studies of its agronomic properties revealed the interesting "seed source effect": via a still unknown mechanism, seed maturation of this genotype in a tropical environment (Puerto Rico) resulted in negative effects on subsequent germination, emergence and yield not observed when seed maturation occurred in a temperate environment (Iowa) [87]. However, a common bean low-phytate mutant conditioned by a mutation in only one of the common bean's two MRP duplications displays little or no negative impact on germination, emergence or subsequent plant growth, performance or yield [62,88]. Furthermore, a sustained breeding effort [79] using the soybean CX1834 line as the source of the low-phytate trait produced two soybean lines whose yields and germination were statistically similar to high-yielding check cultivars (discussed below). These studies of maize and bean mutations in MRP genes, and studies of other *lpa* mutants [84], also illustrate that apart from any agronomic value, *lpa* genetics represent a valuable resource in studies of plant and seed biology and the role that the phytic acid and inositol phosphate pathways may play in it.

But the main point is that yes, one can readily isolate *lpa* alleles homozygosity for which greatly impact performance and yield. If a program isolates *lpa* mutations in a given crop, or engineers a knock-out of a phytic acid pathway gene, and simply evaluates performance of early-generation homozygotes, the chance that they will observe poor performance is fairly high. For example, if nine out of ten randomly-isolated alleles have a clear negative impact, as was the case with maize *lpa*1, then the chance of reporting an *lpa* line with poor yield is pretty high. I think this is exactly what has happened. But with hard work and some breeding, that one-in-ten high-yielding *lpa* line can probably be developed.

Very good examples of this are field trials of barley *lpa* lines, representing variant alleles of at least four genes (Figure 7, [27,89]). Four pairs of near-isogenic lines were included in both rounds of trials and those will be discussed here. These pairs consisted of a wild-type sibling and a sibling homozygous for one of four independent *lpa* mutations: barley *lpa*1-1 conditioning a ~47% reduction in seed phytic acid; barley *lpa*2-1 conditioning a 50% reduction in seed phytic acid; barley *lpa*3-1 conditioning a ~65% reduction in seed phytic acid; barley *lpa*-M955, conditioning a ~95% reduction in seed phytic acid (also illustrated in part in Figure 4). The appropriate comparison in these studies is between the wild-type ("normal phytic acid") sib line versus the homozygous *lpa* sib line in each pair of near-isogenic lines. For the first round of trials in 2002 and 2003, these homozygous mutant or wild-type lines were obtained after only two to four generations of backcrossing to the recurrent parent, the progenitor cultivar "Harrington", followed by self-pollination. Lines for subsequent study were selected via visual inspection for plants displaying nice growth equivalent to the wild-type progenitor cultivar "Harrington". However, no rigorous selection for plant performance or seed weight or quality within the *lpa* class in a given generation of backcrossing was conducted; in each generation seed from essentially all members of a given progeny class (wild-type or mutant) were bulked to plant the next generation.

This first round of field trials were conducted in 2002 and 2003 in four locations in Idaho USA, two of which were irrigated and two of which were only "rain-fed", the latter representing much more stressful and less productive environments (Figure 7A,B). The first important observation is that in the irrigated trials in 2002–2003, we observed no effect of homozygosity for *lpa*1-1 on yield

(Figure 7A, left). However, with larger decreases in grain phytic conditioned by *lpa*2-1, *lpa*3-1, and *lpa*-M955 respectively, a very linear and negative relationship between phytic acid reduction and yield was observed (indicated by the red regression line). The second important observation in the 2002-2003 trials is that the absent or modest reductions in yield in the *lpa* sib lines for the *lpa*1-1, *lpa*2-1, and *lpa*3-1 mutations when assayed in irrigated trials is much more pronounced in the rain-fed trials (Figure 7B). Although not statistically significant in every case, there is clearly a much larger impact of the *lpa* trait on yield, including for the *lpa*1-1 sib line, in this first round of trials in the rain-fed versus irrigated locations. Thus in this first round of field trials: (1) the *lpa*1 line performed best overall and its performance in irrigated locations was excellent; (2) there appeared to be a very linear and negative relationship between the level of phytate reduction and yield in the irrigated trials; (3) this negative impact on yield was very pronounced in the more stressful, non-irrigated locations.

Breeding with these *lpa* mutations continued and a second round of field trials were conducted in 2010 and 2011 (Figure 7C,D; [27]). For this second round seed was obtained following further backcrossing (four or five generations) but again, no rigorous selection for plant performance or seed traits within a given *lpa* progeny class was conducted. When subsequently tested over two years in five locations in Idaho, USA, including locations that again were either irrigated or not, barley *lpa*1-1 had no discernable effect on yield, regardless of location. Furthermore, in the irrigated locations (Figure 7C), the difference in performance between a given wild-type sib line versus its homozygous *lpa* sib was much less pronounced than in 2002 and 2003: yield of *lpa*1-1, *lpa*2-1 and *lpa*3-1 homozygous mutant lines were as good or better than their wild-type sib lines and the loss of yield conditioned by *lpa*-M955, as compared with its wild-type sib, was less dramatic than that observed in 2002 and 2003. While there still was a clear, negative relationship between yield and phytic acid reduction (indicated by the red regression line), the impact of phytic acid reduction on yield was approximately half of that observed in 2002: the regression coefficient was -0.035 in 2002/2003 (Figure 7A) versus -0.018 in 2010/2011 (Figure 7C). One cannot rule out the possibility that this difference in regression of yield against phytic acid reduction is due in whole or part to differences in overall production conditions between 2002/2003 and 2010/2011. For example, overall, yields were lower in the irrigated locations in 2010/2011 as compared with 2002/2003.

While field performance of the barley *lpa* lines was again much more impacted in the far more stressful rain-fed locations (Figure 7D), in 2010 and 2011 *lpa*1-1 still yielded as good as its wild-type sib. The excellent field performance of barley *lpa*1-1 probably is due to the fact that this mutation's phenotype is highly tissue-specific, reducing phytic acid accumulation in the aleurone but not the germ, the two sites of phytic acid accumulation in cereal grains [26,27]. Since the germ of barley *lpa*1-1 is wild-type in terms of phytic acid, the impact of this mutation on germination, emergence, plant growth, and performance is minimal.

Taken together, these data indicate that even a minimal amount of breeding, limited to backcrossing, works! This conclusion is supported by the results of backcrossing the first soybean low-phytate mutations [85] into an elite background in a breeding program at the Univ. or Tennessee (USA, [79]). Two low-phytate soybean lines were obtained after five generations of backcrossing to an elite line. Field trials of these lines found their yield statistically equivalent to the yield of two high-yielding culivars included as checks, and also found no effect of the low-phytate trait on germination. Thus good germination, field performance and yield of low-phytate types has now been documented in two important food legumes, the common bean and the soybean [62,79].

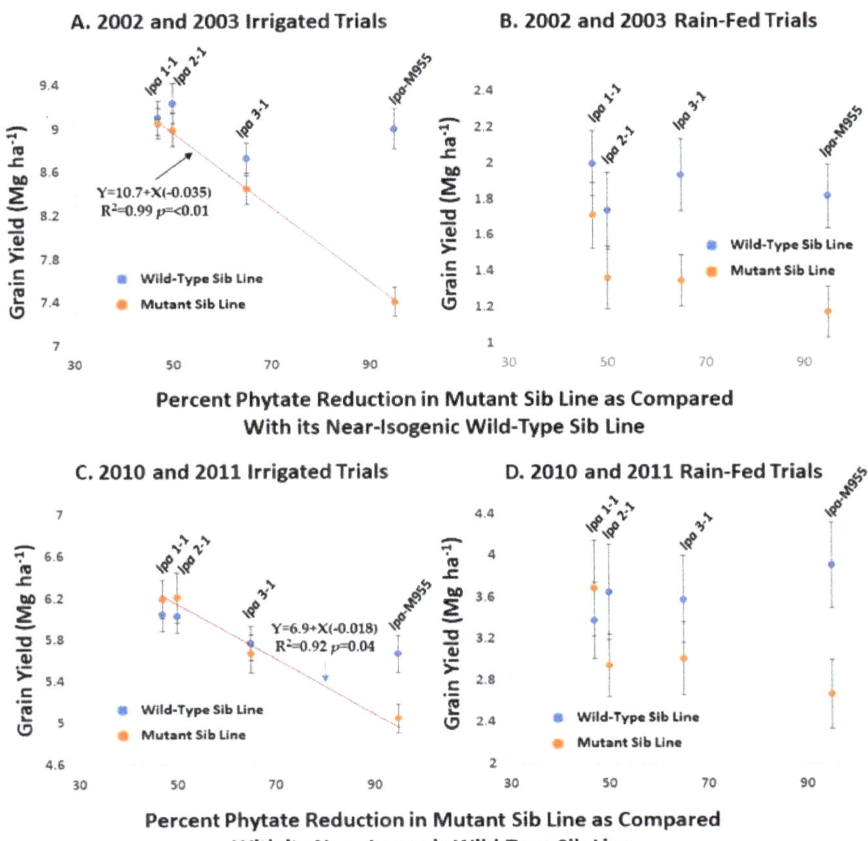

Figure 7. Two rounds of yield trials of barley *low phytic acid* (*lpa*) near-isogenic wild-type (normal phytic acid) lines, conducted in 2002 and 2003 and then in 2010 and 2011. Four pairs of near-isogenic lines, each pair consisting of a wild-type (normal phytic acid) sib and a *lpa* sib, were evaluated in several locations in Idaho (USA). These locations were either irrigated (two locations in 2002/2003 and three locations in 2010/2011) or rain-fed (non-irrigated; two locations in both 2002/2003 and 2010/2011) [27,89]. The homozygous *lpa* member of each isoline pair are: *lpa*1-1, which conditions a ~47% reduction in seed phytic acid; *lpa*2-1, which conditions a ~50% in seed phytic acid; *lpa*3-1, which conditions a ~65% reduction in seed phytic acid; *lpa*-M955, which conditions a ~95% reduction in seed phytic acid. (**A**,**B**): Trials conducted in 2002 and 2003. (**C**,**D**): Trials conducted in 2010 and 2011. The error bars represent the standard deviation of the mean, or standard error, n = 12 to n = 36. Linear regression of yield against % phytate reduction (indicated by red lines) was calculated for the *lpa* isolines in the irrigated trials.

However, what is lacking in these as well as in all other cases of breeding with *lpa* genotypes is recurrent selection within the *lpa* progeny class, over several generations, for seed and plant performance traits. For example, while studies show that recurrent selection might improve the germination and emergence of low-phytate soybean lines [90], the results of such selection have not been reported to date.

There are two additional problems with the viewpoint that the *lpa* trait is obligately associated with reduced yield or seed quality that cannot be overcome with traditional crop breeding methods.

First, the genetics of the *lpa* trait is still in its infancy, as compared with other seed chemistry traits like starch, protein or oil content or chemistry. In addition to the relatively few major *lpa* loci that have been identified to date, to my knowledge there has been only two studies to date documenting secondary loci or allelic variants of genes that have a valuable modifier effect in an *lpa* background (discussed below, [91,92]). Such modifiers are well known for genes that perturb or alter starch, protein and oil content and chemistry, and critically important to breeding elite-performance lines with such traits [93–95]. Consider the genes and loci that modify starch, sugar or carbohydrate content in maize. Not only are already well-known modifiers important in breeding and end-use quality, such as sugary enhancer used to modify *sugary* in sweet corn breeding [93], but additional loci continue to be identified that have modifying effects. These new discoveries continue even after many decades of intensive research, thus the "supply" has not been exhausted, even in the case of truly major seed constituents such as starch, oil and protein.

The first genetic engineering of a gene to serve as a modifier of poor fertility and growth of an *lpa* mutant targeted the *Arabidopsis* "Gle mRNA export factor" [91]. The negative effects on vegetative growth and fertility observed in low-seed phytate *Arabidopsis* lines conditioned by knock-out of its *Ipk1* (inositol pentaki*s*phosphate 2-kinase) gene were rescued by engineering allelic variants of the Gle mRNA export factor for elevated Ins P_6 sensitivity [91].

A sustained classical breeding effort can also overcome seed quality issues associated with the low-phytate trait via selection for beneficial modifiers. An excellent example of this concerns efforts to combine the low-phytate and low-saturated fat traits in soybean. Low-saturated fat represents an important health-beneficial trait for soybean oil. The U.S. Food and Drug Administration set an upper threshold of 89 g saturated fats per kg product in order to refer to a product as "low saturated fat" [96]. A breeding program targeted developing low-phyate/low saturated fat soybean cultivars [92,97]. The low-phytate donor line was the original low-phytate soybean mutant CX1834 [85], and its seed had a relatively high saturated fatty acid (palmitate + stearate) concentration ranging from ~100 to 174 g per kg seed [97]. This line was crossed with a "low saturate donor parent", a line homozygous for alleles of two genes that confer the low-saturated fat seed phenotype, and whose seed saturated fatty acid concentration was 71 g per kg seed [97]. Normal-phytate and low-phytate progeny classes were identified following one generation of backcrossing to the low-saturate parent (followed by self pollination), and amongst these two classes, 20 normal-phytate and 20 low-phytate progenies were further identified that produced seed with less than 50 g palmitate per kg. These lines were then field tested and the normal-phyate and low-phytate progenies were found to produce seed that had 77 and 83 g palmitate+stearate per kg seed, respectively. This difference was not statistically significant and both levels are below the FDA's threshold, but the level of saturated fatty acids in seed of the low-phytate lines was still consistently higher than that observed in the normal-phytate progenies and higher than that observed in the standard low-saturated fat parent. These lines produced seed that was deemed to close to the FDA threshold to warrant recommendation for commercial production [97].

How this group's work illustrates the value of a sustained effort and identification of beneficial modifier alleles is that following further crossing and selection, low-phyate soybean lines with saturated fatty acid levels less than 70 g per kg seed were successfully identified. It was concluded that this was accomplished via selection for "favorable modifying genes for low saturate concentration" [92].

A problem associated with rice *lpa*1 is that it is associated with grain "chalkiness", an undesirable characteristic [98]. This is reminiscent of the case of opaque 2/high-lysine maize. Perhaps the most important and relevant example of the problem of field performance for a nutritionally-enhanced major crop is the case of opaque 2/high-lysine maize. If anyone well versed in the history of efforts to breed nutritionally-enhanced staple crops read that there is concern with the agronomic performance of the early generations of *lpa* crops, that person might think of the story of opaque2/high-lysine corn [94]. After its initial identification, excitement led to disappointment; homozygosity for *o*2 resulted in undesirable grain characteristics including chalkiness, and impaired field performance and yield. But addressing the amino acid deficiencies such as lysine deficiency of standard maize is critically important to nations like Mexico where maize is such an important staple food. As a result, there has been substantial government and institutional support for this work in several nations. Sustained breeding and research over several decades has yielded high-performance high-lysine, "Quality Protein" maize with excellent yields and good grain characteristics. Elite performance of any line or hybrid, regardless of specific end-use or quality traits they might contain, reflects favorable combinations of alleles at large numbers of dispersed loci that only result from sustained and ongoing breeding over many generations. As discussed above, reports of recurrent selection for performance within a *lpa* line, or in fact any sustained effort at *lpa* breeding, has been very limited. In a parallel to the *o2* story, perhaps recurrent selection might correct the problem of "chalkiness" in rice *lpa*1.

Second, recurrent selection within a *lpa* line for field performance or yield probably would not only select for favorable genetic alleles, in the sense of sequence-differing variants, but also will very likely select for favorable epigenetic variants [99]. This latter type of variation, and its role in breeding crop lines with elite performance, is not yet well understood in terms of science nor is its importance appreciated in the crop breeding community. I predict that that will change greatly over the next decade.

This discussion so far has only addressed "conventional" approaches to breeding high-yielding low-phytate crops. Clearly, genetic engineering for the trait provides an additional and powerful approach to developing low-phytate crops with good performance characteristics [33]. There are a growing number of targets for such engineering [3–5,31–33,36,91]. It is interesting that in maize, one of the worlds' most important food and feed crops, both the first use of zinc-finger nucleases and more recently the methods of TALENS and CRISPR/Cas for sequence-targeted gene engineering, targeted genes involved in the low-phytate trait [100,101]. Perhaps due to the significant potential benefits of the trait, developing high-yielding stress-tolerant low-phytate crops has become something of a holy grail for crop genetic engineering. Perhaps the most powerful approach might turn out to be engineering crops with seed tissue-targeted phytase overexpression. This has been accomplished with a growing number of crops including maize [102], soybean [103] and barley [104]. This approach can result in quantitative conversion of phytate P to inorganic P in mature seed [103], but would also provide active phytase action against any source of phytate in a food or feed.

4.2. Is Comparing the Yield of a Low-Phytate Type Against a Wild-Type Like "Comparing Apples and Oranges", or More Appropriately, Comparing Yields of Sweet Corn with Field (Starchy) Corn?

One would not insist that sweet corn yields compare favorably with field or starchy corn yields. But is not that exactly what people have been doing with the low-phytate trait? The problem is that in the context of crop production for mainstream poultry and swine production in the U.S., often *lpa* crops must compete with standard commodity crops. In the industry's primarily horizontal structure, *lpa* crops have to yield close to standard crops and would also have to be handled as a "specialty-use crop", where they are segregated in both storage and shipment. In light of the fact that a practical alternative to the problem that *lpa* crops address in this context already exists, the feed additive phytase, the "yield issue" and need for segregated production and handling present barriers to widespread production. But should *lpa* types be compared directly with standard types? For example, in the case of maize, should not the *lpa* type simply be considered a "specialty use" corn or perhaps represent a new maize commodity classification? For example, in the case of maize, we have "dent corn", "flint corn", "sweet corn", "waxy corn", and "amylomaize". To this we should add low-phytate maize. Would it be appropriate to compare the yield of waxy corn, or sweet corn, with a starchy field corn? Absolutely not. In light of the numerous benefits of low-phytate types as human foods or animal feeds, should one not grow a nutritionally-enhanced crop variant that perhaps has 5% to 10% less yield than a standard variant but one that is substantially more nutritious?

5. Conclusions

The *lpa* or "low-phytate" seed trait can provide numerous potential benefits to the nutritional quality of foods and feeds and to the sustainability of agricultural production. These include enhanced phosphorus management contributing to enhanced sustainability in non-ruminant (poultry, swine, and fish) production; reduced environmental impact via reduced waste P in non-ruminant production; enhanced "global" bioavailability of minerals (iron, zinc, calcium, magnesium) for both humans and non-ruminant animals; altered distribution of minerals in cereal grains potentially resulting in enhanced mineral contents of milled products; enhancement of animal health, productivity and the quality of animal products; potential enhancement of food and feed protein and starch utilization; development of "low seed total P" crops which also can enhance management of P in agricultural production and contribute to its sustainability. Evaluations of this trait by industry, and by advocates of biofortification via breeding for enhanced mineral density, have been too short term and too narrowly focused. Arguments against breeding for the low-phytate trait overstate the negatives such as potentially reduced yields and field performance or possible reductions in phytic acid's health benefits. Progress in breeding or genetically-engineering high-yielding stress-tolerant low-phytate crops continues. While there are widely available and efficacious alternative approaches to deal with the problems posed by seed-derived dietary phytic acid, such as use of the enzyme phytase as a feed additive, or biofortification breeding, if there were an interest in developing low-phytate crops with good field performance or good seed-quality, it could be accomplished given adequate time and support. Perhaps due to the potential benefits of the low-phytate trait, the challenge of developing high-yielding, stress-tolerant low-phytate crops has become something of a holy grail for crop genetic engineering. Even with a moderate reduction in yield, in light of the numerous benefits of low-phytate types as human foods or animal feeds, should one not grow a nutritionally-enhanced crop variant that perhaps has 5% to 10% less yield than a standard variant but one that is substantially more nutritious? Such crops would be a benefit to human nutrition especially in populations at risk for iron and zinc deficiency, and a benefit to the sustainability of agricultural production.

Funding: This research received no external funding.

Conflicts of Interest: The author declares no conflict of interest.

References

1. Raboy, V. myo-Inositol-1, 2, 3, 4, 5, 6-hexakisphosphate. *Phytochemistry* **2003**, *64*, 1033–1043. [CrossRef]
2. Raboy, V.; Gerbasi, P.F.; Young, K.A.; Stoneberg, S.D.; Pickett, S.G.; Bauman, A.T.; Murthy, P.P.N.; Sheridan, W.F.; Ertl, D.S. Origin and seed phenotype of maize low phytic acid 1-1 and low phytic acid 2-1. *Plant Physiol.* **2000**, *124*, 355–368. [CrossRef] [PubMed]
3. Shi, J.; Wang, H.; Hazebroek, J.; Ertl, D.S.; Harp, T. The maize low-phytic acid 3 encodes a myo-inositol kinase that plays a role in phytic acid biosynthesis in developing seeds. *Plant J.* **2005**, *42*, 708–719. [CrossRef] [PubMed]
4. Stevenson-Paulik, J.; Bastidas, R.J.; Chiou, S.-T.; Frye, R.A.; York., J.D. Generation of phytate-free seeds in Arabidopsis through disruption of inositol polyphosphate kinases. *Proc. Natl. Acad. Sci. USA* **2005**, *102*, 12612–12617. [CrossRef] [PubMed]
5. Kim, S.I.; Tai, T.H. Identification of genes necessary for wild-type levels of seed phytic acid in Arabidopsis thaliana using a reverse genetics approach. *Mol. Genet. Genom.* **2011**, *286*, 119–133. [CrossRef] [PubMed]
6. Cichy, K.; Raboy, V. Evaluation and development of low-phytate crops. In *Modification of Seed Composition to Promote Health and Nutrition*; Krishnan, H., Ed.; American Society of Agronomy: Madison, WI, USA, 2009; pp. 177–200.
7. El-Hack, M.E.A.; Alagawany, M.; Arif, M.; Emam, M.; Saeed, M.; Arain, M.A.; Siyal, F.A.; Patra, A.; Elnesr, S.S.; Khan, R.U. The uses of microbial phytase as a feed additive in poultry nutrition—A review. *Ann. Anim. Sci.* **2018**, *18*, 639–658. [CrossRef]

8.   Selle, P.H.; Ravindran, V. Microbial phytase in poultry nutrition. *Anim. Feed Sci. Technol.* **2007**, *135*, 1–41. [CrossRef]

9.   Bouis, H.E.; Welch, R.M. Biofortification—A sustainable agricultural strategy for reducing micronutrient malnutrition in the global south. *Crop Sci.* **2010**, *50*, S20–S32. [CrossRef]

10.   Common, R.H. Phytic acid in mineral metabolism in poultry. *Nature* **1939**, *143*, 379–380. [CrossRef]

11.   Nelson, T.S. The utilization of phytate phosphorus by poultry—A review. *Poult. Sci.* **1967**, *46*, 862–871. [CrossRef]

12.   Cromwell, G.L. Biological availability of phosphorus for pigs. *Feedstuffs* **1980**, *52*, 38–42.

13.   Holt, R.F.; Timmons, D.R.; Latterell, J.J. Accumulation of phosphates in water. *J. Agric. Food Chem.* **1970**, *18*, 781–784. [CrossRef] [PubMed]

14.   McCance, R.A.; Widdowson, R.M. Phytin in human nutrition. *Biochem. J.* **1935**, *29*, 2694–2699. [CrossRef] [PubMed]

15.   Erdman, J.W., Jr. Bioavailability of trace minerals from cereals and legumes. *Cereal Chem.* **1981**, *58*, 21–26.

16.   Brown, K.H.; Solomons, N.W. Nutritional problems in developing countries. *Infect. Disease Clin. N. Am.* **1991**, *5*, 297–317.

17.   Kennedy, G.; Nantel, G.; Shetty, P. The scourge of "hidden hunger": Global dimensions of micronutrient deficiencies. *Food Nutr. Agric.* **2003**, *32*, 8–16.

18.   Hitz, W.D.; Carlson, T.J.; Kerr, P.S.; Sebastian, S.A. Biochemical and molecular characterization of a mutation that confers a decreased raffinosaccharide and phytic acid phenotype on soybean seeds. *Plant Physiol.* **2002**, *128*, 650–660. [CrossRef]

19.   Spencer, J.D.; Allee, G.L.; Sauber, T.E. Growing-finishing performance and carcass characteristics of pigs fed normal and genetically modified low-phytate corn. *J. Anim. Sci.* **2000**, *78*, 1529–1536. [CrossRef]

20.   Stilborn, H.L.; Crum, R.C.; Rice, D.W.; Hinds, M.A.; Ertl, D.S.; Beach, L.R.; Huff, W.E.; Kleese, R.A.; Pioneer Hi-Bred International Inc.; US Department of Agriculture. Method of Reducing Cholesterol in Eggs. U.S. Patent 6,391,348, 21 May 2002.

21.   Lott, J.N.; Ockenden, I.; Raboy, V.; Batten, G.D. Phytic acid and phosphorus in crop seeds and fruits: A global estimate. *Seed Sci. Res.* **2000**, *10*, 11–33. [CrossRef]

22.   Oliveira, M.; Machado, A.V. The role of phosphorus on eutrophication: A historical review and future perspectives. *Environ. Technol. Rev.* **2013**, *2*, 117–127. [CrossRef]

23.   European Environmental Agency. *Nutrient Enrichment and Eutrophication in Europe's Seas: Moving towards a Healthy Marine Environment*; EEA Report; Publication Office of the European Union: Luxembourg, 2019; Volume 14. [CrossRef]

24.   Rose, T.; Liu, L.; Wissuwa, M. Improving phosphorus efficiency in cereal crops: Is breeding for reduced grain phosphorus concentration part of the solution? *Front. Plant Sci.* **2013**, *4*. [CrossRef] [PubMed]

25.   Dorsch, J.A.; Cook, A.; Young, K.A.; Anderson, J.M.; Bauman, A.T.; Volkmann, C.J.; Murthy, P.P.; Raboy, V. Seed phosphorus and inositol phosphate phenotype of barley low phytic acid genotypes. *Phytochemistry* **2003**, *62*, 691–706. [CrossRef]

26.   Raboy, V.; Cichy, K.; Peterson, K.; Reichman, S.; Sompong, U.; Srinives, P.; Saneoka, H. Barley (*Hordeum vulgare* L.) low phytic acid 1-1: An endosperm-specific, filial determinant of seed total phosphorus. *J. Hered.* **2014**, *105*, 656–665. [CrossRef] [PubMed]

27.   Raboy, V.; Peterson, K.; Jackson, C.; Marshall, J.; Hu, G.; Saneoka, H.; Bregitzer, P. A substantial fraction of barley (*Hordeum vulgare* L.) low phytic acid mutations have little or no effect on yield across diverse production environments. *Plants* **2015**, *4*, 225–239. [CrossRef]

28.   Wang, F.; Rose, T.; Jeong, K.; Kretzschmar, T.; Wissuwa, M. The knowns and unknowns of phosphorus loading into grains, and implications for phosphorus efficiency in cropping systems. *J. Exp. Bot.* **2015**, *67*, 1221–1229. [CrossRef]

29.   Bregitzer, P.; Raboy, V.; Obert, D.E.; Windes, J.M.; Whitmore, J.C. Registration of 'Herald' barley. *Crop Sci.* **2007**, *47*, 441–442. [CrossRef]

30.   Bregitzer, P.; Raboy, V.; Obert, D.E.; Windes, J.; Whitmore, J.C. Registration of 'Clearwater' low-phytate hulless spring barley. *J. Plant Regist.* **2008**, *2*, 1–4. [CrossRef]

31.   Ye, H.; Zhang, X.Q.; Broughton, S.; Westcott, S.; Wu, D.; Lance, R.; Li, C. A nonsense mutation in a putative sulphate transporter gene results in low phytic acid in barley. *Funct. Integr. Genom.* **2011**, *11*, 103–110. [CrossRef]

32. Yamaji, N.; Takemoto, Y.; Miyaji, T.; Mitani-Ueno, N.; Yoshida, K.T.; Ma, J.F. Reducing phosphorus accumulation in rice grains with an impaired transporter in the node. *Nature* **2017**, *541*, 92. [CrossRef]

33. Shi, J.; Wang, H.; Schellin, K.; Li, B.; Faller, M.; Stoop, J.M.; Meeley, R.B.; Ertl, D.S.; Ranch, J.P.; Glassman, K. Embryo-specific silencing of a transporter reduces phytic acid content of maize and soybean seeds. *Nat. Biotechnol.* **2007**, *25*, 930–937. [CrossRef]

34. Sasakawa, N.; Sharif, M.; Hanley, M.R. Metabolism and biological activities of inositol pentakisphosphate and inositol hexakisphosphate. *Biochem. Pharmacol.* **1995**, *50*, 137–146. [CrossRef]

35. Tsui, M.M.; York, J.D. Roles of inositol phosphates and inositol pyrophosphates in development, cell signaling and nuclear processes. *Adv. Enzym. Regul.* **2010**, *50*, 324–337. [CrossRef] [PubMed]

36. Kim, S.I.; Andaya, C.B.; Goyal, S.S.; Tai, T.H. The rice OsLpa1 gene encodes a novel protein involved in phytic acid metabolism. *Theor. Appl. Genet.* **2008**, *117*, 769–779. [CrossRef] [PubMed]

37. Desai, M.; Rangarajan, P.; Donahue, J.L.; Williams, S.P.; Land, E.S.; Mandal, M.K.; Phillippy, B.Q.; Perera, I.Y.; Raboy, V.; Gillaspy, G.E. Two inositol hexakisphosphate kinases drive inositol pyrophosphate synthesis in plants. *Plant J.* **2014**, *80*, 642–653. [CrossRef]

38. Huff, W.E.; Moore, P.A., Jr.; Waldroup, P.W.; Waldroup, A.L.; Balog, J.M.; Huff, G.R.; Rath, N.C.; Daniel, T.C.; Raboy, V. Effect of dietary phytase and high available phosphorus corn on broiler chicken performance. *Poult. Sci.* **1998**, *77*, 1899–1904. [CrossRef]

39. Jong, S. Commercializing a disruptive technology. *Nat. Biotechnol.* **2011**, *29*, 685–688. [CrossRef]

40. Dean, J. Pricing Policies for New Products. *Harv. Bus. Rev.* **1976**, *54*, 141–153.

41. Green, J.M. Evolution of glyphosate-resistant crop technology. *Weed Sci.* **2009**, *57*, 108–117. [CrossRef]

42. Parish, A.L. Farmer Willingness to Adopt High Available Phosphorus (HAP) Corn. Ph.D. Thesis, University of Delaware, Newark, DE, USA, 2007.

43. Pesek, J.D., Jr.; Bernard, J.C.; Gupta, M. Consumer interest in environmentally beneficial chicken feeds: Comparing high available phosphorus corn and other varieties. *J. Agric. Appl. Econ.* **2011**, *43*, 591–605. [CrossRef]

44. Veum, T.L.; Ledoux, D.R.; Raboy, V. Low-phytate barley cultivars improve the utilization of phosphorus, calcium, nitrogen, energy, and dry matter in diets fed to young swine. *J. Anim. Sci.* **2007**, *85*, 961–971. [CrossRef]

45. Overturf, K.; Raboy, V.; Cheng, Z.J.; Hardy, R.W. Mineral availability from barley low phytic acid grains in rainbow trout (*Oncorhynchus mykiss*) diets. *Aquacul. Nutr.* **2003**, *9*, 239–246. [CrossRef]

46. Hambidge, K.M.; Krebs, N.F.; Westcott, J.L.; Sian, L.; Miller, L.V.; Peterson, K.L.; Raboy, V. Absorption of calcium from tortilla meals prepared from low-phytate maize. *Am. J. Clin. Nutr.* **2005**, *82*, 84–87. [CrossRef] [PubMed]

47. Hurrell, R.F. Influence of vegetable protein sources on trace element and mineral bioavailability. *J. Nutr.* **2003**, *133*, 2973S–2977S. [CrossRef] [PubMed]

48. Mendoza, C.; Viteri, F.E.; Lonnerdal, B.; Young, K.A.; Raboy, V.; Brown, K.H. Effect of genetically modified, low-phytic acid maize on absorption of iron from tortillas. *Am. J. Clin. Nutr.* **1998**, *68*, 1123–1127. [CrossRef] [PubMed]

49. Morris, E.R.; Ellis, R. Effect of dietary phytate/zinc molar ratio on growth and bone zinc response of rats fed semipurified diets. *J. Nutr.* **1980**, *110*, 1037–1045. [CrossRef]

50. Hambidge, K.M.; Miller, L.V.; Westcott, J.E.; Krebs, N.F. Dietary reference intakes for zinc may require adjustment for phytate intake based upon model predictions. *J. Nutr.* **2008**, *138*, 2363–2366. [CrossRef]

51. Adams, C.L.; Hambidge, M.; Raboy, V.; Dorsch, J.A.; Sian, L.; Westcott, J.L.; Krebs, N.F. Zinc absorption from a low–phytic acid maize. *Am. J. Clin. Nutr.* **2002**, *76*, 556–559. [CrossRef]

52. Hambidge, K.M.; Huffer, J.W.; Raboy, V.; Grunwald, G.K.; Westcott, J.L.; Sian, L.; Miller, L.V.; Dorsch, J.A.; Krebs, N.F. Zinc absorption from low-phytate hybrids of maize and their wild-type isohybrids. *Am. J. Clin. Nutr.* **2004**, *79*, 1053–1059. [CrossRef]

53. Graham, R.; Welch, R.; Bouis, H. Addressing micronutrient malnutrition through the nutritional quality of staple foods: Principles, perspectives, and knowledge gaps. *Adv. Agron.* **2001**, *70*, 77–142.

54. Bouis, H.E.; Saltzman, A. Improving nutrition through biofortification: A review of evidence from HarvestPlus, 2003 through 2016. *Glob. Food Secur.* **2017**, *12*, 49–58. [CrossRef]

55. Petry, N.; Egli, I.; Gahutu, J.B.; Tugirimana, P.L.; Boy, E.; Hurrell, R. Stable iron isotope studies in Rwandese women indicate that the common bean has limited potential as a vehicle for iron biofortification. *J. Nutr.* **2012**, *142*, 492–497. [CrossRef] [PubMed]

56. White, P.J.; Broadley, M.R. Biofortification of crops with seven mineral elements often lacking in human diets–iron, zinc, copper, calcium, magnesium, selenium and iodine. *New Phytol.* **2009**, *182*, 49–84. [CrossRef] [PubMed]

57. Hambidge, K.M.; Miller, L.V.; Westcott, J.E.; Sheng, X.; Krebs, N.F. Zinc bioavailability and homeostasis. *Am. J. Clin. Nutr.* **2010**, *91*, 1478S–1483S. [CrossRef] [PubMed]

58. Linares, L.B.; Broomhead, J.N.; Guaiume, E.A.; Ledoux, D.R.; Veum, T.L.; Raboy, V. Effects of low phytate barley (*Hordeum vulgare* L.) on zinc utilization in young broiler chicks. *Poult. Sci.* **2007**, *86*, 299–308. [CrossRef] [PubMed]

59. Chomba, E.; Westcott, C.M.; Westcott, J.E.; Mpabalwani, E.M.; Krebs, N.F.; Patinkin, Z.W.; Palacios, N.; Hambidge, K.M. Zinc absorption from biofortified maize meets the requirements of young rural Zambian children. *J. Nutri.* **2015**, *145*, 514–519. [CrossRef]

60. Petry, N.; Egli, I.; Campion, B.; Nielsen, E.; Hurrell, R. Genetic reduction of phytate in common bean (*Phaseolus vulgaris* L.) seeds increases iron absorption in young women. *J. Nutr.* **2013**, *143*, 1219–1224. [CrossRef]

61. Petry, N.; Boy, E.; Wirth, J.P.; Hurrell, R.F. The potential of the common bean (*Phaseolus vulgaris*) as a vehicle for iron biofortification. *Nutrients* **2015**, *7*, 1144–1173. [CrossRef]

62. Campion, B.; Sparvoli, F.; Doria, E.; Tagliabue, G.; Galasso, I.; Fileppi, M.; Bollini, R.; Nielsen, E. Isolation and characterisation of an lpa (low phytic acid) mutant in common bean (*Phaseolus vulgaris* L.). *Theor. Appl. Genet.* **2009**, *118*, 1211–1221. [CrossRef]

63. Welch, R.M.; Graham, R.D. Breeding for micronutrients in staple food crops from a human nutrition perspective. *J. Exp. Bot.* **2004**, *55*, 353–364. [CrossRef]

64. Gibson, R.S.; King, J.C.; Lowe, N. A review of dietary zinc recommendations. *Food Nutr. Bull.* **2016**, *37*, 443–460. [CrossRef]

65. Miller, L.V.; Krebs, N.E.; Hambidge, K.M. A mathematical mode of zinc absorption in humans as a function of dietary zinc and phytate. *J. Nutr.* **2007**, *137*, 135–141. [CrossRef] [PubMed]

66. European Food Safety Authority. Scientific opinion on dietary reference values for zinc. EFSA Panel on Dietetic Products, Nutrition and Allergies (NDA). *ESFA J.* **2014**, *12*. [CrossRef]

67. Finkelstein, J.L.; Haas, J.D.; Mehta, S. Iron-biofortified staple food crops for improving iron status: A review of the current evidence. *Curr. Opin. Biotechnol.* **2017**, *44*, 138–145. [CrossRef] [PubMed]

68. Phillippy, B.Q. Transport of calcium across caco-2 cells in the presence of inositol hexakisphosphate. *Nurt. Res.* **2006**, *26*, 146–149. [CrossRef]

69. Han, O.; Failla, M.L.; Hill, A.D.; Morris, E.R.; Smith, J.C. Inositol phosphates inhibit uptake and transport of iron and zinc by a human intestinal-cell line. *J. Nutr.* **1994**, *124*, 580–587. [CrossRef]

70. Glahn, R.; Tako, E.; Gore, M.A. The germ fraction inhibits iron bioavailability of maize: Identification of an approach to enhance maize nutritional quality via processing and breeding. *Nutrients* **2019**, *11*, 833. [CrossRef]

71. Juliano, B.O. Rice bran. In *Rice Chemistry and Technology*; Juliano, B.O., Ed.; AACC International: Saint Paul, MN, USA, 1994; pp. 647–687.

72. Goloran, J.B.; Johnson-Beebout, S.E.; Morete, M.J.; Impa, S.M.; Kirk, G.J.D.; Wissuwa, M. Grain Zn concentrations and yield of Zn-biofortified versus Zn-efficient rice genotypes under contrasting growth conditions. *Field Crops Res.* **2019**, *234*, 26–32. [CrossRef]

73. Zhao, F.J.; Su, Y.H.; Dunham, S.J.; Rakszegi, M.; Bedo, Z.; McGrath, S.P.; Shewry, P.R. Variation in mineral micronutrient concentrations in grain of wheat lines of diverse origin. *J.Cereal Sci.* **2009**, *49*, 290–295. [CrossRef]

74. Funes, J.E.; Benson, T.; Sun, L.; Sedano, F.; Birol, E. The impact of iron biofortified beans on yields and farmers' incomes: The case of Rwanda. In Proceedings of the Agricultural and Applied Economics Association 2019 Annual Meeting, Atlanta, GA, USA, 21–23 July 2019.

75. Holmes, R.A.; Boston, R.S.; Payne, G.A. Diverse inhibitors of aflatoxin biosynthesis. *Appl. Microbiol. Biotechnol.* **2008**, *78*, 559–572. [CrossRef]

76. Silva, E.O.; Bracarense, A.P.F. Phytic acid: From antinutritional to multiple protection factor of organic systems. *J. Food Sci.* **2016**, *81*, R1357–R1362. [CrossRef]

77. Wilson, M.S.; Bulley, S.J.; Pisani, F.; Irvine, R.F.; Saiardi, A. A novel method for the purification of inositol phosphates from biological samples reveals that no phytate is present in human plasma or urine. *Open Biol.* **2015**, *5*, 150014. [CrossRef]

78. Sakamoto, K.; Vucenik, I.; Shamsuddin, A.M. [3H] Phytic acid (inositol hexaphosphate) is absorbed and distributed to various tissues in rats. *J. Nutr.* **1993**, *123*, 713–720. [CrossRef]

79. Boehm, J.D.; Walker, F.R.; Bhandari, H.S.; Kopsell, D.; Pantalone, V.R. Seed inorganic phosphorus stability and agronomic performance of two low-phytate soybean lines evaluated across six southeastern US environments. *Crop Sci.* **2017**, *57*, 2555–2563. [CrossRef]

80. Kishor, D.S.; Lee, C.; Lee, D.; Venkatesh, J.; Seo, J.; Chin, J.H.; Jin, Z.; Hong, S.K.; Ham, J.K.; Koh, H.J. Novel allelic variant of Lpa1 gene associated with a significant reduction in seed phytic acid content in rice (Oryza sativa L.). *PLoS ONE* **2019**, *14*, e0209636. [CrossRef] [PubMed]

81. Ertl, D.S.; Young, K.A.; Raboy, V. Plant genetic approaches to phosphorus management in agricultural production. *J. Environ. Qual.* **1998**, *27*, 299–304. [CrossRef]

82. Raboy, V.; Young, K.A.; Dorsch, J.A.; Cook, A. Genetics and breeding of seed phosphorus and phytic acid. *J. Plant Physiol.* **2001**, *158*, 489–497. [CrossRef]

83. Doria, E.; Galleschi, L.; Calucci, L.; Pinzino, C.; Pilu, R.; Cassani, E.; Nielsen, E. Phytic acid prevents oxidative stress in seeds: Evidence from a maize (Zea mays L.) low phytic acid mutant. *J. Exp. Bot.* **2009**, *60*, 967–978. [CrossRef]

84. Zhou, L.; Ye, Y.; Zhao, Q.; Du, X.; Zakari, S.A.; Su, D.; Pan, G.; Cheng, F. Suppression of ROS generation mediated by higher InsP 3 level is critical for the delay of seed germination in lpa rice. *Plant Growth Reg.* **2018**, *85*, 411–424. [CrossRef]

85. Wilcox, J.R.; Premachandra, G.S.; Young, K.A.; Raboy, V. Isolation of high seed inorganic P, low-phytate soybean mutants. *Crop Sci.* **2000**, *40*, 1601–1605. [CrossRef]

86. Gillman, J.D.; Pantalone, V.R.; Bilyeu, K. The low phytic acid phenotype in soybean line CX1834 is due to mutations in two homologs of the maize low phytic acid gene. *Plant Genome* **2009**, *2*, 179–190. [CrossRef]

87. Meis, S.J.; Fehr, W.R.; Schnebly, S.R. Seed source effect on field emergence of soybean lines with reduced phytate and raffinose saccharides. *Crop Sci.* **2003**, *43*, 1336–1339. [CrossRef]

88. Cominelli, E.; Confalonieri, M.; Carlessi, M.; Cortinovis, G.; Daminati, M.G.; Porch, T.G.; Losa, A.; Sparvoli, F. Phytic acid transport in *Phaseolus vulgaris*: A new low phytic acid mutant in the PvMRP1 gene and study of the PvMRPs promoters in two different plant systems. *Plant Sci.* **2018**, *270*, 1–12. [CrossRef] [PubMed]

89. Bregitzer, P.; Raboy, V. Effects of four independent Low-Phytate mutations on barley agronomic performance. *Crop Sci.* **2006**, *46*, 1318–1322. [CrossRef]

90. Oltmans, S.E.; Fehr, W.R.; Welke, G.A.; Raboy, V.; Peterson, K.L. Agronomic and seed traits of soybean lines with low–phytate phosphorus. *Crop Sci.* **2005**, *45*, 593–598. [CrossRef]

91. Lee, H.S.; Lee, D.H.; Cho, H.K.; Kim, S.H.; Auh, J.H.; Pai, H.S. InsP6-sensitive variants of the Gle1 mRNA export factor rescue growth and fertility defects of the ipk1 low-phytic-acid mutation in Arabidopsis. *Plant Cell* **2015**, *27*, 417–431. [CrossRef]

92. Brace, R.C.; Fehr, W.R. Modifying genes for palmitate and stearate concentration impacts selection for low-phytate, low-saturate soybean lines. *Crop Sci.* **2012**, *52*, 664–668. [CrossRef]

93. Ferguson, J.E.; Rhodes, A.M.; Dickinson, D.B. The genetics of sugary enhancer (se), an independent modifier of sweet corn (su). *J. Hered.* **1978**, *69*, 377–380. [CrossRef]

94. Vasal, S.K. The quality protein maize story. *Food Nutr. Bull.* **2000**, *21*, 445–450. [CrossRef]

95. Holding, D.R.; Hunter, B.G.; Klingler, J.P.; Wu, S.; Guo, X.; Gibbon, B.C.; Wu, R.; Schulze, J.M.; Jung, R.; Larkins, B.A. Characterization of opaque2 modifier QTLs and candidate genes in recombinant inbred lines derived from the K0326Y quality protein maize inbred. *Theor. Appl. Genet.* **2011**, *122*, 783–794. [CrossRef]

96. US Food and Drug Administration. *A Food Labeling Guide*; US Dep. of Health and Human Serv. Public Health Serv.: Washington, DC, USA, 2009.

97. Hulke, B.S.; Fehr, W.R.; Welke, G.A. Agronomic and seed characteristics of soybean with reduced phytate and palmitate. *Crop Sci.* **2004**, *44*, 2027–2031. [CrossRef]

98. Edwards, J.D.; Jackson, A.K.; McClung, A.M. Genetic architecture of grain chalk in rice and interactions with a low phytic acid locus. *Field Crop. Res.* **2017**, *205*, 116–123. [CrossRef]

99. Springer, N.M. Epigenetics and crop improvement. *Trends Genet.* **2013**, *29*, 241–247. [CrossRef] [PubMed]

100. Shukla, V.K.; Doyon, Y.; Miller, J.C.; DeKelver, R.C.; Moehle, E.A.; Worden, S.E.; Mitchell, J.C.; Arnold, N.L.; Gopalan, S.; Meng, X.; et al. Precise genome modification in the crop species Zea mays using zinc-finger nucleases. *Nature* **2009**, *459*, 437–441. [CrossRef] [PubMed]
101. Liang, Z.; Zhang, K.; Chen, K.; Gao, C. Targeted mutagenesis in Zea mays using TALENs and the CRISPR/Cas system. *J. Genet. Genom.* **2014**, *41*, 63–68. [CrossRef] [PubMed]
102. Chen, R.; Xue, G.; Chen, P.; Yao, B.; Yang, W.; Ma, Q.; Fan, Y.; Zhao, Z.; Tarczynski, M.C.; Shi, J. Transgenic maize plants expressing a fungal phytase gene. *Transgenic Res.* **2008**, *17*, 633–643. [CrossRef]
103. Bilyeu, K.D.; Zeng, P.; Coello, P.; Zhang, Z.J.; Krishnan, H.B.; Bailey, A.; Beuselinck, P.R.; Polacco, J.C. Quantitative conversion of phytate to inorganic phosphorus in soybean seeds expressing a bacterial phytase. *Plant Physiol.* **2008**, *146*, 468–477. [CrossRef] [PubMed]
104. Holme, I.B.; Dionisio, G.; Madsen, C.K.; Brinch-Pedersen, H. Barley *HvPAPhy_a* as transgene provides high and stable phytase activities in mature barley straw and in grains. *Plant Biotechnol. J.* **2017**, *15*, 415–422. [CrossRef]

© 2020 by the author. Licensee MDPI, Basel, Switzerland. This article is an open access article distributed under the terms and conditions of the Creative Commons Attribution (CC BY) license (http://creativecommons.org/licenses/by/4.0/).

Review

Phytic Acid and Transporters: What Can We Learn from *low phytic acid* Mutants?

Eleonora Cominelli [1],*, Roberto Pilu [2] and Francesca Sparvoli [1]

[1]   Institute of Agricultural Biology and Biotechnology, Consiglio Nazionale delle Ricerche, Via E. Bassini 15, 20133 Milan, Italy; sparvoli@ibba.cnr.it

[2]   Department of Agricultural and Environmental Sciences—Production Landscape, Agroenergy Università degli Studi di Milano, Via G. Celoria 2, 20133 Milan, Italy; salvatore.pilu@unimi.it

*   Correspondence: cominelli@ibba.cnr.it; Tel.: +39-022-369-9421

Received: 22 November 2019; Accepted: 1 January 2020; Published: 5 January 2020

Abstract: Phytic acid has two main roles in plant tissues: Storage of phosphorus and regulation of different cellular processes. From a nutritional point of view, it is considered an antinutritional compound because, being a cation chelator, its presence reduces mineral bioavailability from the diet. In recent decades, the development of low phytic acid (*lpa*) mutants has been an important goal for nutritional seed quality improvement, mainly in cereals and legumes. Different *lpa* mutations affect phytic acid biosynthetic genes. However, other *lpa* mutations isolated so far, affect genes coding for three classes of transporters: A specific group of ABCC type vacuolar transporters, putative sulfate transporters, and phosphate transporters. In the present review, we summarize advances in the characterization of these transporters in cereals and legumes. Particularly, we describe genes, proteins, and mutants for these different transporters, and we report data of in silico analysis aimed at identifying the putative orthologs in some other cereal and legume species. Finally, we comment on the advantage of using such types of mutants for crop biofortification and on their possible utility to unravel links between phosphorus and sulfur metabolism (phosphate and sulfate homeostasis crosstalk).

Keywords: phytic acid; low phytic acid mutants; MRP transporter; ABCC transporter; SULTR transporter; Pht; phosphate transporter; sulfate transporter

---

1. Introduction

Phytic acid (PA), chemically *myo*-inositol-1,2,3,4,5,5-hexakisphosphate, is the major form of phosphorus (P) storage in seeds (up to 85% of total P) and in other plant organs, such as pollen, roots, tubers, and turions. However, PA is not only an important molecule for P storage but, together with its precursors (lower InsPs and *myo*-inositol) and its derivative molecules (InsP7 and InsP8 inositol pyrophosphates), it is involved in the regulation of different cell signaling and plant processes in vegetative tissues, such as abiotic and biotic stress response, storage and polar transport of auxin, P homeostasis, photomorphogenesis, chromatin modification, and remodeling and mRNA nuclear export [1].

In seeds, where P amounts may even be 1000-fold higher than those detected in vegetative tissues, PA is accumulated during development, reaching a plateau at the end of the "cell expansion phase" [2,3]. PA is synthesized in the cytosol through two different routes: (i) The lipid-independent pathway, the most used in the seed, consisting of the sequential phosphorylation of the 6-carbon *myo*-inositol and soluble inositol phosphates (InsPs), and (ii) the lipid-dependent pathway, using precursors that include phosphatidylinositol (PtdIns) and PtdIns phosphates. PA is transferred from the cytosol to the vacuole where it is accumulated into globoids, spherical inclusions found within protein bodies [4–7]. Interestingly, the amount and distribution of PA in different seed/grain portions vary among different species. In cereals, there are differences between *Zea mays* L. (maize) kernels, where PA is mainly present in the embryo and scutellum, and the small grains of *Hordeum vulgare* L. (barley), different *Triticum* (wheat) species and *Oryza sativa* L. (rice), where 80% of PA is stored in the aleurone and bran (maternal teguments) and only a limited amount accumulates in the embryo [8]. However, in legumes, more than 95% seed PA is accumulated in the cotyledons [9]. During germination, phytases degrade PA and in this way, P is remobilized to support seedling growth [10]. Due to its high negative charge at physiological pH (~6–7), PA easily precipitates in the form of phytate salts binding cations, such as iron, zinc, potassium, calcium, magnesium, some of them (mainly iron and zinc) important from a nutritional point of view, in this way reducing their bioavailability. Only ruminants are able to degrade PA, due to the presence of microbial phytases in their digestive tracts, while for monogastric animals, including humans, mainly in those populations whose diet is largely based on staple crops, the presence of PA decreases seeds' nutritional value [11,12]. Moreover, as undigested PA is excreted by non-ruminants, such as swine, fowl, and fish, the supplementation of feed with nutrient P is a common practice, in order to provide for an animal's nutritional requirement. In this way P concentrations increase in manure, consequently in soils, finally contributing to P pollution in runoff water [13]. Hence, PA is considered an antinutrient and in recent decades, many efforts were spent to isolate and develop low phytic acid (*lpa*) crops, in which a 45–90% reduction of PA was achieved [1]. Unfortunately, it was shown that the reduction in PA content may affect plant growth, plant stress response and seed development and germination, thus limiting the efficacy of the introgression of the *lpa* trait into breeding programs [14]. The negative pleiotropic effects of the *lpa* mutations depend on the previously mentioned important roles PA has in different regulatory processes.

Hence, it is very important to identify the best strategy in order to specifically decrease PA content in the seeds without affecting plant and seed performance and possibly contribute to reducing the environmental impact. The *lpa* mutations so far isolated can be classified into three classes, depending on the step of the biosynthetic pathway or transport they affect: (i) Mutations altering the MIPS activity, the first steps of the biosynthetic pathway (from glucose 6-P to *myo*-inositol[3]-monophosphate), (ii) mutations affecting the following phosphorylation of the InsP6 pathway (from *myo*-inositol[3]-monophosphate to PA), (iii) mutations perturbing the final transport of PA.

Only mutants belonging to class (ii) accumulate InsPs intermediates. The mutations belonging to the (i) and (iii) classes induce a decrease of PA amount, accompanied by a molar equivalent increase of inorganic phosphate (P_i) in the homozygous mutants. Moreover, they are usually perturbed in different branches of the biosynthetic pathway common to PA and other compounds (e.g., galactinol, raffinose, stachyose, and ascorbic acid). Mutants in three classes of transporters have been characterized for their *lpa* phenotype, affected in: (i) A specific group of ABCC type vacuolar transporters [15], orthologues to the *Arabidopsis thaliana* (L.) Heinh AtMRP5 (also referred to as AtABCC5) [16,17], also known as multidrug resistance-associated proteins (MRPs), (ii) putative sulfate transporters, orthologues of the Arabidopsis AtSULTR3;3 [18,19] and AtSULTR3;4 proteins [20]; (iii) the rice OsPht1;4 phosphate transporter [21,22]. Only in the case of an ABCC transporter was it shown that the protein is able to actually transport PA [16].

In this review we will discuss the advances in the characterization of PA-MRP, PA-SULTR, and OsPht1;4 transporters, and of the corresponding mutants described so far in cereals and legumes. Particular emphasis will be given to the reported differences among cereals and legumes of *lpa* mutant phenotypes in the PA-MRP genes, depending on the presence of one or more partially redundant copies of these genes and to their tissue-specific expression. Moreover, we identified in silico the putative orthologs of PA-SULTR in species of interest for the isolation of *lpa* mutants. We will also discuss the advantages of these mutants for crop biofortification. Furthermore, we will highlight how the study of these mutants may help to elucidate phosphate and sulfur metabolism, and the possible roles that the transporters described here may play.

2. PA-MRP Transporters

MRP proteins are transmembrane transporters involved in several functions, such as organic ions transport, xenobiotic detoxification, oxidative stress tolerance, and transpiration control [23]. The first evidence of the involvement of an MRP-type ABC transporter in PA transport was reported for the maize ZmMRP4 protein from the analysis of the insertional *lpa1* mutant [24]. *ZmMRP4* gene is orthologous to *AtMRP5*, which had already been characterized some years ago as an anion transporter involved in root growth, lateral root formation, regulation of stomatal movement, guard cell hormonal signaling, and water use efficiency [25–27], aspects not immediately attributable to PA transport. The biochemical demonstration that PA transport was dependent on the presence of a functional MRP transporter in an ATP-dependent manner was given for AtMRP5, showing a very high affinity for PA (maximum reaction velocity -V_{max}- values of about 1.6–2.5 μmol min^{-1} mg^{-1} and Michaelis—Mentent constant -K_m- ranging between 263 and 310 nM) and a vacuolar subcellular localization [16].

As summarized in Table 1, other *PA-MRP* genes and the corresponding mutants/transgenics have been hereafter characterized in rice, *Glycine max* (L.) Merr. (soybean), *Phaseolus vulgaris* L. (common bean) and *Triticum aestivum* L. (soft wheat) [28–33]. Moreover, the putative *Pennisetum glaucum* (L.) R.Br. (pearl millet) *PA-MRP* gene has recently been described [34].

The main difference between cereals and legumes for which PA-MRPs have been characterized so far is the gene number: While only one gene is present in diploid maize, pearl millet, and rice genomes, and three copies in the hexaploid *Triticum aestivum*, two and three paralogues are present in common bean and soybean, respectively [17,32,33]. The presence of more than one member of the *PA-MRP* genes seems to be a common feature of legumes, for example also in *Medicago truncatula* it is possible to predict two *PA-MRP* genes [35], unlike the situation in other dicotyledons, such as Arabidopsis and *Solanum lycopersicum* L. (tomato) in which only one *PA-MRP* gene was described [16,36]. As discussed below, the gene copy number has a significant influence on the *lpa* mutant phenotypes.

The gene structure of PA-MRP transporters is very highly conserved: All analyzed genes in the present study have 11 exons and 10 introns with very similar lengths, only some differences can be found in the lengths of some introns between cereals and legumes. In Figure 1a, the rice *OsMRP5* and the soybean *GmABCC1* gene schematic representations are given as examples.

PA-MRP proteins are full-length ABC transporters (length from 1501 aa of TaABCC13–4B to 1539 aa of GmABCC1) with three membrane-spanning domains (TMD0, containing five transmembrane α-helices and TMD1, TMD2, each with six α-helices) and two cytosolic nucleotide-binding domains (NBD1 and NBD2, containing the Walker A and B motifs), arranged in the TMD0-TMD1-NBD1_TMD2-NBD2 so-called forward orientation (Figure 1b), as previously described [17]. Although it is not known which specific amino acids are involved in PA transport, a conserved lysine residues stretch, located in the cytosolic loop linking NBD1 and TMD2 and a number of charged amino acid residues (mostly lysine and arginine) found in other conserved stretches in TMD1 and TMD2 have been suggested to be involved in PA transport [17].

PA-MRP protein sequences are phylogenetically very highly conserved among different species, mainly in the TMD and NBD domains, but also outside, particularly among cereals or legumes (Figure 1 and Figure S1). As shown in Figure 1c, the degree of aminoacid identity among different PA-MRP proteins belonging to different species is very high, also between cereals and legumes where it ranges from 67 to 71.5% (similarity between cereals and legumes ranges from 86 to 89%, data not shown).

Table 1. Described cereal and legume transporters involved in PA accumulation and corresponding mutations affecting PA seed/grain content. EMS stands for ethyl methanesulfonate. * Indicates incomplete sequence.

Class of Transporters	Species	Gene	Phytozome/Genbank/Ensembl Accession Number	Origin of Mutation	Mutation	Reference
MRP	*Zea mays*	*ZmMRP4/ZmABCC4*	EF586878	EMS	*lpa1-1*	[37]
					lpa1-241	[38]
					lpa1-7	[39]
				T-DNA insertion	*lpa1-mum1*	[24]
				Embryo specific:RNAi	*Ole::MRP4 Glb::MRP4*	[24]
	Oryza sativa	*OsMRP5/OsABCC13*	LOC_Os03g04920	γ rays + sodium azide	*Os-lpa-XS110-2*	[40]
					Os-lpa-XS110-3	[41]
				T-DNA insertion	*4A-02500*	[41]
				Embryo specific amiRNA	*Ami-MRP5*	[42]
	Triticum aestivum	*TaABCC13-4B* *TaABCC13-4D* *TaABCC13-5A**	TraesCS4B02G343800 TraesCS4D02G339000 TraesCS5A02G512500	Constitutive RNAi	*TaABCC13 RNAi*	[33]
	Glycine max	*GmMRP3/GmABCC1* *GmMRP19/GmABCC2* *GmMRP13/GmABCC3*	Glyma.03G167800 Glyma.19G169000 Glyma.13G127500	EMS no reported mutant	*CX1834* no reported mutant	[29–31,43] [32]
	Phaseolus vulgaris	*PvMRP1/PvABCC1*	Phvul.001G165500	EMS	*lpa1* *lpa1²*	[32,44] [35]
		PvMRP2/PvABCC2	Phvul.007G153800	no reported mutant	no reported mutant	[32]
SULTR	*Oryza sativa*	*OsSULTR3;3*	LOC_Os04g55800	γ rays	*Oslpa-MH86-1* *Os-lpa-Z9B-1*	[28] [19]
		OsSULTR3;4	LOC_Os06g05160	retrotransposon *Tos-17* insertion	*spdt-1, spdt-2, spdt-3*	[20]
	Hordeum vulgare	*Hvst*	HORVU2Hr1G113050	sodium azide	*lpa1-1(M422)*	[45] [19]
Pht	*Oryza sativa*	*OsPht1;4*	LOC_Os04g10750	retrotransposon *Tos-17* insertion	*ospt4-1 (NE1260) ospt4-2 (SHIP_ZSF6267)* RNAi	[21] [22]

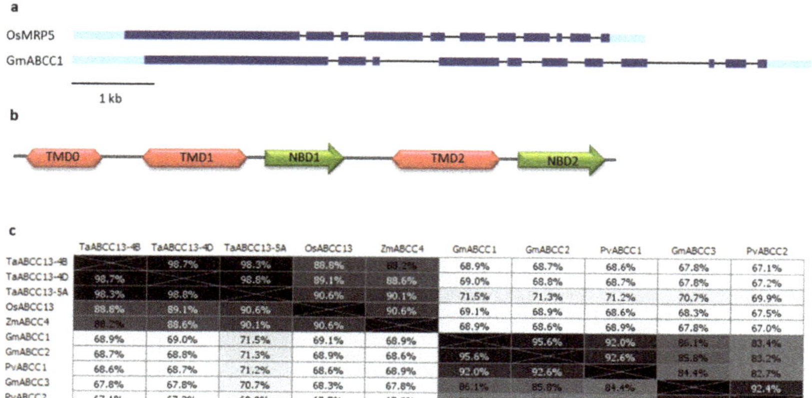

	TaABCC13-4B	TaABCC13-4D	TaABCC13-5A	OsABCC13	ZmABCC4	GmABCC1	GmABCC2	PvABCC1	GmABCC3	PvABCC2
TaABCC13-4B		98.7%	98.3%	88.8%	88.2%	68.9%	68.7%	68.6%	67.8%	67.1%
TaABCC13-4D	98.7%		98.8%	89.1%	88.6%	69.0%	68.8%	68.7%	67.8%	67.2%
TaABCC13-5A	98.3%	98.8%		90.6%	90.1%	71.5%	71.3%	71.2%	70.7%	69.9%
OsABCC13	88.8%	89.1%	90.6%		90.6%	69.1%	68.9%	68.6%	68.3%	67.5%
ZmABCC4	88.2%	88.6%	90.1%	90.6%		68.9%	68.6%	68.9%	67.8%	67.0%
GmABCC1	68.9%	69.0%	71.5%	69.1%	68.9%		95.6%	92.0%	86.1%	83.4%
GmABCC2	68.7%	68.8%	71.3%	68.9%	68.6%	95.6%		92.6%	85.8%	83.2%
PvABCC1	68.6%	68.7%	71.2%	68.6%	68.9%	92.0%	92.6%		84.4%	82.7%
GmABCC3	67.8%	67.8%	70.7%	68.3%	67.8%	86.1%	85.8%	84.4%		92.4%
PvABCC2	67.1%	67.2%	69.9%	67.5%	67.0%	83.4%	83.2%	82.7%	92.4%	

Figure 1. (**a**) Gene structure of *OsMRP5* and *GmABCC1* genes, as an example of a cereal and a legume *PA-MRP* gene, respectively. Light and dark blue rectangles represent UTRs and coding exons, respectively, the black bars correspond to introns. Gene Structure Display Server [46] was used; (**b**) Predicted domains of the PA-MRP protein. The transmembrane domains (TMD) and the nucleotide-binding domains (NBD) are represented in red and green, respectively. The structure of the PA-MRP proteins was previously described [17]; (**c**) Distances between PA-MRP proteins, expressed as a percentage of identity. Phylogenies were constructed with the Geneious Tree Builder tool, using the Jukes—Cantor distance model, neighbor-joining tree build method.

In Figure 2, pictographic representations of the different organs' expression patterns of the *OsMRP5* and the soybean *PA-MRP* genes, taken as examples for cereals and legumes, are reported from the rice and soybean eFP Browsers [47]. The rice gene is expressed at high levels in different organs including the caryopsis, as previously reported [41]. The maize ortholog shows a similar expression pattern [24]. The *TaABCC13* genes are expressed in different plant organs, preferentially during grain developmental stages, with the transcript accumulation derived from the B genome the highest one, mainly at 14 days after anthesis [33,48].

The expression pattern of the different legume genes varies. As shown in Figure 2, *GmMRP3* and *GmMRP19* genes are expressed in different organs and highly expressed in seed, particularly at the late stage of development, while the *GmMRP13* gene is mainly expressed in root and flower and at a very low level in seed. As discussed below, when both *GmMRP3* and *GmMRP19* are mutated an *lpa* seed phenotype occurs [30]. It suggests that *GmMRP3* and *GmMRP19* have an important role in PA accumulation and their function is redundant, while *GmMRP13* is not active in the seed.

A similar diversified expression pattern was observed in common bean, where the *PvMRP1* gene, coding for a protein more similar to GmMRP3 and GmMRP19 (Figure 1c and Figure S1), is highly expressed in cotyledons, where its transcript levels continue to increase during seed development, reaching the highest levels at 28 days after flowering (DAF) with a similar kinetics to that reported for the accumulation of PA in the same organ. The *PvMRP2* gene, ortholog of *GmMRP13*, is expressed similarly to *PvMRP1* in vegetative organs, but at no appreciable level in cotyledons. Interestingly, both genes are expressed in root nodules, organs specialized in symbiosis with nitrogen-fixing bacteria, in which the role of PA is still unknown [35]. Recently, a detailed analysis was reported of GUS activity in *Arabidopsis thaliana* and *Medicago truncatula* plants, harboring a promoter sequence of *PvMRP1* and *PvMRP2* genes, fused upstream of the *GUS* reporter gene. The strongest GUS activity, driven by both constructs, in organs other than the seeds was present in the vascular tissues [35]. Similar patterns of reporter gene activity were previously shown in transgenic plants harboring the *AtMRP5* promoter [25] and promoters of different Arabidopsis genes coding for enzymes involved in different steps in PA pathway [49–56]. These data suggest that vascular tissues are an important site for synthesis and transport of PA involved in the regulation of different cellular processes, the so-called "signaling PA" [57].

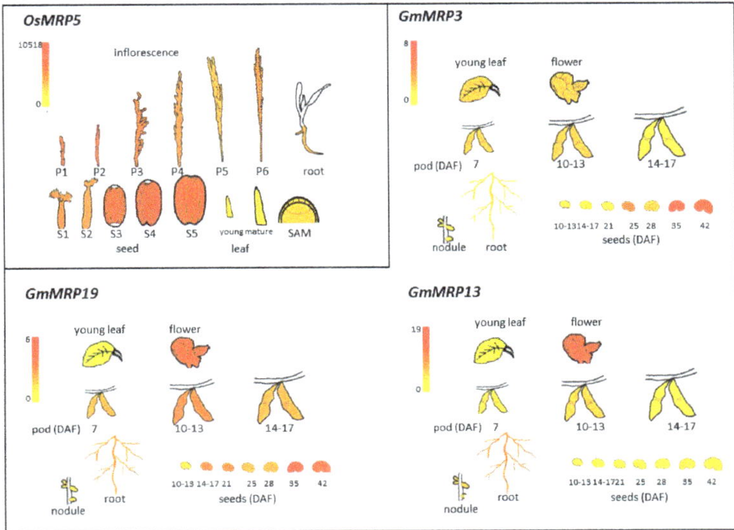

Figure 2. PA-MRP gene expression data in various rice and soybean organs and tissues were obtained from the rice and soybean eFP Browsers [47]. Rice MAS and soybean Severin data sources were used. For rice, the default signal threshold was used, while for the three soybean genes the signal threshold was arbitrarily put to the same value (8.00) in order to compare expression data between different genes.

lpa Mutants in PA-MRP Transporters

As shown in Table 1, the majority of the *lpa* mutations affecting transporters concern mutations in PA-MRP proteins. Differences exist between cereal and legume *lpa* mutants, with cereal mutants generally affected by more pronounced negative pleiotropic effects mainly due to: (i) The different accumulation of PA in seed/caryopsis compartments, as previously mentioned, (ii) the presence of only one gene coding for a PA-MRP transporter in cereal genomes and more than one in legumes [17,37,38]. As previously discussed, there are similarities between mutants affected in PA biosynthetic genes and in PA transport in the reduction of PA content, accompanied by a molar equivalent increase of P_i and the absence of accumulation of InsPs intermediates. For this reason, the first efforts to map the maize *lpa1* mutation suggested that the *myo*-inositol 3-phosphate synthase (*MIPS*) gene coding for the first enzyme of the pathway was mutated [37,38]. This was also corroborated by mapping and expression data, since in maize the *ZmMIPS1S* and the *ZmMRP4* genes map very closely on chromosome 1S, and in mutants affecting *ZmMRP4* the expression of *ZmMIPS1S* is reduced [24,37,38,58]. However, transposon mutagenesis tagging experiments conducted by Shi et al. (2007) demonstrated that *lpa1* gene encodes a multidrug-associated-protein (MRP) named ZmMRP4 (accession number EF586878). As shown in Table 1, different *lpa* mutations were isolated in the maize ZmMRP4 and the rice OsMRP5 PA transporters [4,24,28,37,38,41,58–63]. Due to the previously mentioned important roles of PA in different regulatory processes and due to the fact that in these species only a *PA-MRP* gene is present, these mutants display negative pleiotropic effects on plants (stunted vegetative growth) and seeds, such as reduced seed development and weight, low germination rates, making these mutants of limited value to breeders [37,59,64–66]. In maize, the most studied model species, four different mutants affecting the *ZmMRP4* locus were isolated: *lpa1-1*, consisting of a point mutation that causes an A1432V substitution in the NBD2 region [24,37], *lpa1-241*, a paramutagenic allele [60] that causes a remarkable variability of expression with a different degree of negative pleiotropic effects depending on its strength [59], *lpa1-7*, whose molecular feature is not known, although the nature of a paramutagenic allele can be excluded [39] and *lpa1-5525*, not yet fully characterized [67]. In the *lpa1-1* mutant, kernel PA was reduced by 66% [37], whilst *lpa1-241* and *lpa1-7* mutants showed the highest reduction in PA with more than 80% [39,59]. All these mutants do not perturb the total P present but are characterized by a five- to ten-fold increase in the amount of free phosphate in the kernel [37–39].

In rice, the *Os-XS-lpa2-1* and *Os-XS-lpa2-2* mutations have been isolated at the *OsMRP5* locus [28]. The *Os-XS-lpa2-1* mutant shows a grain PA reduction of about 20% caused by a single base pair substitution mutation in the transmembrane domain TMD2 [41]. In the case of *Os-XS-lpa2-2* the PA reduction is more than 90% due to a 5-bp deletion determining a frame shift causing a premature stop codon at aa 474. The same phenotype was observed in a T-DNA knock outline (4A-02500), demonstrating the important involvement of this gene in PA transport [41]. Unfortunately, in these maize and rice mutants, there is a correlation between the severity of the negative pleiotropic effects and the PA content. In fact, the strongest maize *lpa1-241* and *lpa1-7* mutants and rice *Os-XS-lpa2-2* and *4A-02500* are lethal in the homozygous state, while the other milder mutants (*lpa1-1* and *Os-XS-lpa2-1*) are viable, although showing yield losses compared to wild type [37,39,41,59]. The incapacity to germinate is probably due to the impaired embryo development, mainly because of the displacement of the root primordium and the consequent asymmetry in the body plan, as shown in the maize *lpa1-241* mutant [39,59]. Furthermore, the maize *lpa1-1* mutant and barley *lpa* mutants, such as *Hvlpa1*, *Hvlpa2*, *Hvlpa3*, and *Hv-M955* mutants affected in other genes, are more sensitive to drought stress in the field [68]. The negative pleiotropic effects could be associated with an alteration of the mature root system, as demonstrated in the case of the maize *lpa1-7* mutant [39]. In the latter mutant, other pleiotropic effects associated with the *lpa* mutation have been described, such as reduced carotenoid and chlorophyll content and increased length and trichome density compared with wild type sibling leaves [39].

Another explanation for the lethal phenotype due to the strongest mutations was proposed by Doria and colleagues [62]: They showed that whole *lpa1-241* mutant kernels contained about 50% more free iron associated with a higher content of free radicals than the wild type control. Furthermore, higher production of hydrogen peroxide was found in the embryo of *lpa1-241* grains, particularly in the ones artificially aged. Taken together, these results confirmed that PA is involved in the prevention of oxidative stress in grains, previously only suggested [69–71] and considered to be important for the maintenance of the viability of grains [37,72]. Another hypothesis to explain the negative pleiotropic effect associated with mutations affecting the multidrug-associated-protein (MRP) in *lpa* mutants could be that this protein is involved directly or indirectly in the transport of other molecules in addition to PA. In fact, it was observed that the *lpa1-241* mutation, in a genetic background capable of accumulating anthocyanins in the scutellum (embryo tissue), conferred a bluish color in comparison to the reddish wild type control. This alteration was attributed to a defect in the pigment transport in the vacuole, causing a mislocalized accumulation of these pigments in the cytosol, suggesting that ZmMRP4 could have a direct or indirect role in anthocyanin transport [62].

To overcome the negative pleiotropic effects present in maize and rice, *lpa* mutants affected in *ZmMRP4* and *OsMRP5* genes, respectively, seed-specific silencing of both *MRP* genes was undertaken [24,42]. Transgenic lines expressing an antisense sequence for a fragment of the cDNA for the ZmMRP4 transporter under the control of the embryo-specific *Ole16* and *Glb* promoters produced *lpa*, high P_i grains that germinated normally and did not have any significant reduction in grain dry weight, revealing the potential of this approach in maize nutritional quality improvement [24]. On the other hand, plants silenced in the *OsMRP5* gene through the artificial microRNA (amiRNA) technology, under the control of the *Ole18* promoter, active in the embryo and aleurone, produced *lpa* grains (PA reduced by 35.8–71.9% with increased levels of P_i of up to 7.5 times). Although no consistent significant differences of plant height or number of tillers per plant were observed, significantly lower grain weights (up to 17.8% reduction) and reduced seed germination were observed, suggesting that this strategy is not successful for practical application in rice breeding. The different results obtained in maize and rice may depend on the different promoters used, with the rice ones also being active in aleurone and endosperm beyond the embryo [42]. A similar approach was also used in hexaploid wheat, where the three copies of the *TaABCC13* gene, previously shown to encode a protein able to transport cadmium [73] were silenced through RNA-interference (RNAi). In transgenic lines, a reduction in PA content of 34–22% was observed. Moreover, these lines were characterized by reduced grain filling, reduced numbers of spikelets, reduced kernel viability, delayed germination, early emergence of lateral roots, and defects in metal uptake and development of lateral roots in the presence of cadmium stress, compared to non-transgenic lines. These data show that TaABCC13 is important for several other aspects of growth as well as for grain nutritional quality and for root development and detoxification of heavy metals [33].

Mutations in PA-MRP transporters have also been reported in soybean and common bean, two of the most relevant legume crops worldwide [43,44]. Following EMS mutagenesis of the soybean breeding line *CX1515-4*, the two independent *M153* and *M766* mutant lines were isolated, with the *M153* line displaying a stronger PA reduction compared to the *M766* one (80% vs. 76.3%, respectively) [43]. However, the content of PA drops to 94% of that of the parental line when the double mutant is produced [31,43,74,75]. Although at the beginning it was hypothesized that a mutation in the *MIPS* gene could be responsible for the *lpa* phenotype of these lines [29,76], genetic and fine-mapping studies revealed that the trait was under the control of two loci, named *lpa1* and *lpa2* [74,77]. These contained independent but interactive recessive alleles coding for PA-MRP transporters, GmMRP3/GmABCC1, and GmMRP19/GmABCC2, respectively [30,31] (Table 1). It was shown that the *lpa1-a* allele (line M153) carries a nonsense mutation at R893, which results in a truncated protein [29,30], while in the case of the *lpa1-b* allele (M766 line) a single T > A SNP 7 bp upstream of the start of exon 10 was identified, which introduced an alternative splicing site producing five additional base pairs from the intron sequence and a frame shift starting at exon 10. Concerning the second locus, an R1039K change was identified in the *lpa2-a* allele (*M153* line), while in the *lpa2-b* allele (*M766* line) a single base change at position 1039 causes a premature termination [31].

A number of agronomic analyses have been performed on the soybean breeding line *CX1834-1-6* (derived from the mutant lines *M153*), and in different studies, a reduction in seedling emergence (about 22–30% less than wt) has been reported [75,78–80]. In particular, Anderson et al. (2008) demonstrated that the environment of reproduction of the *lpa* plants has important implications for seedlings' field emergence. In fact, *lpa* seeds harvested in Puerto Rico (tropical environment) displayed decreased germination, compared to those harvested in Iowa (temperate environment). However, genetic improvement through advanced backcrossing was successful and *lpa* lines with normal seedling emergence were obtained [79].

In common bean, two *lpa* mutants in the PA-MRP transporter have been isolated in two different backgrounds [35,44]. In the *lpa1* mutant, a highly conserved Glu changed to Lys at position 1155, in the transmembrane domain TMD2, while in the *lpa1²* mutant a single base pair change in the first exon caused a non-sense mutation (R500Stop) leading to a truncated protein. Reduction of PA accumulation was about 90% and 75% compared with the wt parent, for the *lpa1* and *lpa1²* mutants, respectively, suggesting a highly critical functional role of the conserved Glu_{1155} residue. In the *lpa1* mutant, it has also been demonstrated that PA accumulation is accompanied by a decrease of raffinose-containing sugars by 25% and *myo*-inositol by 30% [32,44], thus indicating metabolic rearrangements of derived pathways. Despite the strong PA reduction in the seed, the different bean *lpa1* mutant lines showed that seedling emergence, seed yield, and plant growth were not statistically different from those of wt and parental genotypes [81]. Furthermore, germination of *lpa1* seeds in stressful conditions: By the accelerated aging test (AAT) and the stress integrated germination test (SIGT) showed that there was equal (SIGT) or even better (AAT) germination performance of *lpa1* seeds compared to the wt ones [44]. The finding that in common bean a second gene, *PvMRP2*, paralog of *PvMRP1*, is present, indicates that most likely it is able to complement the absence of a functional PvMRP1 in tissues and organs other than the seed. Another interesting feature of the *lpa1* mutant is that its seeds when germinated in the presence of ABA were hypersensitive to the presence of this phytohormone [26] a result contrasting with the finding of Klein and coworkers who reported that the Arabidopsis *mrp5* mutant has reduced sensitivity to ABA during germination [26]. Since the sensitivity of seed germination to ABA has been reported to correlate negatively to seed *myo*-inositol content [32,50,82,83], it is possible that *myo*-inositol levels are not reduced or may even be increased in the Arabidopsis *mrp5* mutant.

3. SULTR3.3 and SULTR3.4 Transporters Involved in PA Metabolism

Two *lpa* mutants isolated in barley and rice are affected in *HvST* and *OsSULTR3;3* genes, respectively [19,28,45,84], coding for two putative sulfate transporters, belonging to the SULTR3;3 class [18]. Recently, another rice *lpa* mutant, affected in the OsSULTR3;4 putative sulfate transporter, also called SULTR-like Phosphorus Distribution Transporter (SPDT), was isolated [20].

Here, we present an in silico analysis of *SULTR3;3* and *SULTR3;4* genes, including the ones already described (Table 1) and also putative *SULTR3;3* and *SULTR3;4* orthologs from other cereal and legume crops for which interest in the isolation of *lpa* mutants is considered an important challenge, such as maize, barley, common bean and soybean (Table 2). A phylogenetic tree with all the SULTR3;3 and SULTR3;4 proteins of cereals and legumes analyzed in the present work is shown in Figure S3.

Table 2. Putative orthologous genes of *OsSULTR3;3/HvSULTR3;3* and *OsSULTR3;4*, identified in maize, common bean, and soybean by in silico analysis.

SULTR Group.	Species	Gene Name	Phytozome Accession Number
SULTR3;3	*Zea mays*	*ZmSULTR3;3*	GRMZM2G395114
	Phaseolus vulgaris	*PvSULTR3;3*	Phvul.002G095300
	Glycine max	*GmSULTR3;3a*	Glyma.20G017100
		GmSULTR3;3b	Glyma.07G218900
SULTR3;4	*Zea mays*	*ZmSULTR3;4*	GRMZM2G444801
	Phaseolus vulgaris	*PvSULTR3;4a*	Phvul.005G171800
		PvSULTR3;4b	Phvul.010G151000
	Glycine max	*GmSULTR3;4a*	Glyma.07G006500
		GmSULTR3;4b	Glyma.08G207100
		GmSULTR3;4c	Glyma.13G360000
		GmSULTR3;4d	Glyma.15G014000

3.1. SULTR3;3

In all analyzed species, one putative ortholog belonging to the SULTR3;3 group was found by BLAST analysis of HvST or OsSULTR3;3 against the different genomes, except for soybean, in which two different genes have been identified (Table 2, Figures S3 and S4). Indeed, this is not unexpected, as soybean underwent an ancient event of genome duplication [85]. The SULTR3;3 gene structure is quite conserved among species and consists of 13 exons in the majority of the genes, with the exceptions of barley and maize with only 12 exons. All genomic sequences are characterized by the presence of a long fourth or fifth intron, as reported in Figure 3a, where the structure of the characterized OsSULTR3;3 and HvST and of PvSULTR3;3 is shown as an example.

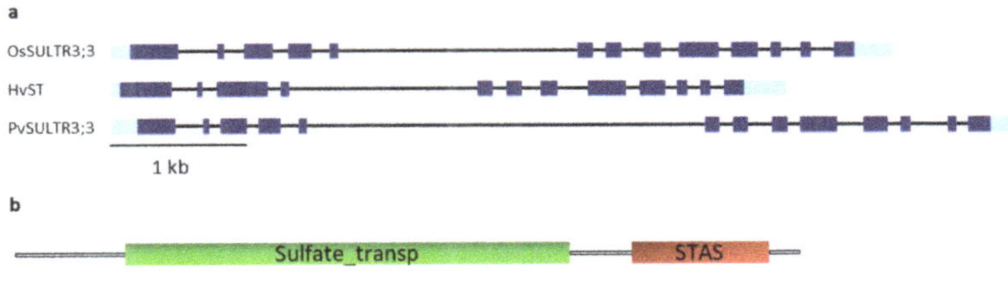

	ZmSULTR3;3	OsSULTR3;3	HvST	GmSULTR3;3b	GmSULTR3;3a	PvSULTR3;3
ZmSULTR3;3		86.5%	84.1%	68.3%	67.8%	68.0%
OsSULTR3;3	86.5%		86.1%	68.7%	68.0%	69.2%
HvST	84.1%	86.1%		68.1%	67.5%	68.5%
GmSULTR3;3b	68.3%	68.7%	68.1%		97.9%	91.7%
GmSULTR3;3a	67.8%	68.0%	67.5%	97.9%		90.9%
PvSULTR3;3	68.0%	69.2%	68.5%	91.7%	90.9%	

Figure 3. (**a**) Gene structure of *OsSULTR3;3, HvST* and putative *PvSULTR3;3* genes. Light and dark blue rectangles represent UTRs and coding exons, respectively, the black bars correspond to introns. Table 1. a legend. (**b**) Predicted domains of the SULTR3;3 protein. The sulfate transporter and the anti-sigma factor antagonist (STAS) domains are represented in red and green, respectively. Picture reproduced from [84]. (**c**) Distances between SULTR3;3 proteins, expressed as the percentage of identity. Phylogenies were constructed as described in Figure 1c.

Predicted domains of SULTR3;3 proteins are represented in Figure 3b and correspond to a sulfate transporter domain and an anti-sigma factor antagonist (STAS) domain, as previously reported [84]. Protein length varies from 647 aa of PvSULTR3;3 to 661 aa of OsSULTR3;3 (Figure S4).

Protein identity is generally very high among different species, ranging from 84.1% to 86.5% among the considered cereals and from 90.9% to 91.7% among legumes, and at 97.9% in the two paralogs of soybean, as shown in the Figure 3c diagram.

In the case of the OsSULTR3;3 gene detailed expression analysis was reported: Transgenic lines harboring the promoter of this gene fused to the GUS reporter gene revealed that a strong GUS activity was present in vascular bundles of shoots, leaves, flowers, and grains, where it was mainly detected in the scutellum. Moreover, the subcellular localization was defined to be in the endoplasmic reticulum [84]. Interestingly, both GmSULTR3;3a and GmSULTR3;3b are expressed in leaves and flowers, while only GmSULTR3;3a was expressed in the seed, with an increasing expression during seed development with a peak at 35 DAF (data not shown, in silico analysis performed using the soybean eFP Browser [47]).

The exact function of this family of proteins is still unknown and in the case of OsSULTR3;3, which was the only one analyzed in detail, no activity was revealed for the transport of phosphate, sulfate, inositol or inositol 1,4,5 triphosphate by heterologous expression in either yeast or Xenopus oocytes [84].

3.2. SULTR3;4

In the case of the SULTR3;4 group of transporters a similar situation to the one previously described for MRP proteins is present: In cereals, only one protein for each species can be found by BLAST analysis of OsSULTR3;4 against the different genomes, while in legumes, two or four paralogous proteins are present in common bean and soybean, respectively (Table 2 and Figures S3 and S5). The gene structure differs between cereals with 10 exons (the barley sequence present in the Phytozome database is incomplete with only eight exons) and legumes with 13 exons and also in this case, the fourth (the fifth in maize) intron is quite long. In Figure 4a the structures of the characterized OsSULTR3;4 and of putative ZmSULTR3;4 and PvSULTR3;4a genes are given as examples.

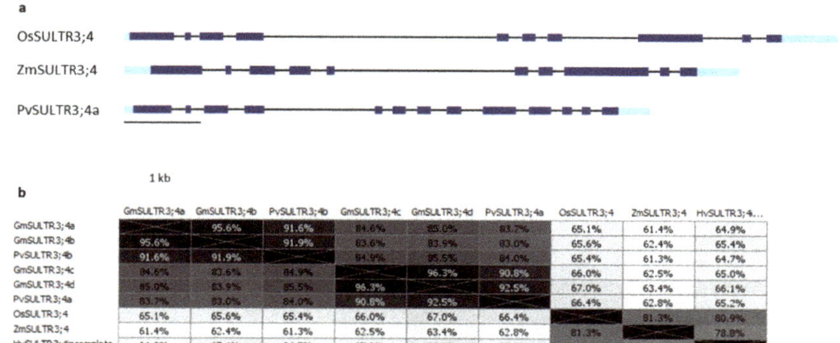

b

	GmSULTR3;4a	GmSULTR3;4b	PvSULTR3;4b	GmSULTR3;4c	GmSULTR3;4d	PvSULTR3;4a	OsSULTR3;4	ZmSULTR3;4	HvSULTR3;4...
GmSULTR3;4a		95.6%	91.6%	84.6%	85.0%	83.7%	65.1%	61.4%	64.9%
GmSULTR3;4b	95.6%		91.9%	83.6%	83.9%	83.0%	65.6%	62.4%	65.4%
PvSULTR3;4b	91.6%	91.9%		84.9%	85.5%	84.0%	65.4%	61.3%	64.7%
GmSULTR3;4c	84.6%	83.6%	84.9%		96.3%	90.8%	66.0%	62.5%	65.0%
GmSULTR3;4d	85.0%	83.9%	85.5%	96.3%		92.5%	67.0%	63.4%	66.1%
PvSULTR3;4a	83.7%	83.0%	84.0%	90.8%	92.5%		66.4%	62.8%	65.2%
OsSULTR3;4	65.1%	65.6%	65.4%	66.0%	67.0%	66.4%		81.3%	80.9%
ZmSULTR3;4	61.4%	62.4%	61.3%	62.5%	63.4%	62.8%	81.3%		78.8%
HvSULTR3;4incomplete	64.9%	65.4%	64.7%	65.0%	66.1%	65.2%	80.9%	78.8%	

Figure 4. (**a**) Gene structure of *OsSULTR3;4* and putative *ZmSULTR3;4* and *PvSULTR3;4a* genes. Light and dark blue rectangles represent UTRs and coding exons, respectively, the black bars correspond to introns. The gene structure was obtained as described in Figure 1a legend. (**b**) Distances between SULTR3;4 proteins, expressed as the percentage of identity. Phylogenies were constructed as described in Figure 1c.

Predicted domains are the same as those already described for SULTR3;3, represented in Figure 3b. Protein length varies from 648 aa of the soybean protein to 670 aa of the rice one (Figure S5).

Also for SULTR3;4 proteins, identity is quite high ranging from 78.8% to 81.3% among cereals, from 83% to 91.9% among legumes and from 61.3% to 67% between cereals and legumes, as shown in the diagram in Figure 4b.

Analysis of the phylogenetic tree (Figure S3) clearly shows a separation between monocotyledons and dicotyledons. Furthermore, in the two legume species, the gene is duplicated, with soybean carrying four genes arising from an ancient event of genome duplication [85].

qRT-PCR expression analysis of OsSULTR3;4 gene revealed that during grain filling it was mainly expressed in node I, a very important hub for mineral distribution to upper node and panicle in Poaceae [86]. Moreover, immunostaining against GFP in lines harboring OsSULTR3;4 promoter fused to GFP, showed the highest staining in the xylem region of both enlarged- and diffuse-vascular bundles of the basal node and in node I, as well as in the parenchyma tissues between them, but not in the phloem region [20]. The activity of OsSULTR3;4 as an influx plasma-membrane localized H+/P$_i$ symporter was shown in proteoliposomes as well as in Xenopus oocytes. Particularly, it was found that OsSULTR3;4 is involved in the intervascular transfer of P at the nodes, unloading P from xylem towards phloem [20].

3.3. lpa Mutants in SULTR Transporters

The first mutants affected in SULTR3;3 and SULTR3;4 genes were described in *Arabidopsis thaliana*. They have been characterized for phenotypic alterations related to sulfate translocation between seed compartments [87]. Moreover, using the quintuple mutant defective in all SULTR3 subfamily members, it was recently shown that they have functional redundancy in chloroplast sulfate uptake and consequent influence on Cys, glutathione, and ABA biosynthesis, with the resulting growth retardation and altered stress responses in the multiple mutants [88]. Otherwise, no evidence of the involvement of these Arabidopsis genes in PA metabolism has been reported so far, with the only exception of AtSULTR3;4 for which contrasting results have been reported. In fact, very recently, Ding and co-workers [89] demonstrated that AtSULTR3;4/SPDT functions as a high-affinity P$_i$ transporter, being able to mediate P$_i$ uptake when injected in the Xenopus oocyte. Furthermore, it has been shown to localize to the plasma membrane, while Chen et al. reported a chloroplast localization [89]. On the other hand, these data are in agreement with those reported below on mutations affecting the HvST,

OsSULTR3;3 and OsSULTR3;4 genes which confer grain *lpa* phenotype and in which the relevant proteins are localized in the endoplasmic reticulum and plasma membranes, respectively [19,20,84].

3.3.1. Mutants Affected in the *SULTR3;3* Genes

In the case of *HvST* a nonsense mutation (*M422*) in the last exon of the barley *lpa1-1* gene was isolated from a sodium azide mutagenized population [19], and in the case of *OsSULTR3;3* the two different *Os-lpa-Z9B-1* and *Os-lpa-MH86-1* mutations were a 6 bp deletion in the first exon and a 1 bp deletion in the 12th exon, identified through screening of a gamma-ray irradiation mutagenized population [28,45,84]. These barley and rice mutants exhibit a decrease in phytic acid-P like other *lpa* mutants, but also a decrease in total P in the seed (about 15% in barley mutant and 27.5–18.9% in rice mutants) [19,84,90], differently from *lpa* mutants affected in biosynthetic or *MRP*-transporter genes. Particularly, an endosperm-specific total P reduction [90] was reported that is not due to a reduction in the uptake of P in the maternal plant, suggesting that HvST functions as a seed-specific or filial determinant of barley endosperm total P. Moreover, OsSULTR3;3 disruption dramatically alters the grain metabolite profile. In fact, an increase was observed in the concentration of sugars involved in the close biosynthetic pathway leading to PA, sugar alcohols, free fatty acids, organic acids, biogenic amine GABA, serine, and lysine. However, the concentration of cysteine was decreased [84]. These traits were also stably maintained in the homozygous *lpa* progeny of generations F4 to F7 of crosses between the original *Os-lpaMH86-1* mutant with a commercial rice cultivar [91]. In addition, the metabolic profiles of the *lpa* progenies were strongly influenced by the lipid profiles of the wild type cultivar used as the crossing parent [92].

The OsSULTR3;3 mutants also show a significant increase in seed total sulfur and in sulfate concentration in embryo and pericarp/aleuronic layers. The mutations also increase root and leaf P and P_i concentrations and decrease root and leaf sulfate concentration in comparison to their corresponding wild type parents. Moreover, the analysis performed on developing seeds of the *MH86* mutant showed that the expression of genes coding for the last steps of PA biosynthesis was altered: Generally, an up-regulation was shown, and the expression of genes for sulfur metabolism and sulfate transport was different in the mutant compared to the corresponding wild type. However, the most dramatic effects on gene expression concern several genes involved in P signaling and homeostasis [84]. A redistribution of P_i in endosperm and a reduction of lysophospholipid content were also observed in the rice mutant [93].

As previously mentioned, the role of the SULTR3;3 transporter is not clear, as, when expressed in heterologous systems, such as yeast or Xenopus oocytes, it is unable to transport either sulfate, or phosphate, or PA precursors [84]. However, it cannot be excluded that in plant systems OsSULTR3;3 may transport these molecules as well as PA. In plant cells OsSULTR3;3 is ER-localized. Previous studies have suggested that the final steps of PA synthesis (from InsP3 to InsP6) take place in the ER [94]. Zhao and collaborators suggested that OsSULTR3;3 may have a specific role in the existing cross-talk between sulfate and phosphate homeostasis and/or signaling, as it has effects on phosphate as well as on sulfate concentrations in both vegetative tissues and grain [84].

Unfortunately, from an agronomic point of view, these mutants show some negative pleiotropic effects. In the rice mutant, grain weight reduction and yield per plant reductions have been shown [45]. In barley, only in rain-fed locations and not in irrigated ones, the mutation is associated with reduced test weight and percentage of plump kernels [95].

Interestingly, the mutant barley straws, although not showing significant differences in terms of fiber composition, compared to the wild type, after an acidic pre-treatment, showed increased fiber hydrolysibility, thus representing a promising material for cellulosic ethanol production [96].

3.3.2. The *spdt* Mutants

The rice *spdt* mutants, affected in the *OsSULTR3;4/SPDT* gene are retrotransposon *Tos-17* insertion lines (the transposon is in the fourth exon in *spdt1* and in the eighth exon in *spdt-2* and *spdt-3*).

The analysis of these mutants, grown under field conditions, revealed that the distribution of P in different organs was greatly altered, with a reduction by 20% of P concentration and P content in the seeds, without a significant penalty on grain yield, and a comparable 20% increase of P in the straw. Moreover, in the mutant seeds, a reduction in the concentration of PA by 25–32% was observed, compared to the wild type. However, neither the seed germination rate nor the early growth was affected by the reduced phytate content. P in the grain comes from re-translocation from old leaves or from node-based distribution of P newly taken up after the flowering stage. The reported results indicate that SPDT, localized in the nodes, especially in the uppermost node I, functions as a switch for P distribution to the grains. Indeed, another meaning of the acronym SPDT, used by Yamaji et al. (2017), is "single-pole, double throw", corresponding to a type of two-way electrical switch.

As knockout of *SPDT* resulted in a 20% reduction of total P and about 30% of PA in the grain without an obvious penalty of grain yield, and in increased P in the straw, the use of these mutants may present some advantages: As straw will be returned to the field after harvest, less P will be removed from the field, reducing the requirement for P fertilizer input. Their *lpa* phenotype may increase mineral bioavailability and lower the risk of eutrophication of waterways [20]. Very recently, the *atsultr3;4* mutant of Arabidopsis has been characterized and demonstrated to be a high-affinity P_i transporter that mediates xylem to phloem transfer of phosphate. In particular, it has been shown that, like the *OsSULTR3;4/SPDT* mutant, *atsultr3;4* seeds accumulate less P (about 15%) than the wt ones. This decrease is accompanied by a P increase in the shoot, indicating a role of AtSULTR3;4/SPDT in mediating P allocation to the seeds [89].

4. OsPht1;4 Phosphate Transporter

The OsPht1;4 (or OsPT4, corresponding to LOC_Os04g10750) phosphate transporter, belonging to the Pht1 family, was described as influencing grain PA content, as the corresponding mutant produces *lpa* grains [21]. As the identification of putative orthologs of this protein in other species is not so obvious, due to the high number of Pht1 genes and to their sequence similarity, (in rice there are 26 [97]), in the present review we limit our consideration to OsPT4.

The genomic sequence is characterized by the presence of a single exon (Figure 5a) and the protein, 538 aa long, by a major facilitator superfamily domain, characteristic of different transporters, including phosphate transporters (Figure 5b).

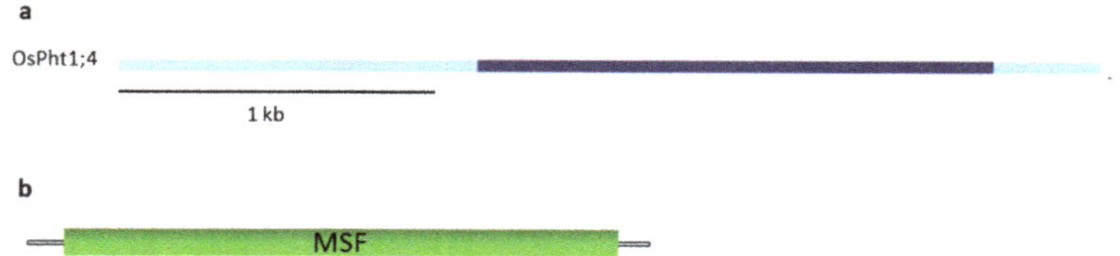

Figure 5. (a) Gene structure of OsPHT1;4. Light and dark blue rectangles represent UTRs and coding exons, respectively, the black bars correspond to introns. The gene structure was obtained as described in Figure 1a legend. (b) Predicted domain of the OsPHT1;4 protein by PFAM [98] software. The major facilitator superfamily (MFS) domain is represented.

The *OsPT4* gene is mainly expressed in roots, flag leaves and embryos, and its expression is increased in response to prolonged P starvation conditions in shoots and roots, where the signal is specifically localized to the exodermis. The protein is localized to the plasma membrane, as shown in the protoplast system and it is a functional P_i influx transporter, able to complement a yeast mutant defective in P_i uptake and to facilitate the increased accumulation of P_i in Xenopus oocytes.

Ospt4 Mutants, OsPT4 RNAi and Overexpression Lines

The OsPT4 functional characterization was performed using transposon insertional mutants, knockdown lines harboring the *OsPT4-RNAi* construct and overexpression lines. P_i and total P concentration is strongly reduced in mutant lines, attenuated in RNAi lines and increased in overexpression lines, in roots as well as in shoots. Moreover, a dramatic reduction in P_i uptake in mutants, a small reduction in RNAi lines, and an increase in overexpression lines were observed [21,22]. There was altered expression of genes that regulate Pi absorption and homeostasis, such as *OsPHO2*, *OsPHR2*, and *OsSPX1*. A detailed analysis of the grains revealed a decrease in total P concentration in the embryo and in one line also in the endosperm, attenuated and increased effects in RNAi lines and overexpression lines, respectively. Moreover, a decrease of 32–22% in PA concentration was observed in *ospt4-1* and RNAi lines' grains and an increase of 10% in overexpression lines. The alteration in mutant and RNAi embryos correlates with a reduction in the transcript levels of *OsRINO* (coding for *myo*-inositol 3-phosphate synthase—MIPS) and of *OsIPK1* (coding for 1,3,5,6-pentakisphosphate 2-kinase), coding for enzymes catalyzing the first and the last step of the PA biosynthetic pathway, respectively. Both genes' transcript levels were also significantly reduced in the endosperm of mutant grains and only partially reduced in RNAi lines [21]. From all these data, it is clear that OsPT4 has an important role in acquisition and mobilization of P_i and also during embryogenesis and seed development, so it is a good candidate to improve P efficiency, although alterations in panicle robustness, grain-setting rates, grain weight, grain yield per plant and seed germination registered in *ospt4* and *OsPt4* RNAi lines need breeding actions to ensure acceptable agronomic performance and avoid yield penalties.

5. Conclusions

Strategies for controlling the accumulation of specific metabolites are commonly based on switching off structural or regulatory genes of the biosynthetic pathway. However, in order to avoid or reduce downstream effects on derived pathways, a different approach is to interfere with compound transport to the site of accumulation (organ, cell type, subcellular compartment). In this review we reported data showing how PA reduction can be achieved with mutations in different types of transporters that control PA transport to the vacuole (MRP), or by modifying P_i availability for PA synthesis through mutations in transporters involved in Pi loading and organ/intracellular distribution (SULTR) or by P_i acquisition and mobilization during seed development (PHT1;4) (Figure 6).

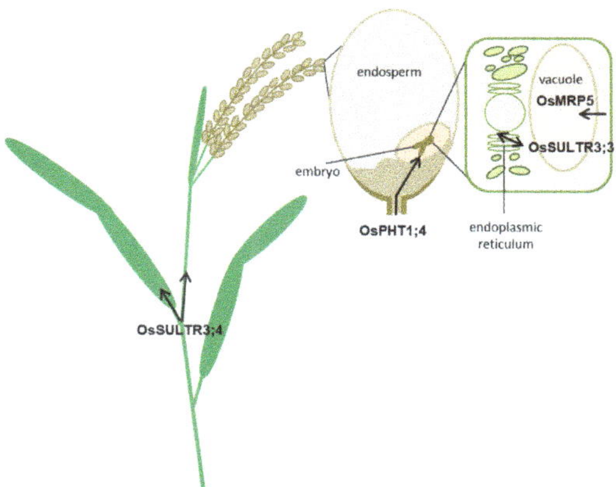

Figure 6. Rice transporters identified to modulate PA homeostasis. Modified from [99].

These types of *lpa* mutants are potentially advantageous over other *lpa* mutants in structural biosynthetic genes to achieve a triple goal: (i) Increased bioavailability of mineral cations, which are no longer chelated by PA, (ii) less PA is excreted into the environment with manures, and hence there is reduced impact on water eutrophication, (iii) increased P use efficiency as the seeds are loaded with less P which remains in the straw and may potentially contribute to reducing the demand for P fertilizers, hence increasing crop sustainability [99].

From an agronomic point of view, mutants in *MRP* and *PHT1;4* genes show compromised yields, seed setting, seed germination, and/or seedling growth, etc., so that they are not very attractive for breeders [21,22,37]. However, good field performance has been demonstrated for *MRP* mutants of common bean, which is due to the presence of duplicated *MRP* gene(s) able to complement the mutated seed-specific copy [32]. So far, *sultr3;3* and *sultr3;4/spdt* mutants, besides Arabidopsis, have been isolated and described only in rice and barley. Interestingly, these mutants do not display negative pleiotropic effects as observed for the other transporters described above, hence they potentially represent a valuable tool to simultaneously achieve seed biofortification and a more sustainable crop while reducing the environmental impact of the crop cultivation. It would be very interesting to verify the role and function of *sultr3;3* and *sultr3;4/spdt* genes in legume crops as well, since contrasting data on function and subcellular localization have been reported for Arabidopsis mutants and a possible species-specific and/or development-specific behavior has been proposed [100].

Finally, the finding that both phosphate and putative sulfate transporters produce similar *lpa* phenotypes suggests the "existence of a multilevel coordination in the regulation of the two ions in which currently unidentified key elements are actively cross-talking between the two signaling pathways" [101,102]. The availability of such mutants from different crops may help towards understanding this cross-talk and identifying new players.

Supplementary Materials: The following are available online at http://www.mdpi.com/2223-7747/9/1/69/s1. Figure S1. PA-MRP proteins alignment. See Table 1 for the correspondence with genes accession numbers. The Clustal W alignment (cost matrix Blosum, gap open cost 10, gap extend cost 0.1) of Geneious 11.0.2 software was used. Figure S2. Phylogenetic tree of characterized crop PA-MRP proteins, listed in Table 1. Phylogenies were constructed with the Geneious Tree Builder tool, using the Jukes-Cantor distance model, Neighbor-Joining tree build method. Figure S3. Phylogenetic tree of SULTR3;3 and SULTR3;4 proteins, listed in Table 2. Phylogenies were constructed as described in Figure S2. Figure S4. SULTR3;3 proteins alignment. See Tables 1 and 2 for the correspondence with genes accession numbers. The method used is described in Figure S1 legend. Figure S5. SULTR3;4 proteins alignment. See Tables 1 and 2 for the correspondence with genes accession numbers. The method used is described in Figure S1 legend.

Funding: This research received no external funding.

Acknowledgments: We thank Fabio Nocito, Università degli Studi di Milano (Milan, Italy), for fruitful discussions.

Conflicts of Interest: The authors declare no conflict of interest.

References

1. Sparvoli, F.; Cominelli, E. Seed biofortification and phytic acid reduction: A conflict of interest for the plant? *Plants* **2015**, *4*, 728–755. [CrossRef] [PubMed]
2. Coelho, C.; Tsai, S.; Vitorello, V. Dynamics of inositol phosphate pools (tris-, tetrakis- and pentakisphosphate) in relation to the rate of phytate synthesis during seed development in common bean (i). *J. Plant Physiol.* **2005**, *162*, 1–9. [CrossRef] [PubMed]
3. Hatzack, F.; Johansen, K.; Rasmussen, S. Nutritionally relevant parameters in low-phytate barley (*Hordeum vulgare* L.) grain mutants. *J. Agric. Food Chem.* **2000**, *48*, 6074–6080. [CrossRef] [PubMed]
4. Lin, L.; Ockenden, I.; Lott, J. The concentrations and distribution of phytic acid-phosphorus and other mineral nutrients in wild-type and low phytic acid1-1 (lpa1-1) corn (*Zea mays* L.) grains and grain parts. *Can. J. Bot.* **2005**, *83*, 131–141. [CrossRef]
5. Ockenden, I.; Dorsch, J.; Reid, M.; Lin, L.; Grant, L.; Raboy, V.; Lott, J. Characterization of the storage of phosphorus, inositol phosphate and cations in grain tissues of four barley (*Hordeum vulgare* L.) low phytic acid genotypes. *Plant Sci.* **2004**, *167*, 1131–1142. [CrossRef]

6. Regvar, M.; Eichert, D.; Kaulich, B.; Gianoncelli, A.; Pongrac, P.; Vogel-Mikus, K.; Kreft, I. New insights into globoids of protein storage vacuoles in wheat aleurone using synchrotron soft X-ray microscopy. *J. Exp. Bot.* **2011**, *62*, 3929–3939. [CrossRef]

7. Krishnan, H. Preparative procedures markedly influence the appearance and structural integrity of protein storage vacuoles in soybean seeds. *J. Agric. Food Chem.* **2008**, *56*, 2907–2912. [CrossRef]

8. O'Dell, B.L.; de Boland, A.R.; Koirtyohann, S.T. Distribution of phytate and nutritionally important elements among the morphological components of cereal grains. *J. Agric. Food Chem.* **1972**, *20*, 718–721. [CrossRef]

9. Ariza-Nieto, M.; Blair, M.; Welch, R.; Glahn, R. Screening of iron bioavailability patterns in eight bean (*Phaseolus vulgaris* L.) genotypes using the caco-2 cell in vitro model. *J. Agric. Food Chem.* **2007**, *55*, 7950–7956. [CrossRef]

10. Raboy, V. myo-Inositol-1,2,3,4,5,6-hexakisphosphate. *Phytochemistry* **2003**, *64*, 1033–1043. [CrossRef]

11. Raboy, V. Seeds for a better future: 'low phytate', grains help to overcome malnutrition and reduce pollution. *Trends Plant Sci.* **2001**, *6*, 458–462. [CrossRef]

12. Schlemmer, U.; Frølich, W.; Prieto, R.M.; Grases, F. Phytate in foods and significance for humans: Food sources, intake, processing, bioavailability, protective role and analysis. *Mol. Nutr. Food Res.* **2009**, *53* (Suppl. 2), S330–S375. [CrossRef] [PubMed]

13. Leytem, A.B.; Maguire, R.O. Environmental implications of inositol phosphates in animal manures. In *Inositol Phosphates: Linking Agriculture and the Environment*; Turner, B.L., Richardson, A.E., Mullaney, E.J., Eds.; CAB International: Wallingford, CT, USA; Oxfordshire, UK, 2007; pp. 150–168.

14. Raboy, V. Approaches and challenges to engineering seed phytate and total phosphorus. *Plant Sci.* **2009**, *177*, 281–296. [CrossRef]

15. Martinoia, E. Vacuolar transporters—Companions on a longtime journey. *Plant Physiol.* **2018**, *176*, 1384–1407. [CrossRef]

16. Nagy, R.; Grob, H.; Weder, B.; Green, P.; Klein, M.; Frelet-Barrand, A.; Schjoerring, J.; Brearley, C.; Martinoia, E. The Arabidopsis ATP-binding cassette protein AtMRP5/AtABCC5 is a high affinity inositol hexakisphosphate transporter involved in guard cell signaling and phytate storage. *J. Biol. Chem.* **2009**, *284*, 33614–33622. [CrossRef]

17. Sparvoli, F.; Cominelli, E. Phytate Transport by MRPs. In *Plant ABC Transporters*; Geisler, M., Ed.; Springer: Cham, Switzerland, 2014; pp. 19–38.

18. Takahashi, H.; Buchner, P.; Yoshimoto, N.; Hawkesford, M.J.; Shiu, S.H. Evolutionary relationships and functional diversity of plant sulfate transporters. *Front. Plant Sci.* **2011**, *2*, 119. [CrossRef] [PubMed]

19. Ye, H.; Zhang, X.; Broughton, S.; Westcott, S.; Wu, D.; Lance, R.; Li, C. A nonsense mutation in a putative sulphate transporter gene results in low phytic acid in barley. *Funct. Integr. Genom.* **2011**, *11*, 103–110. [CrossRef]

20. Yamaji, N.; Takemoto, Y.; Miyaji, T.; Mitani-Ueno, N.; Yoshida, K.T.; Ma, J.F. Reducing phosphorus accumulation in rice grains with an impaired transporter in the node. *Nature* **2017**, *541*, 92–95. [CrossRef]

21. Zhang, F.; Sun, Y.; Pei, W.; Jain, A.; Sun, R.; Cao, Y.; Wu, X.; Jiang, T.; Zhang, L.; Fan, X.; et al. Involvement of OsPht1;4 in phosphate acquisition and mobilization facilitates embryo development in rice. *Plant J.* **2015**, *82*, 556–569. [CrossRef]

22. Ye, Y.; Yuan, J.; Chang, X.; Yang, M.; Zhang, L.; Lu, K.; Lian, X. The phosphate transporter gene OsPht1;4 is involved in phosphate homeostasis in rice. *PLoS ONE* **2015**, *10*, e0126186. [CrossRef]

23. Hwang, J.U.; Song, W.Y.; Hong, D.; Ko, D.; Yamaoka, Y.; Jang, S.; Yim, S.; Lee, E.; Khare, D.; Kim, K.; et al. Plant ABC transporters enable many unique aspects of a terrestrial plant's lifestyle. *Mol. Plant* **2016**, *9*, 338–355. [CrossRef] [PubMed]

24. Shi, J.; Wang, H.; Schellin, K.; Li, B.; Faller, M.; Stoop, J.; Meeley, R.; Ertl, D.; Ranch, J.; Glassman, K. Embryo-specific silencing of a transporter reduces phytic acid content of maize and soybean seeds. *Nat. Biotechnol.* **2007**, *25*, 930–937. [CrossRef] [PubMed]

25. Gaedeke, N.; Klein, M.; Kolukisaoglu, U.; Forestier, C.; Muller, A.; Ansorge, M.; Becker, D.; Mamnun, Y.; Kuchler, K.; Schulz, B.; et al. The Arabidopsis thaliana ABC transporter AtMRP5 controls root development and stomata movement. *EMBO J.* **2001**, *20*, 1875–1887. [CrossRef]

26. Klein, M.; Perfus-Barbeoch, L.; Frelet, A.; Gaedeke, N.; Reinhardt, D.; Mueller-Roeber, B.; Martinoia, E.; Forestier, C. The plant multidrug resistance ABC transporter AtMRP5 is involved in guard cell hormonal signalling and water use. *Plant J.* **2003**, *33*, 119–129. [CrossRef]

27. Suh, S.J.; Wang, Y.F.; Frelet, A.; Leonhardt, N.; Klein, M.; Forestier, C.; Mueller-Roeber, B.; Cho, M.H.; Martinoia, E.; Schroeder, J.I. The ATP binding cassette transporter AtMRP5 modulates anion and calcium channel activities in Arabidopsis guard cells. *J. Biol. Chem.* **2007**, *282*, 1916–1924. [CrossRef]

28. Liu, Q.; Xu, X.; Ren, X.; Fu, H.; Wu, D.; Shu, Q. Generation and characterization of low phytic acid germplasm in rice (*Oryza sativa* L.). *Theor. Appl. Genet.* **2007**, *114*, 803–814. [CrossRef]

29. Maroof, M.; Glover, N.; Biyashev, R.; Buss, G.; Grabau, E. Genetic basis of the low-phytate trait in the soybean line CX1834. *Crop Sci.* **2009**, *49*, 69–76. [CrossRef]

30. Gillman, J.; Pantalone, V.; Bilyeu, K. The low phytic acid phenotype in soybean line CX1834 is due to mutations in two homologs of the maize low phytic acid gene. *Plant Genome* **2009**, *2*, 179–190. [CrossRef]

31. Gillman, J.; Baxter, I.; Bilyeu, K. Phosphorus partitioning of soybean lines containing different mutant alleles of two soybean seed-specific adenosine triphosphate-binding cassette phytic acid transporter paralogs. *Plant Genome* **2013**, *6*. [CrossRef]

32. Panzeri, D.; Cassani, E.; Doria, E.; Tagliabue, G.; Forti, L.; Campion, B.; Bollini, R.; Brearley, C.A.; Pilu, R.; Nielsen, E.; et al. A defective ABC transporter of the MRP family, responsible for the bean lpa1 mutation, affects the regulation of the phytic acid pathway, reduces seed myo-inositol and alters ABA sensitivity. *New Phytol.* **2011**, *191*, 70–83. [CrossRef]

33. Bhati, K.K.; Alok, A.; Kumar, A.; Kaur, J.; Tiwari, S.; Pandey, A.K. Silencing of ABCC13 transporter in wheat reveals its involvement in grain development, phytic acid accumulation and lateral root formation. *J. Exp. Bot.* **2016**, *67*, 4379–4389. [CrossRef] [PubMed]

34. Boncompagni, E.; Orozco-Arroyo, G.; Cominelli, E.; Gangashetty, P.I.; Grando, S.; Kwaku Zu, T.T.; Daminati, M.G.; Nielsen, E.; Sparvoli, F. Antinutritional factors in pearl millet grains: Phytate and goitrogens content variability and molecular characterization of genes involved in their pathways. *PLoS ONE* **2018**, *13*, e0198394. [CrossRef] [PubMed]

35. Cominelli, E.; Confalonieri, M.; Carlessi, M.; Cortinovis, G.; Daminati, M.G.; Porch, T.G.; Losa, A.; Sparvoli, F. Phytic acid transport in Phaseolus vulgaris: A new low phytic acid mutant in the PvMRP1 gene and study of the PvMRPs promoters in two different plant systems. *Plant Sci.* **2018**, *270*, 1–12. [CrossRef] [PubMed]

36. Ofori, P.A.; Mizuno, A.; Suzuki, M.; Martinoia, E.; Reuscher, S.; Aoki, K.; Shibata, D.; Otagaki, S.; Matsumoto, S.; Shiratake, K. Genome-wide analysis of ATP binding cassette (ABC) transporters in tomato. *PLoS ONE* **2018**, *13*, e0200854. [CrossRef]

37. Raboy, V.; Gerbasi, P.F.; Young, K.A.; Stoneberg, S.D.; Pickett, S.G.; Bauman, A.T.; Murthy, P.P.; Sheridan, W.F.; Ertl, D.S. Origin and seed phenotype of maize low phytic acid 1-1 and low phytic acid 2-1. *Plant Physiol.* **2000**, *124*, 355–368. [CrossRef]

38. Pilu, R.; Panzeri, D.; Gavazzi, G.; Rasmussen, S.K.; Consonni, G.; Nielsen, E. Phenotypic, genetic and molecular characterization of a maize low phytic acid mutant (lpa241). *Theor. Appl. Genet.* **2003**, *107*, 980–987. [CrossRef]

39. Cerino Badone, F.; Amelotti, M.; Cassani, E.; Pilu, R. Study of Low Phytic Acid1-7 (lpa1-7), a New ZmMRP4 Mutation in Maize. *J. Hered.* **2012**, *103*, 598–605. [CrossRef]

40. Liu, K.; Peterson, K.; Raboy, V. Comparison of the phosphorus and mineral concentrations in bran and abraded kernel fractions of a normal barley (Hordeum vulgare) cultivar versus four low phytic acid isolines. *J. Agric. Food Chem.* **2007**, *55*, 4453–4460. [CrossRef]

41. Xu, X.; Zhao, H.; Liu, Q.; Frank, T.; Engel, K.; An, G.; Shu, Q. Mutations of the multi-drug resistance-associated protein ABC transporter gene 5 result in reduction of phytic acid in rice seeds. *Theor. Appl. Genet.* **2009**, *119*, 75–83. [CrossRef]

42. Li, W.; Zhao, H.; Pang, W.; Cui, H.; Poirier, Y.; Shu, Q. Seed-specific silencing of OsMRP5 reduces seed phytic acid and weight in rice. *Transgenic Res.* **2014**, *23*, 585–599. [CrossRef]

43. Wilcox, J.; Premachandra, G.; Young, K.; Raboy, V. Isolation of high seed inorganic P, low-phytate soybean mutants. *Crop Sci.* **2000**, *40*, 1601–1605. [CrossRef]

44. Campion, B.; Sparvoli, F.; Doria, E.; Tagliabue, G.; Galasso, I.; Fileppi, M.; Bollini, R.; Nielsen, E. Isolation and characterisation of an lpa (low phytic acid) mutant in common bean (*Phaseolus vulgaris* L.). *Theor. Appl. Genet.* **2009**, *118*, 1211–1221. [CrossRef] [PubMed]

45. Zhao, H.; Liu, Q.; Ren, X.; Wu, D.; Shu, Q. Gene identification and allele-specific marker development for two allelic low phytic acid mutations in rice (*Oryza sativa* L.). *Mol. Breed.* **2008**, *22*, 603–612. [CrossRef]

46. Gene Structure Display Server. Available online: http://gsds.cbi.pku.edu.cn/ (accessed on 20 November 2019).

47. Patel, R.; Nahal, H.; Breit, R.; Provart, N. BAR expressolog identification: Expression profile similarity ranking of homologous genes in plant species. *Plant J.* **2012**, *71*, 1038–1050. [CrossRef]

48. Bhati, K.; Aggarwal, S.; Sharma, S.; Mantri, S.; Singh, S.; Bhalla, S.; Kaur, J.; Tiwari, S.; Roy, J.; Tuli, R.; et al. Differential expression of structural genes for the late phase of phytic acid biosynthesis in developing seeds of wheat (*Triticum aestivum* L.). *Plant Sci.* **2014**, *224*, 74–85. [CrossRef]

49. Latrasse, D.; Jegu, T.; Meng, P.; Mazubert, C.; Hudik, E.; Delarue, M.; Charon, C.; Crespi, M.; Hirt, H.; Raynaud, C.; et al. Dual function of MIPS1 as a metabolic enzyme and transcriptional regulator. *Nucleic Acids Res.* **2013**, *41*, 2907–2917. [CrossRef]

50. Donahue, J.; Alford, S.; Torabinejad, J.; Kerwin, R.; Nourbakhsh, A.; Ray, W.; Hernick, M.; Huang, X.; Lyons, B.; Hein, P.; et al. The Arabidopsis thaliana myo-inositol 1-phosphate synthase1 gene is required for myo-inositol synthesis and suppression of cell death. *Plant Cell* **2010**, *22*, 888–903. [CrossRef]

51. Chen, H.; Xiong, L. myo-inositol-1-phosphate synthase is required for polar auxin transport and organ development. *J. Biol. Chem.* **2010**, *285*, 24238–24247. [CrossRef]

52. Sato, Y.; Yazawa, K.; Yoshida, S.; Tamaoki, M.; Nakajima, N.; Iwai, H.; Ishii, T.; Satoh, S. Expression and functions of myo-inositol monophosphatase family genes in seed development of Arabidopsis. *J. Plant Res.* **2011**, *124*, 385–394. [CrossRef]

53. Nourbakhsh, A.; Collakova, E.; Gillaspy, G.E. Characterization of the inositol monophosphatase gene family in Arabidopsis. *Front. Plant Sci.* **2014**, *5*, 725. [CrossRef]

54. Sweetman, D.; Stavridou, I.; Johnson, S.; Green, P.; Caddick, S.; Brearley, C. Arabidopsis thaliana inositol 1,3,4-trisphosphate 5/6-kinase 4 (AtITPK4) is an outlier to a family of ATP-grasp fold proteins from Arabidopsis. *FEBS Lett.* **2007**, *581*, 4165–4171. [CrossRef] [PubMed]

55. Xia, H.; Brearley, C.; Elge, S.; Kaplan, B.; Fromm, H.; Mueller-Roeber, B. Arabidopsis inositol polyphosphate 6-/3-kinase is a nuclear protein that complements a yeast mutant lacking a functional ArgR-Mcm1 transcription complex. *Plant Cell* **2003**, *15*, 449–463. [CrossRef]

56. Zhang, Z.B.; Yang, G.; Arana, F.; Chen, Z.; Li, Y.; Xia, H.J. Arabidopsis inositol polyphosphate 6-/3-kinase (AtIpk2beta) is involved in axillary shoot branching via auxin signaling. *Plant Physiol.* **2007**, *144*, 942–951. [CrossRef] [PubMed]

57. Munnik, T.; Vermeer, J. Osmotic stress-induced phosphoinositide and inositol phosphate signalling in plants. *Plant Cell Environ.* **2010**, *33*, 655–669. [CrossRef] [PubMed]

58. Shukla, S.; VanToai, T.; Pratt, R. Expression and nucleotide sequence of an INS (3) P-1 synthase gene associated with low-phytate kernels in maize (*Zea mays* L.). *J. Agric. Food Chem.* **2004**, *52*, 4565–4570. [CrossRef] [PubMed]

59. Pilu, R.; Landoni, M.; Cassani, E.; Doria, E.; Nielsen, E. The maize lpa241 mutation causes a remarkable variability of expression and some pleiotropic effects. *Crop Sci.* **2005**, *45*, 2096–2105. [CrossRef]

60. Pilu, R.; Panzeri, D.; Cassani, E.; Badone, F.C.; Landoni, M.; Nielsen, E. A paramutation phenomenon is involved in the genetics of maize low phytic acid1-241 (lpa1-241) trait. *Heredity* **2009**, *102*, 236–245. [CrossRef]

61. Doria, E.; Galleschi, L.; Calucci, L.; Pinzino, C.; Pilu, R.; Cassani, E.; Nielsen, E. Phytic acid prevents oxidative stress in seeds: Evidence from a maize (*Zea mays* L.) low phytic acid mutant. *J. Exp. Bot.* **2009**, *60*, 967–978. [CrossRef]

62. Cerino Badone, F.; Cassani, E.; Landoni, M.; Doria, E.; Panzeri, D.; Lago, C.; Mesiti, F.; Nielsen, E.; Pilu, R. The low phytic acid 1-241 (lpa1-241) maize mutation alters the accumulation of anthocyanin pigment in the kernel. *Planta* **2010**, *231*, 1189–1199. [CrossRef]

63. Landoni, M.; Badone, F.; Haman, N.; Schiraldi, A.; Fessas, D.; Cesari, V.; Toschi, I.; Cremona, R.; Delogu, C.; Villa, D.; et al. Low phytic acid 1 mutation in maize modifies density, starch properties, cations, and fiber contents in the seed. *J. Agric. Food Chem.* **2013**, *61*, 4622–4630. [CrossRef]

64. Meis, S.; Fehr, W.; Schnebly, S. Seed source effect on field emergence of soybean lines with reduced phytate and raffinose saccharides. *Crop Sci.* **2003**, *43*, 1336–1339. [CrossRef]

65. Bregitzer, P.; Raboy, V. Effects of four independent low-phytate mutations in barley (*Hordeum vulgare* L.) on seed phosphorus characteristics and malting quality. *Cereal Chem.* **2006**, *83*, 460–464. [CrossRef]

66. Guttieri, M.; Peterson, K.; Souza, E. Mineral distributions in milling fractions of low phytic acid wheat. *Crop Sci.* **2006**, *46*, 2692–2698. [CrossRef]

67. Borlini, G.; Rovera, C.; Landoni, M.; Cassani, E.; Pilu, R. lpa1-5525: A new lpa1 mutant isolated in a mutagenized population by a novel non-disrupting screening method. *Plants* **2019**, *8*, 209. [CrossRef]

68. Raboy, V.; Peterson, K.; Jackson, C.; Marshall, J.; Hu, G.; Saneoka, H.; Bregitzer, P. A substantial fraction of barley (*Hordeum vulgare* L.) low phytic acid mutations have little or no effect on yield across diverse production environments. *Plants* **2015**, *4*, 225–239. [CrossRef] [PubMed]

69. Graf, E.; Empson, K.; Eaton, J. Phytic acid—A natural antioxidant. *J. Biol. Chem.* **1987**, *262*, 11647–11650. [PubMed]

70. Graf, E.; Eaton, J.W. Antioxidant functions of phytic acid. *Free Radic. Biol. Med.* **1990**, *8*, 61–69. [CrossRef]

71. Empson, K.; Labuza, T.; Graf, E. Phytic acid as food antioxidant. *J. Food Sci.* **1991**, *56*, 560–563. [CrossRef]

72. Dorsch, J.; Cook, A.; Young, K.; Anderson, J.; Bauman, A.; Volkmann, C.; Murthy, P.; Raboy, V. Seed phosphorus and inositol phosphate phenotype of barley low phytic acid genotypes. *Phytochemistry* **2003**, *62*, 691–706. [CrossRef]

73. Bhati, K.K.; Sharma, S.; Aggarwal, S.; Kaur, M.; Shukla, V.; Kaur, J.; Mantri, S.; Pandey, A.K. Genome-wide identification and expression characterization of ABCC-MRP transporters in hexaploid wheat. *Front. Plant Sci.* **2015**, *6*, 488. [CrossRef]

74. Oltmans, S.; Fehr, W.; Welke, G.; Cianzio, S. Inheritance of low-phytate phosphorus in soybean. *Crop Sci.* **2004**, *44*, 433–435. [CrossRef]

75. Oltmans, S.; Fehr, W.; Welke, G.; Raboy, V.; Peterson, K. Agronomic and seed traits of soybean lines with low-phytate phosphorus. *Crop Sci.* **2005**, *45*, 593–598. [CrossRef]

76. Chappell, A.; Scaboo, A.; Wu, X.; Nguyen, H.; Pantalone, V.; Bilyeu, K. Characterization of the MIPS gene family in Glycine max. *Plant Breed.* **2006**, *125*, 493–500. [CrossRef]

77. Scaboo, A.; Pantalone, V.; Walker, D.; Boerma, H.; West, D.; Walker, F.; Sams, C. Confirmation of molecular markers and agronomic traits associated with seed phytate content in two soybean RIL populations. *Crop Sci.* **2009**, *49*, 426–432. [CrossRef]

78. Hulke, B.; Fehr, W.; Welke, G. Agronomic and seed characteristics of soybean with reduced phytate and palmitate. *Crop Sci.* **2004**, *44*, 2027–2031. [CrossRef]

79. Spear, J.; Fehr, W. Genetic improvement of seedling emergence of soybean lines with low phytate. *Crop Sci.* **2007**, *47*, 1354–1360. [CrossRef]

80. Anderson, B.; Fehr, W. Seed source affects field emergence of low-phytate soybean lines. *Crop Sci.* **2008**, *48*, 929–932. [CrossRef]

81. Campion, B.; Glahn, R.; Tava, A.; Perrone, D.; Doria, E.; Sparvoli, F.; Cecotti, R.; Dani, V.; Nielsen, E. Genetic reduction of antinutrients in common bean (*Phaseolus vulgaris* L.) seed, increases nutrients and in vitro iron bioavailability without depressing main agronomic traits. *Field Crops Res.* **2013**, *141*, 27–37. [CrossRef]

82. Zhang, W.; Gruszewski, H.; Chevone, B.; Nessler, C. An Arabidopsis purple acid phosphatase with phytase activity increases foliar ascorbate. *Plant Physiol.* **2008**, *146*, 431–440. [CrossRef]

83. Torabinejad, J.; Donahue, J.; Gunesekera, B.; Allen-Daniels, M.; Gillaspy, G. VTC4 is a bifunctional enzyme that affects myoinositol and ascorbate biosynthesis in plants. *Plant Physiol.* **2009**, *150*, 951–961. [CrossRef]

84. Zhao, H.; Frank, T.; Tan, Y.; Zhou, C.; Jabnoune, M.; Arpat, A.B.; Cui, H.; Huang, J.; He, Z.; Poirier, Y.; et al. Disruption of OsSULTR3;3 reduces phytate and phosphorus concentrations and alters the metabolite profile in rice grains. *New Phytol.* **2016**, *211*, 926–939. [CrossRef] [PubMed]

85. Shoemaker, R.C.; Polzin, K.; Labate, J.; Specht, J.; Brummer, E.C.; Olson, T.; Young, N.; Concibido, V.; Wilcox, J.; Tamulonis, J.P.; et al. Genome duplication in soybean (*Glycine subgenus* soja). *Genetics* **1996**, *144*, 329–338. [PubMed]

86. Yamaji, N.; Ma, J.F. Node-controlled allocation of mineral elements in Poaceae. *Curr. Opin. Plant Biol.* **2017**, *39*, 18–24. [CrossRef] [PubMed]

87. Zuber, H.; Davidian, J.C.; Aubert, G.; Aimé, D.; Belghazi, M.; Lugan, R.; Heintz, D.; Wirtz, M.; Hell, R.; Thompson, R.; et al. The seed composition of Arabidopsis mutants for the group 3 sulfate transporters indicates a role in sulfate translocation within developing seeds. *Plant Physiol.* **2010**, *154*, 913–926. [CrossRef]

88. Chen, Z.; Zhao, P.X.; Miao, Z.Q.; Qi, G.F.; Wang, Z.; Yuan, Y.; Ahmad, N.; Cao, M.J.; Hell, R.; Wirtz, M.; et al. SULTR3s function in chloroplast sulfate uptake and affect ABA biosynthesis and the stress response. *Plant Physiol.* **2019**, *180*, 593–604. [CrossRef]

89. Ding, G.; Lei, G.J.; Yamaji, N.; Yokosho, K.; Mitani-Ueno, N.; Huang, S.; Ma, J.F. Vascular cambium-localized AtSPDT mediates xylem-to-phloem transfer of phosphorus for its preferential distribution in Arabidopsis. *Mol. Plant* **2019**. [CrossRef]

90. Raboy, V.; Cichy, K.; Peterson, K.; Reichman, S.; Sompong, U.; Srinives, P.; Saneoka, H. Barley (*Hordeum vulgare* L.) Low phytic acid 1-1: An endosperm-specific, filial determinant of seed total phosphorus. *J. Hered.* **2014**, *105*, 656–665. [CrossRef]

91. Zhou, C.; Tan, Y.; Goßner, S.; Li, Y.; Shu, Q.; Engel, K.H. Stability of the metabolite signature resulting from the OsSULTR3;3 mutation in low phytic acid rice (*Oryza sativa* L.) seeds upon cross-breeding. *J. Agric. Food Chem.* **2018**, *66*, 9366–9376. [CrossRef]

92. Zhou, C.; Tan, Y.; Goßner, S.; Li, Y.; Shu, Q.; Engel, K.H. Impact of crossing parent and environment on the metabolite profiles of progenies generated from a low phytic acid rice (*Oryza sativa* L.) mutant. *J. Agric. Food Chem.* **2019**, *67*, 2396–2407. [CrossRef]

93. Tong, C.; Chen, Y.; Tan, Y.; Liu, L.; Waters, D.L.E.; Rose, T.J.; Shu, Q.; Bao, J. Analysis of lysophospholipid content in low phytate rice mutants. *J. Agric. Food Chem.* **2017**, *65*, 5435–5441. [CrossRef]

94. Otegui, M.; Capp, R.; Staehelin, L. Developing seeds of Arabidopsis store different minerals in two types of vacuoles and in the endoplasmic reticulum. *Plant Cell* **2002**, *14*, 1311–1327. [CrossRef] [PubMed]

95. Bregitzer, P.; Raboy, V. Effects of four independent low-phytate mutations on barley agronomic performance. *Crop Sci.* **2006**, *46*, 1318–1322. [CrossRef]

96. Li, Z.; Liu, Y.; Liao, W.; Chen, S.; Zemetra, R. Bioethanol production using genetically modified and mutant wheat and barley straws. *Biomass Bioenergy* **2011**, *35*, 542–548. [CrossRef]

97. Liu, F.; Chang, X.; Ye, Y.; Xie, W.; Wu, P.; Lian, X. Comprehensive sequence and whole-life-cycle expression profile analysis of the phosphate transporter gene family in rice. *Mol. Plant* **2011**, *4*, 1105–1122. [CrossRef] [PubMed]

98. PFAM Software. Available online: http://pfam.xfam.org/ (accessed on 20 November 2019).

99. Kopriva, S.; Chu, C. Are we ready to improve phosphorus homeostasis in rice? *J. Exp. Bot.* **2018**, *69*, 3515–3522. [CrossRef] [PubMed]

100. Takahashi, H. Sulfate transport systems in plants: Functional diversity and molecular mechanisms underlying regulatory coordination. *J. Exp. Bot.* **2019**, *70*, 4075–4087. [CrossRef]

101. Rouached, H. Multilevel coordination of phosphate and sulfate homeostasis in plants. *Plant Signal. Behav.* **2011**, *6*, 952–955. [CrossRef]

102. Sacchi, G.A.; Nocito, F.F. Plant sulfate transporters in the *low phytic acid* network: Some educated guesses. *Plants* **2019**, *8*, 616. [CrossRef]

© 2020 by the authors. Licensee MDPI, Basel, Switzerland. This article is an open access article distributed under the terms and conditions of the Creative Commons Attribution (CC BY) license (http://creativecommons.org/licenses/by/4.0/).

 *plants*

Opinion

Plant Sulfate Transporters in the Low Phytic Acid Network: Some Educated Guesses

Gian Attilio Sacchi and Fabio Francesco Nocito *

Dipartimento di Scienze Agrarie e Ambientali—Produzione, Territorio, Agroenergia,
Università degli Studi di Milano, 20133 Milano, Italy; gianattilio.sacchi@unimi.it
* Correspondence: fabio.nocito@unimi.it

Received: 29 November 2019; Accepted: 16 December 2019; Published: 17 December 2019

Abstract: A few new papers report that mutations in some genes belonging to the group 3 of plant sulfate transporter family result in low phytic acid phenotypes, drawing novel strategies and approaches for engineering the low-phytate trait in cereal grains. Here, we shortly review the current knowledge on phosphorus/sulfur interplay and sulfate transport regulation in plants, to critically discuss some hypotheses that could help in unveiling the physiological links between sulfate transport and phosphorus accumulation in seeds.

Keywords: sulfate transporters; phytic acid; sulfur; phosphorous

---

1. Background

Phytic acid (PA)—the major phosphorus (P) store in seeds—cannot be digested by humans and monogastric animals who lack the digestive enzyme phytase. For this reason, almost 90% of phytate consumed by humans is excreted, contributing to eutrophication of rivers, lakes, and oceans [1]. Furthermore, high levels of PA largely prevent the absorption of essential metals in the intestine, thus reducing further the nutritional value of the seeds [2,3].

In the last decades, several approaches have been proposed to solve the seed PA-related problems, including the engineering of crops for high phytase activity in seeds, or the selection of suitable *low phytic acid* (*lpa*) genotypes for crop breeding [4]. Today, numerous *lpa* genotypes have been identified and studied in several major crops, including maize, barley, wheat, rice, soybean, and common bean, reveling several mutations and alleles that could be potentially useful for breeding. However, a large part of the *lpa* phenotypes is caused by mutations in genes involved in PA biosynthesis or compartmentalization and often results in undesirable pleiotropic effects on yield-related traits and agronomic performances, since PA and inositol phosphates play pivotal roles in a plethora of developmental and signaling processes [4,5]. As a result, the use of these genetic resources to engineer seed PA content has proven to be challenging. Most recent advances in this research topic revealed that mutations in some members of the sulfate transporter gene family might result in *lpa* phenotypes. Unfortunately, little data are available to explain such effects fully or to develop new strategies for engineering seed PA content. Trying to fill this gap, here, we shortly review the current knowledge on plant sulfate transporters, trying to provide a glimpse into the complex and, in many respects, unexpected connections among the regulatory layers of sulfur (S) and P homeostasis in plants.

2. Sulfate Transporters: A Short Overview

S is an essential nutrient for plants. It is found in the amino acids cysteine and methionine, which are essential components of proteins and peptides, in vitamins and cofactors, and in a plethora of secondary compounds. S plays important and critical roles in a wide variety of cellular processes involved in plant development and response to environmental changes [6–9].

Sulfate (SO_4^{2-}) ions in the rhizosphere are the major source of S for plants. They are absorbed by roots and then allocated to different sinks by mean of specific sulfate transporters (SULTRs). The oxidized S atom in SO_4^{2-} is then reduced and assimilated into cysteine, before entering other metabolic pathways, or directly used for sulfation reactions [9–11]. SULTRs are classified as H^+/SO_4^{2-} co-transporters, are integrated into membranes by 12 membrane-spanning domains, and contain a carboxyl-terminal region, named STAS (Sulfate Transporter/AntiSigma-factor), which is thought to be critical for both activity and stability of the transporters, as well as for their interaction with other proteins [9,12–14].

A multigene family encodes plant SULTRs. In the best-characterized species—*Arabidopsis thaliana* and, to a lesser extent, rice (*Oryza sativa* L.)—12 *SULTR* genes have been reported [14,15]. SULTRs can be divided into four functional groups or subfamilies, according to their amino acid sequences. The members of each group have specialized functions for SO_4^{2-} uptake and distribution within the cells and among plant organs, as indicated by their different tissue and subcellular localization, and regulation pathways.

Group 1 of the family encodes high-affinity SULTRs. Two members of this group, SULTR1;1 and SULTR1;2, are mainly expressed in the outermost cell layers of the root (root hairs, epidermis, and cortex), where they largely contribute in determining the rate of SO_4^{2-} uptake. Arabidopsis *sultr1;1sultr1;2* double-knockout lines are severely impaired in growth and unable to take up SO_4^{2-} at low external concentrations [16–19]. Although these transporters seem to share the same function, they are differently regulated to fulfill the plant demand for S-containing compounds under different SO_4^{2-} availabilities or soil conditions. In the currently accepted model, SULTR1;2 is thought to be the major component of the SO_4^{2-} uptake system under normal S supply, whereas SULTR1;1 should play a most significant role under S deficiency or during other stresses [16,17,20–22].

Sulfate ions absorbed by root are translocated to shoot throughout the xylem and then distributed to different sink organs and tissues. It has been proposed that SULTR2;1, a low-affinity SULTR expressed in pericycle and xylem parenchyma, may play a pivotal role in controlling the amount of SO_4^{2-} available to be loaded into the xylem, by acting as a scavenger reabsorbing the excess of the anion in the apoplastic space inside the root stele. Under S starvation, the increase in the transcript level of *SULTR2;1* could help in maintaining adequate fluxes of SO_4^{2-} directed to the xylem [16,23]. It is important to note that a local expression of *SULTR2;1* has also been observed in the xylem parenchyma and phloem cells of the leaves, and that it is not possible to rule out that *SULTR2;1* transcript is also expressed below detection levels in the phloem companion cells of the root [16,24].

An interesting regulatory circuit controls SO_4^{2-} translocation and partitioning at the post-transcriptional level (Figure 1). The *SULTR2;1* mRNA is targeted and degraded by the miRNA-395 (miR395), which accumulates under S deficiency mainly in the companion cells of the phloem of both root and shoot [24]. The induction of miR395 is, in turn, activated by *SLIM1/EIL3* (*SULFUR LIMITATION 1/ETHYLENE-INSENSITIVE3-LIKE3*), a major regulator gene belonging to the EIL family transcription factors, which controls the expression of several S-responsive genes [25,26]. The mechanism by which miR395 controls *SULTR2;1* transcript level is not conventional, since the accumulation of both miR395 and *SULTR2;1* mRNA is induced under S starvation. However, the non-overlapping spatial expression domains of the two transcripts allows miR395 to restrict the expression of *SULTR2;1* to the xylem parenchyma cells of the root, thus inhibiting long-distance SO_4^{2-} transport to sink tissues via the phloem and facilitating, at the same time, xylem SO_4^{2-} translocation to the leaves [24,25].

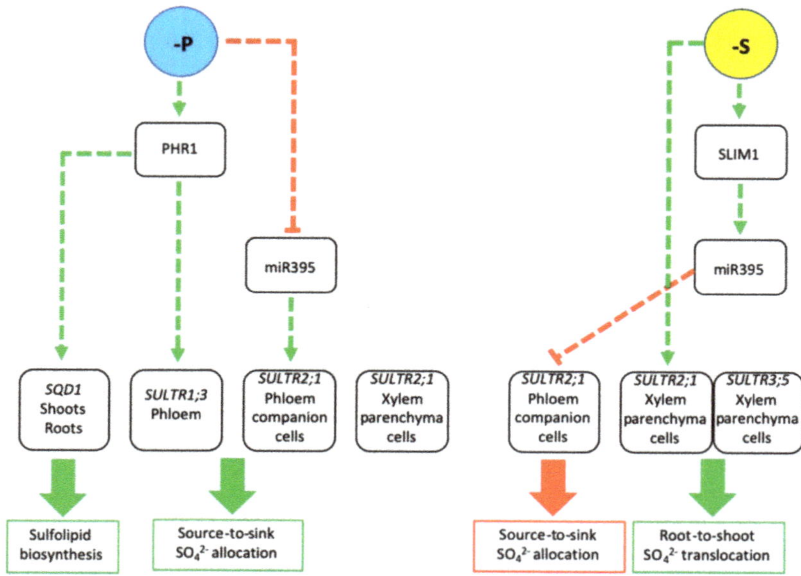

Figure 1. Main regulatory circuits controlling SO_4^{2-} distribution in response to P or S status. Under S deficiency, the induction of *SULTR2;1*, in xylem parenchyma cells, and miR395, in phloem companion cells, enhances root-to-shoot SO_4^{2-} translocation. In this condition, the co-expression of *SULTR3,5* could help the activity of SULTR2;1 in reabsorbing the excess of SO_4^{2-} in the apoplastic space of the root. Under P deficiency, an extra regulatory circuit involving *PHR1* allows changes in SO_4^{2-} to support sulfolipids biosynthesis.

Another low-affinity SULTR belonging to group 2, SULTR2;2, seems to be involved in controlling the source-to-sink distribution of SO_4^{2-} inside the plant. Localization analyses indicate that SULTR2;2 may play a role in the transport of SO_4^{2-} via root phloem, as well as in the distribution of the anion from leaf vasculature to the leaf palisade and mesophyll, which are thought to be the primary sites for SO_4^{2-} assimilation [16]. Finally, long-distance transport of SO_4^{2-} from source to sink organs could also involve SULTR1;3, a high-affinity SULTR of group 1, as indicated by the peculiar expression of this transporter in sieve elements and companion cells of the phloem [27].

Inside the cells, SO_4^{2-} is further transported into the vacuole and chloroplast/plastid, where it is compartmentalized as S store or reduced and assimilated into cysteine for further metabolic processes, respectively. To date, tonoplast proteins mediating vacuolar SO_4^{2-} influx have not been identified. On the other hand, SULTR4;1 and SULTR4;2 are known to be involved in downloading SO_4^{2-} from the vacuoles under S limiting conditions [28].

Recently, all five members of group 3 have been indicated as redundantly involved in SO_4^{2-} uptake across the chloroplast envelope membrane [29,30]. However, these observations do not appear to be conclusive, since several other functions could be postulated for these transporters on the base of observations that are crucial for our dissertation about the hypothetical links between *SULTRs* and *lpa* phenotypes. It is important to note that if, on the one hand, reasonable uncertainties about the capacity of both SULTR1s and SULTR2s to selectively move SO_4^{2-} do not exist, on the other, no direct evidence has been provided about the actual SO_4^{2-} transport activity of most of the SULTR3 subfamily members [14]. A few papers indeed indicate that both substrate preference and subcellular localization of some SULTR3s could be different than expected.

Kataoka et al. [31] reported that SULTR3;5 is expressed in the root vasculature of Arabidopsis—showing the same expression domain of the low-affinity SULTR2;1—and subcellular localizes on the plasma membrane. The heterologous expression of *SULTR3;5* in yeasts defective for SO_4^{2-} uptake shows that this protein does not transport SO_4^{2-} itself, whereas it enhances the SO_4^{2-} uptake capacity of SULTR2;1 when co-expressed in the same yeast mutant. These results, along with the observation that the Arabidopsis *sultr3;5* mutant retains more SO_4^{2-} in the root under S starvation, strongly suggest that SULTR3;5 may have the function to help SULTR2;1 in retrieval apoplastic SO_4^{2-}, contributing in this way to root-to-shoot SO_4^{2-} translocation (Figure 1).

SULTR3;4 from rice and Arabidopsis have been recently indicated as SULTR-like phosphorus distribution transporters (SPDTs) playing essential roles in controlling the allocation of phosphate to grains and developing tissues, respectively [32,33]. Tissue-specific expression analyses show that SULTR3;4/SPDT of rice is expressed in the xylem region of both enlarged- and diffuse-vascular bundles of nodes [32]. The Arabidopsis ortholog gene shows a more complex expression pattern, since it is mainly expressed in the fascicular cambium between the xylem and phloem and in the interfascicular cambium of lower stem, as well as in the cambial zone of the leaf petiole, rosette basal region, hypocotyl, and in the parenchyma cells of both xylem and phloem surrounding the cambial zone [33]. Moreover, the SULTR3;4/SPDTs are localized at the plasma membrane, show proton-dependent transport activities for phosphate, do not transport SO_4^{2-}, and are up-regulated by phosphate deficiency but not under S starvation [32,33]. Mutations in *OsSULTR3;4/SPDT* alter the distribution of P in rice plants, decreasing both total P (−20%) and phytate (−30%) in the brown de-husked grains, without affecting yield, seed germination, and seedling vigor.

Another member of group 3, SULTR3;3, has been indicated as implicated in PA accumulation in barley and rice grains. Zhao et al. [34] recently reported that disruptions in rice *SULTR3;3* gene are the casual events of two interesting allelic mutations, previously described as *lpa-MH86-1* and *Os-lpa-Z9B-1*, since they produce grains with a reduced concentration of both PA and total P [35]. Tissue-specific expression analyses reveal that OsSULTR3;3 is expressed in the vascular bundles of shoots, leaves, flowers, and seeds, but not in the roots. This protein seems to be localized in the endoplasmic reticulum, when expressed in onion epidermal cells, and it does not show any transport activity for both SO_4^{2-} and phosphate when heterologously expressed in yeast mutant strains defective for SO_4^{2-} or phosphate uptake, or *Xenopus* oocytes. However—as underlined by Zhao et al. [34]—the lack of transport activity for SO_4^{2-} or phosphate in heterologous systems does not necessarily mean that OsSULTR3;3 does not have a role in SO_4^{2-} of phosphate transport, since its activity may depend on other proteins or post-translational modifications not present in non-plant hosts. Moreover, *OsSULTR3;3* mutations affect the concentrations of total P and phosphate of both root and shoot—which result higher in the mutants than in the wild type—but also reduce the concentrations of SO_4^{2-} in the same organs. Finally, transcriptional analyses performed on developing grains reveal that *OsSULTR3;3* disruptions are associated with significant changes in the transcript level of genes involved in S and P homeostasis, suggesting a possible role of this gene in the cross-talk between the two nutrients [34]. Interestingly, a single base pair substitution in the last exon of an ortholog gene of *OsSULTR3;3* (designed as *HvST*) has also been identified as the causal event for the *low phytic acid* phenotype of the *lpa1-1* barley mutants [36].

Taken as a whole, these findings strongly indicate that expression domains and subcellular localizations, as well as substrate preferences of the SULTR3 subfamily members are variable, and may depend on plant species, development stage, or experimental approaches used to study their functions. Further efforts will be necessary to understand better whether this variability could play a role in the regulation of SO_4^{2-} fluxes under different environmental conditions, also concerning the level of other essential mineral nutrients.

3. Sulfur and Phosphorous Interplay

Similar to S, P is also an essential macronutrient for plants. P is found as phosphate ester in the majority of the molecular constituents essential for plant cell functions, including nucleic acids, proteins, phospholipids, sugars, ATP, and NADPH. Important aspects related to P acquisition and homeostasis in plants have been recently reviewed elsewhere [37–39]. Here, we mainly focus our attention on S and P interplay by analyzing specific aspects related to SO_4^{2-} transport and distribution inside the plants.

Although it is clear that S or P deficiencies have diverse phenotypic effects on plant growth, development, and productivity, intriguing interconnected responses to the internal levels of these two nutrients have been described at metabolic and transcriptional levels, suggesting the existence of coordination between S and P homeostasis. Rouached [40] pointed out that deficiency or surplus of only one of the two nutrients often results in changes in the expression levels of genes specifically involved in controlling the homeostasis of the other nutrient and underlined as comparable molecular mechanisms regulate both SO_4^{2-} and phosphate transport in plants.

At the metabolic level, one of the most evident relationships between S and P is linked to membrane composition. It is known that cells can replace sulfolipids by phospholipids under S starvation, as well as they are able to replace phospholipids by sulfolipids and/or galactolipids under P starvation [41–46]. In Arabidopsis, the synthesis of sulfolipids is catalyzed by two enzymes, SQD1 and SQD2, whose expressions are increased by P starvation [42,43]. Although lipid shifts could be interpreted as adaptive mechanisms for plant survival under different nutrient availabilities, the physiological and biochemical consequences of phospholipids-sulfolipids substitutions on plant membrane functions are still unclear. Reprogramming membrane compositions under nutrient deficiency could have profound impacts on both S and P availability for plant metabolism. Moreover, recent studies have shown that the lipid environment and lipid-protein interactions may have crucial roles in modulating the functions as well as the conformational dynamics of membrane transporters [47].

Unfortunately, our basic knowledge about the interactions between P metabolism and SO_4^{2-} transport is limited. A few papers show that P deficiency or perturbations in P metabolism may impact the SO_4^{2-} allocation inside the plants. It has been reported that SO_4^{2-} concentration increases in roots and decreases in shoots of Arabidopsis as a consequence of reduced phosphate availabilities in the growing medium [48]. Transcriptional analysis of the main *SULTR* genes implicated in long-distance SO_4^{2-} transport reveals that, under P starvation, the transcript of *AtSULTR1;3* accumulates in both roots and shoots, whereas that of *AtSULTR2;1* weakly accumulates only in the roots. In the same conditions, *AtSQD1* transcript increases in both roots and shoots, indicating that adaptive modulations of SULTRs controlling the inter-organ distribution of SO_4^{2-} are required for the replacement of phospholipid by sulfolipids induced by P starvation [48]. Most of these responses seem to be dependent on *PHR1* (*PHOSPHATE RESPONSE1*), a gene encoding a protein belonging to the MYB-CC family transcription factors involved in the activation of several phosphate starvation-induced genes (PSI) [38,49]. PHR1 binds to an imperfect palindromic motif, named P1BS, which is prevalent in the promoter of the PSI genes [49,50]. Interestingly *cis*-regulatory motifs for PHR1-dependent gene activation are also present in the promoters of both *AtSULTR1;3* and *AtSQD1* genes, whose expressions are coherently reduced in the Arabidopsis *phr1* mutant grown under P starvation [48,51,52]. Interestingly, other evidence indicates *PHR1* as the convergent point for the cross-talk between P and other essential nutrients, such as zinc and iron [53,54]. *AtSULTR2;1*, is up-regulated by P starvation in a PHR1-independent manner, since *AtSULTR2;1* transcript further accumulates in *phr1* P deficient plants [48]. In this context, the observation that the expression of miR395—the microRNA that mainly controls the spatial expression of *SULTR2;1* in vascular tissues—is suppressed under P deficiency, allows us to speculate about the existence of an extra regulatory circuit which controls the inter-organ distribution of SO_4^{2-} under P starvation [55]. In this circuit (Figure 1): (i) the suppression of miR395 should allow SULTR2;1 to control root-to-shoot SO_4^{2-} translocation via the xylem route, as well as the source-to-sink SO_4^{2-} re-allocation via the phloem; (ii) PHR1 activates the expression of SULTR1;3 increasing further the capacity of the plants to move SO_4^{2-} from source to sink tissues. Unfortunately, no other evidence is available to support this extra regulatory circuit further and to fully appreciate its possible physiological impact on S metabolism in P deficient plants. Finally, the observation that Arabidopsis lines engineered for low PA content show alterations in SO_4^{2-} distribution and changes in expression of some *SULTRs* suggests the existence of another level of complexity in the cross-talk between S and P, which directly involves PA [56].

4. SULTRs as Novel Elements in the *lpa* Network

As mentioned above, genetic lesions in some genes putatively involved in sulfate transport result in *lpa* phenotypes in rice and barley [32,34,36]. Interestingly, all the mutations described thus far affect putative SULTR genes belonging to the SULTR3 subfamily, which includes elements whose functions are still objects of debate. For a detailed description of the *SULTR3/lpa* alleles, readers are referred to Cominelli et al. [57].

Differently from other SULTRs, whose capability to transport SO_4^{2-} has mostly been proven using yeast mutants as heterologous expression systems, the function of the SULTR3s as SO_4^{2-} transporters has only been hypothesized on the base of their sequence homologies with other SULTRs. Moreover, species-specific differences could explain the variability observed for the subcellular membrane localization of SULTR3 subfamily members.

Mineral nutrients required for plant growth are absorbed by the roots from the soil solution and then released to the xylem to be translocated to different tissues together with the transpiration flow. However, transpiration cannot be considered as the sole driving force for the root-to-shoot movement of nutrients, since developing organs such as new leaves and seeds are not photosynthetically active. Recently, nodes of gramineous plants have been identified as the main actors controlling nutrient delivery to developing tissues in a transpiration-independent way [58,59]. Several rice transporters involved in the intervascular transfer of nutrients from enlarged vascular bundles to diffuse vascular bundles of nodes seem to be essential to ensure this process [59]. Among these, *OsSULTR3;4/SPDT* has been indicated as pivotal in controlling phosphate delivery to developing tissues since it shows a proton-dependent transport activity for phosphate (but not for SO_4^{2-}), and it is highly expressed in the node 1 of rice at the reproductive stage [32]. Moreover, *OsSULTR3;4/SPDT* knockout mutants reduce P allocation to new leaves and grains, raveling the essential role of this transporter in switching phosphate toward developing leaves and grains.

The recent finding that the Arabidopsis *SULTR3;4* ortholog gene also controls xylem-to-phloem phosphate transfer, strongly suggests that sequence homology of SULTR3;4 with other SULTRs does not necessarily indicate that they share the same function [33]. All these observations not only show SULTR3;4s as phosphate transporters rather than as SO_4^{2-} transporters, but may also explain the role of these proteins in the *lpa* network.

If, on the one hand, the recent description of SULTR3;4/SPDTs as phosphate transporters seems to leave no room for doubt, on the other, assessment of the role of SULTR3;3s on P allocation still appears challenging. Rice SULTR3;3 is mainly expressed in vascular tissues and does not show any transport activity for SO_4^{2-} and phosphate [34]. Further studies are thus needed to uncover its function and subcellular localization. However, the lack of specific transport activity for SO_4^{2-} has also been indicated for the Arabidopsis SULTR3;5, which has been described as an essential component of the SO_4^{2-} transport system that facilitates the root-to-shoot SO_4^{2-} translocation in the vasculature [31]. Although the mechanisms controlling SO_4^{2-} allocation in rice are still known, it is possible to speculate that also *OsSULT3;3* could have a role in SO_4^{2-} partitioning among organs, by helping the activity of some other vascular transporter, or in SO_4^{2-} transport into the chloroplast, as recently suggested for its ortholog in Arabidopsis [30]. Rice *sultr3;3* mutants show significant alterations in S and P homeostasis, as indicated by the reduced concentration of SO_4^{2-} in both shoots and roots, as well as by the accumulation of transcripts of several S- and P-responsive genes in developing grains. Disruption in *OsSULTR3;3* also affects the concentrations of various grain metabolites not directly involved in PA biosynthesis. In particular, the reduced level of cysteine, along with the accumulation of its precursor serine, seems to indicate an insufficient supply of S during seed differentiation. Interestingly, reduced levels of cysteine have also been observed in the chloroplasts isolated from different Arabidopsis *sultr3* mutants [30]. Thus, the alterations in S homeostasis could be interpreted as the primary physiological event that reduces the accumulation of P in the grains of *sultr3;3* mutants. Finally, since total P and phosphate concentrations in root and shoot are higher in mutants than in the wild type, we may further speculate about the existence of mechanisms that somehow limit the systemic mobility of P in the plant. The analysis of the membrane lipid composition could provide in the next future a possible explanation for this phenomenon since substitution of sulfolipids by phospholipids caused by an insufficient S supply could increase the amount of P immobilized within cell membranes.

5. Conclusions and Perspectives

The implication of SULTRs in seed P accumulation not only provides novel opportunities to design routes for the breading of new *lpa* varieties in important cereal crops but also reveals the existence of a complex network of interactions between S and P homeostasis. The recent finding showing the involvement of *SULTR3;4/SPTDs* in delivering phosphate, and not sulfate, to developing tissues opens new questions about the nature of the other members of the SULTR3 subfamily [32,33]. Further investigations aimed at determining their substrate preference between sulfate and phosphate are then essential to unveil the actual role of these transporters in the control of nutrient homeostasis. In this context, the recent study of Cao et al. [29], suggesting that all the Arabidopsis *SULTR3* homologs may redundantly mediate sulfate import into the chloroplast, needs to be carefully reconsidered since chloroplasts isolated from the *sultr3* quintuple mutant retain about 50% of the sulfate uptake capacity of the wild type. Redundancy versus diversity will be the novel challenge to face.

Author Contributions: G.A.S. and F.F.N. equally contributed in discussing this opinion. F.F.N. wrote the manuscript. G.A.S. and F.F.N. revised and approved the final version of the opinion.

Funding: This work received no external funding.

Acknowledgments: The music of Gustav Mahler's V Symphony inspired this work. We would like to thank Maurizio Cocucci and Silvia Morgutti for their precious support during the writing of the manuscript.

Conflicts of Interest: The authors declare no conflict of interest.

References

1. Bohn, L.; Meyer, A.S.; Rasmussen, S.K. Phytate: Impact on environment and human nutrition. A challenge for molecular breeding. *J. Zhejiang Univ. Sci. B* **2008**, *9*, 165–191. [CrossRef] [PubMed]

2. Raboy, V. Seeds for a better future: "Low phytate", grains help to overcome malnutrition and reduce pollution. *Trends Plant Sci.* **2001**, *6*, 458–462. [CrossRef]

3. Schlemmer, U.; Frølich, W.; Prieto, R.M.; Grases, F. Phytate in foods and significance for humans: Food sources, intake, processing, bioavailability, protective role and analysis. *Mol. Nutr. Food Res.* **2009**, *53*, S330–S375. [CrossRef]

4. Raboy, V. Approaches and challenges to engineering seed phytate and total phosphorus. *Plant Sci.* **2009**, *177*, 281–296. [CrossRef]

5. Sparvoli, F.; Cominelli, E. Seed biofortification and phytic acid reduction: A conflict of interest for the plant? *Plants* **2015**, *4*, 728–755. [CrossRef]

6. Leustek, T.; Martin, M.N.; Bick, J.A.; Davies, J.P. Pathways and regulation of sulfur metabolism revealed through molecular and genetic studies. *Annu. Rev. Plant Physiol. Plant Mol. Biol.* **2000**, *51*, 141–165. [CrossRef]

7. Saito, K. Sulfur assimilatory metabolism. The long and smelling road. *Plant Physiol.* **2004**, *136*, 2443–2450. [CrossRef]

8. Nocito, F.F.; Lancilli, C.; Giacomini, B.; Sacchi, G.A. Sulfur metabolism and cadmium stress in higher plants. *Plant Stress* **2007**, *1*, 142–156.

9. Takahashi, H.; Kopriva, S.; Giordano, M.; Saito, K.; Hell, R. Sulfur assimilation in photosynthetic organisms: Molecular functions and regulations of transporters and assimilatory enzyme. *Annu. Rev. Plant Biol.* **2011**, *62*, 157–184. [CrossRef]

10. Mugford, S.G.; Lee, B.R.; Koprivova, A.; Matthewman, C.; Kopriva, S. Control of sulfur partitioning between primary and secondary metabolism. *Plant J.* **2011**, *65*, 96–105. [CrossRef] [PubMed]

11. Koprivova, A.; Kopriva, S. Sulfation pathways in plants. *Chem. Biol. Interact.* **2016**, *259*, 23–30. [CrossRef] [PubMed]

12. Shibagaki, N.; Grossman, A.R. The role of the STAS domain in the function and biogenesis of a sulfate transporter as probed by random mutagenesis. *J. Biol. Chem.* **2006**, *281*, 22964–22973. [CrossRef] [PubMed]

13. Shibagaki, N.; Grossman, A.R. Binding of cysteine synthase to the STAS domain of sulfate transporter and its regulatory consequences. *J. Biol. Chem.* **2010**, *285*, 25094–25102. [CrossRef] [PubMed]

14. Takahashi, H. Sulfate transport systems in plants: Functional diversity and molecular mechanisms underlying regulatory coordination. *J. Exp. Bot.* **2019**, *70*, 4075–4087. [CrossRef] [PubMed]

15. Kumar, S.; Asif, M.H.; Chakrabarty, D.; Tripathi, R.D.; Trivedi, P.K. Differential expression and alternative splicing of rice sulphate transporter family members regulate sulphur status during plant growth, development and stress conditions. *Funct. Integr. Genom.* **2011**, *11*, 259–273. [CrossRef]

16. Takahashi, H.; Watanabe-Takahashi, A.; Smith, F.W.; Blake-Kalff, M.; Hawkesford, M.J.; Saito, K. The roles of three functional sulphate transporters involved in uptake and translocation of sulphate in *Arabidopsis thaliana*. *Plant J.* **2000**, *23*, 171–182. [CrossRef]

17. Yoshimoto, N.; Takahashi, H.; Smith, F.W.; Yamaya, T.; Saito, K. Two distinct high-affinity sulfate transporters with different inducibilities mediate uptake of sulfate in Arabidopsis roots. *Plant J.* **2002**, *29*, 465–473. [CrossRef]

18. Yoshimoto, N.; Inoue, E.; Watanabe-Takahashi, A.; Saito, K.; Takahashi, H. Posttranscriptional regulation of high-affinity sulfate transporters in Arabidopsis by sulfur nutrition. *Plant Physiol.* **2007**, *145*, 378–388. [CrossRef]

19. Barberon, M.; Berthomieu, P.; Clairotte, M.; Shibagaki, N.; Davidian, J.C.; Gosti, F. Unequal functional redundancy between the two *Arabidopsis thaliana* high-affinity sulphate transporters *SULTR1;1* and *SULTR1;2*. *New Phytol.* **2008**, *180*, 608–619. [CrossRef]

20. El Kassis, E.; Cathala, N.; Rouached, H.; Fourcroy, P.; Berthomieu, P.; Terry, N.; Davidian, J.C. Characterization of a selenate-resistant Arabidopsis mutant. Root growth as a potential target for selenate toxicity. *Plant Physiol.* **2007**, *143*, 1231–1241. [CrossRef]

21. Rouached, H.; Wirtz, M.; Alary, R.; Hell, R.; Arpat, A.B.; Davidian, J.C.; Fourcroy, P.; Berthomieu, P. Differential regulation of the expression of two high-affinity sulfate transporters, SULTR1.1 and SULTR1.2, in Arabidopsis. *Plant Physiol.* **2008**, *147*, 897–911. [CrossRef] [PubMed]

22. Ferri, A.; Lancilli, C.; Maghrebi, M.; Lucchini, G.; Sacchi, G.A.; Nocito, F.F. The sulfate supply maximizing Arabidopsis shoot growth is higher under long- than short-term exposure to cadmium. *Front. Plant Sci.* **2017**, *8*, 854. [CrossRef] [PubMed]

23. Maruyama-Nakashita, A.; Watanabe-Takahashi, A.; Inoue, E.; Yamaya, T.; Saito, K.; Takahashi, H. Sulfur-responsive elements in the 3'-nontranscribed intergenic region are essential for the induction of *SULFATE TRANSPORTER 2;1* gene expression in Arabidopsis roots under sulfur deficiency. *Plant Cell* **2015**, *27*, 1279–1296. [CrossRef] [PubMed]

24. Kawashima, C.G.; Yoshimoto, N.; Maruyama-Nakashita, A.; Tsuchiya, Y.N.; Saito, K.; Takahashi, H.; Dalmay, T. Sulphur starvation induces the expression of microRNA-395 and one of its target genes but in different cell types. *Plant J.* **2009**, *57*, 313–321. [CrossRef]

25. Kawashima, C.G.; Matthewman, C.A.; Huang, S.; Lee, B.R.; Yoshimoto, N.; Koprivova, A.; Rubio-Somoza, I.; Todesco, M.; Rathjen, T.; Saito, K.; et al. Interplay of SLIM1 and miR395 in the regulation of sulfate assimilation in Arabidopsis. *Plant J.* **2011**, *66*, 863–876. [CrossRef]

26. Maruyama-Nakashita, A.; Nakamura, Y.; Tohge, T.; Saito, K.; Takahashi, H. Arabidopsis SLIM1 is a central transcriptional regulator of plant sulfur response and metabolism. *Plant Cell* **2006**, *18*, 3235–3251. [CrossRef]

27. Yoshimoto, N.; Inoue, E.; Saito, K.; Yamaya, T.; Takahashi, H. Phloem-localizing sulfate transporter, Sultr1;3, mediates re-distribution of sulfur from source to sink organs in Arabidopsis. *Plant Physiol.* **2003**, *131*, 1511–1517. [CrossRef]

28. Kataoka, T.; Watanabe-Takahashi, A.; Hayashi, N.; Ohnishi, M.; Mimura, T.; Buchner, P.; Hawkesford, M.J.; Yamaya, T.; Takahashi, H. Vacuolar sulfate transporters are essential determinants controlling internal distribution of sulfate in Arabidopsis. *Plant Cell* **2004**, *16*, 2693–2704. [CrossRef]

29. Cao, M.J.; Wang, Z.; Wirtz, M.; Hell, R.; Oliver, D.J.; Xiang, C.B. SULTR3;1 is a chloroplast-localized sulfate transporter in *Arabidopsis thaliana*. *Plant J.* **2013**, *73*, 607–616. [CrossRef]

30. Chen, Z.; Zhao, P.X.; Miao, Z.Q.; Qi, G.F.; Wang, Z.; Yuan, Y.; Ahmad, N.; Cao, M.J.; Hell, R.; Wirtz, M.; et al. SULTR3s function in chloroplast sulfate uptake and affect ABA biosynthesis and the stress response. *Plant Physiol.* **2019**, *180*, 593–604. [CrossRef]

31. Kataoka, T.; Hayashi, N.; Yamaya, T.; Takahashi, H. Root-to-shoot transport of sulfate in Arabidopsis. Evidence for the role of SULTR3;5 as a component of low-affinity sulfate transport system in the root vasculature. *Plant Physiol.* **2004**, *136*, 4198–4204. [CrossRef] [PubMed]

32. Yamaji, N.; Takemoto, Y.; Miyaji, T.; Mitani-Ueno, N.; Yoshida, K.T.; Ma, J.F. Reducing phosphorus accumulation in rice grains with an impaired transporter in the node. *Nature* **2017**, *541*, 92–95. [CrossRef] [PubMed]

33. Ding, G.; Lei, G.J.; Yamaji, N.; Yokosho, K.; Mitani-Ueno, N.; Huang, S.; Ma, J.F. Vascular cambium-localized AtSPDT mediates xylem-to-phloem transfer of phosphorus for its preferential distribution in *Arabidopsis*. *Mol. Plant* **2019**. [CrossRef] [PubMed]

34. Zhao, H.; Frank, T.; Tan, Y.; Zhou, C.; Mabnoune, M.; Arpat, A.B.; Cui, H.; Huang, J.; He, Z.; Poirier, Y.; et al. Disruption of *OsSULTR3;3* reduces phytate and phosphorus concentrations and alters the metabolite profile in rice grains. *New Phytol.* **2016**, *211*, 926–939. [CrossRef] [PubMed]

35. Liu, Q.L.; Xu, X.H.; Ren, X.L.; Fu, H.W.; Wu, D.X.; Shu, Q.Y. Generation and characterization of low phytic acid germplasm in rice (*Oryza sativa* L.). *Theor. Appl. Genet.* **2007**, *114*, 803–814. [CrossRef] [PubMed]

36. Ye, H.; Zhang, X.Q.; Broughton, S.; Westcott, S.; Wu, D.; Lance, R.; Li, C. A nonsense mutation in a putative sulphate transporter gene results in low phytic acid in barley. *Funct. Integr. Genom.* **2011**, *11*, 103–110. [CrossRef]

37. Wang, F.; Deng, M.; Xu, J.; Zhu, X.; Mao, C. Molecular mechanisms of phosphate transport and signaling in higher plants. *Semin. Cell Dev. Biol.* **2018**, *74*, 114–122. [CrossRef]

38. Ham, B.K.; Chen, J.; Yan, Y.; Lucas, W.J. Insights into plant phosphate sensing and signaling. *Curr. Opin. Biotechnol.* **2018**, *49*, 1–9. [CrossRef]

39. Veneklaas, E.J.; Lambers, H.; Bragg, J.; Finnegan, P.M.; Lovelock, S.E.; Plaxton, W.C.; Price, C.A.; Scheible, W.R.; Shane, M.W.; White, P.J.; et al. Opportunities for improving phosphorus-use efficiency in crop plants. *New Phytol.* **2012**, *195*, 306–320. [CrossRef]

40. Rouached, H. Multilevel coordination of phosphate and sulfate homeostasis in plants. *Plant Signal. Behav.* **2011**, *6*, 952–955. [CrossRef]

41. Härtel, H.; Essigmann, B.; Lokstein, H.; Hoffmann-Benning, S.; Peters-Kotting, M.; Benning, C. The phospholipid-deficient *pho1* mutant of *Arabidopsis thaliana* is affected in the organization, but not in the light acclimation, of the thylakoid membrane. *Biomembranes* **1998**, *1415*, 205–218. [CrossRef]

42. Essigmann, B.; Guler, S.; Narang, R.A.; Linke, D.; Benning, C. Phosphate availability affects the thylakoid lipid composition and the expression of *SQD1*, a gene required for sulfolipid biosynthesis in *Arabidopsis thaliana*. *Proc. Natl. Acad. Sci. USA* **1998**, *95*, 1950–1955. [CrossRef] [PubMed]

43. Yu, B.; Xu, C.; Benning, C. *Arabidopsis* disrupted in *SQD2* encoding sulfolipid synthase is impaired in phosphate-limited growth. *Proc. Natl. Acad. Sci. USA* **2002**, *99*, 5732–5737. [CrossRef] [PubMed]

44. Sugimoto, K.; Sato, N.; Tsuzuki, M. Utilization of a chloroplast membrane sulfolipid as a major internal sulfur source for protein synthesis in the early phase of sulfur starvation in *Chlamydomonas reinhardtii*. *FEBS Lett.* **2007**, *581*, 4519–4522. [CrossRef] [PubMed]

45. Tjellström, H.; Andersson, M.X.; Larsson, K.E.; Sandelius, A.S. Membrane phospholipids as a phosphate reserve: The dynamic nature of phospholipid-to-digalactosyl diacylglycerol exchange in higher plants. *Plant Cell Environ.* **2008**, *31*, 1388–1398. [CrossRef] [PubMed]

46. Sugimoto, K.; Tsuzuki, M.; Sato, N. Regulation of synthesis and degradation of a sulfolipid under sulfur-starved conditions and its physiological significance in *Chlamydomonas reinhardtii*. *New Phytol.* **2010**, *185*, 676–686. [CrossRef]

47. Martens, C.; Shekhar, M.; Borysik, A.J.; Lau, A.M.; Reading, E.; Tajkhorshid, E.; Booth, P.J.; Politis, A. Direct protein-lipid interactions shape the conformational landscape of secondary transporters. *Nat. Commun.* **2018**, *9*, 4151. [CrossRef]

48. Rouached, H.; Secco, D.; Arpat, B.; Poirier, Y. The transcription factor PHR1 plays a key role in the regulation of sulfate shoot-to-root flux upon phosphate starvation in Arabidopsis. *BMC Plant Biol.* **2011**, *11*, 19. [CrossRef]

49. Rubio, V.; Linhares, F.; Solano, R.; Martín, A.C.; Iglesias, J.; Leyva, A.; Paz-Ares, J. A conserved MYB transcription factor involved in phosphate starvation signaling both in vascular plants and in unicellular algae. *Genes Dev.* **2001**, *15*, 2122–2133. [CrossRef]

50. Bustos, R.; Castrillo, G.; Linhares, F.; Puga, M.I.; Rubio, V.; Pérez, J.; Solano, R.; Leyva, A.; Paz-Ares, J. A central regulatory system largely controls transcriptional activation and repression responses to phosphate starvation in Arabidopsis. *PLoS Genet.* **2010**, *6*, e1001102. [CrossRef]

51. Franco-Zorrilla, J.M.; Gonzalez, E.; Bustos, R.; Linhares, F.; Leyva, A.; Paz-Ares, J. The transcriptional control of plant responses to phosphate limitation. *J. Exp. Bot.* **2004**, *55*, 285–293. [CrossRef] [PubMed]

52. Stefanovic, A.; Ribot, C.; Rouached, H.; Wang, Y.; Chong, J.; Belbahri, L.; Delessert, S.; Poirier, Y. Members of the *PHO1* gene family show limited functional redundancy in phosphate transfer to the shoot, and are regulated by phosphate deficiency via distinct pathways. *Plant J.* **2007**, *50*, 982–994. [CrossRef] [PubMed]

53. Khan, G.A.; Bouraine, S.; Wege, S.; Li, Y.; de Carbonnel, M.; Berthomieu, P.; Poirier, Y.; Rouached, H. Coordination between zinc and phosphate homeostasis involves the transcription factor PHR1, the phosphate exporter PHO1, and its homologue PHO1;H3 in Arabidopsis. *J. Exp. Bot.* **2014**, *65*, 871–884. [CrossRef] [PubMed]

54. Bournier, M.; Tissot, N.; Mari, S.; Boucherez, J.; Lacombe, E.; Briat, J.F.; Gaymard, F. *Arabidopsis* ferritin 1 (*AtFer1*) gene regulation by the phosphate starvation response 1 (AtPHR1) transcription factor reveals a direct molecular link between iron and phosphate homeostasis. *J. Biol. Chem.* **2013**, *288*, 22670–22680. [CrossRef] [PubMed]

55. Hsieh, L.; Lin, S.; Shih, A.; Chen, J.; Lin, W.; Tseng, C.; Lin, W.Y.; Tseng, C.Y.; Li, H.Y.; Tzyy-Jen Chiou, T.J. Uncovering small RNA-mediated responses to phosphate-deficiency in Arabidopsis by deep sequencing. *Plant Physiol.* **2009**, *151*, 2120–2132. [CrossRef]

56. Belgaroui, N.; Zaidi, I.; Farhat, A.; Chouayekh, H.; Bouain, N.; Chay, S.; Curie, C.; Mari, S.; Masmoudi, K.; Davidian, J.C.; et al. Over-expression of the bacterial phytase US417 in *Arabidopsis* reduces the concentration of phytic acid and reveals its involvement in the regulation of sulfate and phosphate homeostasis and signaling. *Plant Cell Physiol.* **2004**, *55*, 1912–1924. [CrossRef]

57. Cominelli, E.; Pilu, R.; Sparvoli, F. Phytic acid, transporters and the generation of low phytic acid mutants in grain crops. *Plants* **2019**. submitted for publication.

58. Yamaji, N.; Ma, J.F. The node, a hub for mineral nutrient distribution in graminaceous plants. *Trends Plant Sci.* **2014**, *19*, 556–563. [CrossRef]

59. Yamaji, N.; Ma, J.F. Node-controlled allocation of mineral elements in Poaceae. *Curr. Opin. Plant Biol.* **2017**, *39*, 18–24. [CrossRef]

© 2019 by the authors. Licensee MDPI, Basel, Switzerland. This article is an open access article distributed under the terms and conditions of the Creative Commons Attribution (CC BY) license (http://creativecommons.org/licenses/by/4.0/).

Article

lpa1-5525: A New lpa1 Mutant Isolated in a Mutagenized Population by a Novel Non-Disrupting Screening Method

Giulia Borlini [1], Cesare Rovera [1], Michela Landoni [2], Elena Cassani [1] and Roberto Pilu [1,*]

[1] Department of Agricultural and Environmental Sciences-Production, Landscape, Agroenergy-Università degli Studi di Milano, Via Celoria 2, 20133 Milan, Italy
[2] Department of Biosciences-Università degli Studi di Milano, Via Celoria 26, 20133 Milan, Italy
* Correspondence: roberto.pilu@unimi.it; Tel: +39-02-50316549

Received: 15 June 2019; Accepted: 4 July 2019; Published: 6 July 2019

Abstract: Phytic acid, or myo-inositol 1,2,3,4,5,6-hexakisphosphate, is the main storage form of phosphorus in plants. It is localized in seeds, deposited as mixed salts of mineral cations in protein storage vacuoles; during germination, it is hydrolyzed by phytases to make available P together with all the other cations needed for seed germination. When seeds are used as food or feed, phytic acid and the bound cations are poorly bioavailable for human and monogastric livestock due to their lack of phytase activity. Therefore, reducing the amount of phytic acid is one strategy in breeding programs aimed to improve the nutritional properties of major crops. In this work, we present data on the isolation of a new maize (*Zea mays* L.) low phytic acid 1 (lpa1) mutant allele obtained by transposon tagging mutagenesis with the *Ac* element. We describe the generation of the mutagenized population and the screening to isolate new lpa1 mutants. In particular, we developed a fast, cheap and non-disrupting screening method based on the different density of lpa1 seed compared to the wild type. This assay allowed the isolation of the lpa1-5525 mutant characterized by a new mutation in the *lpa1* locus associated with a lower amount of phytic phosphorus in the seeds in comparison with the wild type.

Keywords: maize; low phytic acid; regional mutagenesis; *Ac* transposon; density assay; free phosphate; Chen's assay; PCR based molecular marker

---

1. Introduction

Phytic acid (*myo*-inositol-1,2,3,4,5,6-hexakisphosphate, or InsP6) is the most common form of phosphate present in cereal kernels as well as in seeds of most plants [1,2]. Cereals' seeds contain on average 10 mg of phytic acid, which account from 65% to 90% of the total phosphorus present inside the seed [3]. Among cereals, maize possesses a great importance not only in human and animal nutrition but also in the use of its derivatives in many industrial sectors. As animal feed, maize can be considered one of the most important staple foods available and, in many cases, is indispensable for the formulation of animal rations. Moreover, maize is a model plant in genetic studies aimed at understanding the role of genes involved in biosynthetic pathways and in plant morphogenesis. For those reasons, maize is considered one of the most important plants studied and used in genetic improvement programs.

Phytic acid is synthesized in the endoplasmic reticulum, and then is deposited in protein bodies organized in specific structures called globoids as a mixture of phytic salts of several cations, such as potassium, iron, zinc and magnesium [4]. In cereals' seeds, the phytates are mainly localized in the embryo (80%) and in the aleurone layer (20%) [1]. During germination, the phytate salts are broken down by phytase activity releasing free phosphate, minerals and myo-inositol, necessary for seedling growth [5]. Furthermore, phytic acid, by firmly chelating iron cations, is able to counteract the formation of reactive oxygen species and is thus involved in the preservation of viability of plant seeds [6,7]. Phytic acid is poorly digested by monogastric animals, and since it is a chelator of cations, it is considered an anti-nutritional factor. Moreover, as it is not assimilated, it is expelled with manure, becoming a pollutant of cultivated land and contributing to the eutrophication of surface waters [8]. It has been estimated that nearly 50% of elemental P used in global agricultural activities is accumulated in the phytic acid fraction. For these reasons, its reduction or elimination in the seeds is a major challenge in genetic improvement programs.

The conventional breeding protocols designed for reducing phytic acid content, rely on the isolation of low phytic acid (*lpa*) mutations impairing the biosynthesis or the storage of phytic acid in the seed; the increased P and mineral cation bioavailability in the lpa mutant seeds so far isolated was confirmed by nutritional trials [9–11]. Low phytic acid mutations in maize can be classified into three categories: type 1 are mutations altering the biosynthetic pathway (MIPS, *myo*-inositol 3-phosphate synthase, catalyzing the step from glucose 6-P to myo-inositol[3]-monophosphate); type 2 are mutations altering the following phosphorylation steps (e.g., ITPK$_S$, inositol tris/tetra kisphosphate kinases); type 3 are mutations altering the transport of phytic acid to the vacuole (MRP, multidrug-resistance-associated protein) [12,13]. In recent years, several breeding programs aimed at selecting new maize varieties with seeds characterized by a lower level of phytic acid compared to traditional cultivars, have been carried out [14,15]. The lpa phenotype has been isolated in different crops, maize [16–18], barley [19–21], wheat [22], rice [23–26] and common bean [27], by physical, chemical or transposon tagging mediated mutagenesis. In maize, there are three different *lpa* mutations (*lpa1, 2* and *3*), with the *lpa1* mutation showing the lowest phytic acid content in the seed [16,17].

Due to the strong pleiotropic effects associated with the *lpa* mutations, the mutants thus far isolated are generally lethal; the challenge will be to isolate new lpa mutants in which the pleiotropic effects will be absent or dramatically reduced, allowing the *lpa* mutation to be viable.

lpa1 mutation does not modify the total amount of seed P but reduces phytic acid content, thus leading to a proportionally increased level of free phosphate [18]. Owing to this, an HIP (high inorganic phosphate) phenotype is diagnostic for the presence of the *lpa1* mutation, making it quite easy to identify the lpa1 phenotype. The screenings so far used to identify the lpa mutants are destructive methods based on the quantification of P in the flour obtained by milling the seeds (as, for example, the Chen method [28]). For this reason, these screenings are not performed on the M1 generation but on the M2 progeny, with the consequent increase of time required to acquire the samples and number of samples to be analyzed.

Transposon tagging mutagenesis experiments [29] demonstrated that *lpa1* gene encodes a multidrug-associated-protein (MRP) named *ZmMRP4* (accession number EF586878). MRP proteins are transmembrane transporters involved in several functions such as organic ions transport, xenobiotic detoxification, oxidative stress tolerance and transpiration control [30,31].

The transposon tagging mutagenesis consists in introducing, by crossing, a known DNA sequence in the genome of a target species. If the insertion event occurs within a gene sequence, altering its expression, a mutant phenotype can be observed [32]. Because of the possibility of the transposon moving within the genome, transposon tagging mutagenesis provides the so-called mutable alleles that cannot be directly used in genetic improvement programs but that may be useful to clone the gene responsible for the mutated phenotype and to isolate stable excision events that lead to stable genomic mutations which can be used in genetic improvement programs.

The *Ac/Ds* transposon system of maize is often used in transposon tagging mutagenesis experiments. In fact, the elements of this transposon family can be transferred into the genome of different species, determining the generation of insertional mutants.

The *Ac/Ds* system is made up with two types of transposons: the autonomous element, which can transpose (the *Activator* element, *Ac*), and the non-autonomous element, which cannot transpose independently (the *Dissociator* element, *Ds*). Sequence analyses have shown that the *Ds* element is derived by deletion from the autonomous element of the family, with the loss of the function of one or more genes required for transposition. Therefore, only the *Ac* element encodes for the transposase, the enzyme required to mobilize both the *Ac* and the non-autonomous *Ds* elements [33]. Studies on *Ac/Ds* transposition have revealed a strong preference for insertion in regions of the genome in close genetic linkage to the donor site. In particular, it has been reported that the majority of *Ac* transpositions were within 10cM from the donor site [34]. This characteristic short-range transposition is used in regional mutagenesis studies to create multiple alleles in a target locus close to an *Ac* donor site.

In this work we present a new non-disrupting, fast and simple method to select lpa1 mutants. We describe the development and the screening of a transposon tagging mutagenized population which enabled the isolation of a new lpa1 mutant, named lpa1-5525.

2. Results

2.1. Transposon Tagging Mutagenesis

In order to isolate new lpa1 mutants, a transposon-mediated mutagenesis experiment was performed. The mutagenized population was generated by crossing the lpa1-1 mutant with a line carrying the *Ac* transposon (Figure 1). We chose an *Ac* line in which the transposon was inserted on the short arm of chromosome 1 in the 1.03/1.04 bin region, where *ZmMRP4* gene maps (Figure 1a).

We used *lpa1-1* homozygous plants as female and the *Ac* line as pollen donor. The mutagenized population, consisting of 4787 F1 seeds, was screened to find new lpa1 mutants.

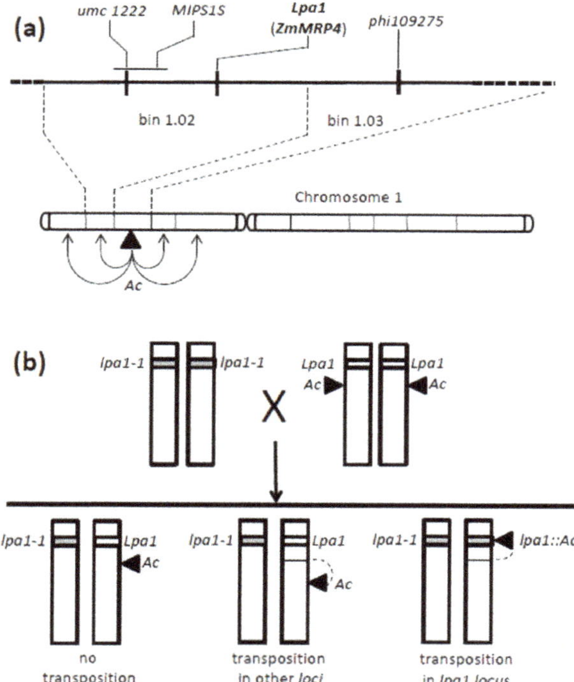

Figure 1. (**a**) Diagram of chromosome 1, the position of the *Ac* element and of *ZmMRP* locus, bin 1.02/1.03 are indicated. (**b**) Genetic scheme used to generate the F1 mutagenized population.

2.2. Density Assay Screening of the Mutagenized Population

As previously reported, the *lpa1* mutation showed the lowest phytic acid content in the seed in comparison to the *lpa* mutations thus far characterized [16,17]. This mutation does not modify the total amount of seed P, but reduces phytic acid content, leading to a proportionally increased level of free phosphate [18]. The screening methods to identify *lpa1* mutations were thus far based on the identification of HIP (high inorganic phosphate) phenotypes by the quantification of the phosphate level in the seeds with disruptive methods [28].

In order to isolate new lpa1 mutants inside the mutagenized populations, we developed a non-disrupting, fast, simple and cheap method. In particular, this assay was able to identify the mutants' seeds because in an highly concentrated sugar solution (1.218–1.222 g/L) the lpa1 seeds, due to their lower density [35], can float, unlike the wild controls that sink and stay on the bottom of the beaker (Figure 2a). Among the 4787 F1 seeds screened with this test, 271 were identified as low density seeds (Table 1).

Table 1. Density assay on mutagenized F1 seeds obtained by crossing *lpa1-1/lpa1-1* (codes R3962 to R3966 and R3893 and R864) with *Ac* line (R3967, R3970 and R916).

Code	Total Seeds Tested	Putative Mutant Seeds Isolated	Sucrose Solution Density (23 °C)
R3962/R3967 (1)	138	0	1.218
R3962/R3967 (2)	271	2	1.218
R3962/R3967 (3)	108	0	1.218
R3962/R3967 (4)	232	4	1.218
R3962/R3967 (5)	246	12	1.218
R3962/R3967 (6)	175	5	1.218
R3962/R3967 (7)	132	3	1.218
R3963/R3967 (1)	159	25	1.218
R3963/R3967 (2)	196	1	1.218
R3963/R3967 (3)	170	1	1.218
R3963/R3967 (4)	101	14	1.218
R3963/R3970 (1)	194	3	1.218
R3963/R3970 (2)	143	7	1.218
R3964/R3967 (1)	232	4	1.218
R3964/R3967 (2)	258	4	1.218
R3964/R3967 (4)	219	27	1.218
R3964/R3967 (5)	232	20	1.218
R3964/R3967 (6)	217	29	1.218
R3965/R3970 (2)	160	23	1.218
R3966/R3970 (8)	114	2	1.218
R3893/R916-300	166	8	1.218
R864/R916/(3)	335	20	1.222
R864/R916 (5)	285	14	1.222
R864/R916 (8)	159	27	1.222
R864/R916 (9)	145	16	1.222
Total	4787	271	

Out of these 271 seeds, 50 were put aside and stored for further analyses, 41 were tested for the HIP phenotype by the Chen method [28] and the remaining 180 were sown in the experimental field (Figure 2b).

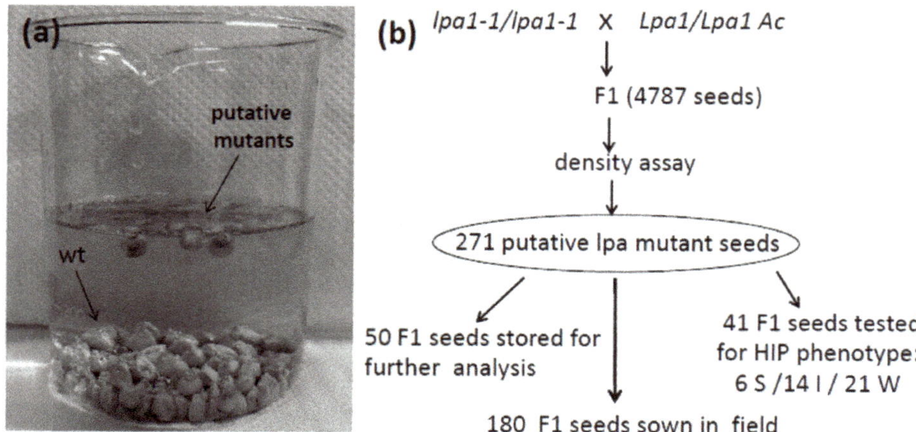

Figure 2. (**a**) Density assay used to isolate new lpa1 mutants: in the sucrose solution lpa1 mutant seeds float and wild type seeds stay on the bottom of the beaker. (**b**) Scheme of the experimental plan used to generate and to screen the F1 mutagenized population.

2.3. Confirmation of the HIP Phenotype of the Low Density Seeds

The density assay we developed was very rapid, cheap and easy, but it also revealed a limitation, i.e., since this screening is based only on the low density of the seeds to be selected, it did not allow the exclusive identification of lpa1 seeds, but it selected all seeds characterized by low density, including seeds affected by any kind of mutations impairing endosperm/embryo development and moldy seeds present inside the mutagenized population. For this reason, 41 out of the 271 seeds selected by the density assay were chosen and tested for the phosphate content (HIP phenotype) in order to verify that the low density character was associated with the lpa1 phenotype (Figure 2b). The determination of the free phosphorus content inside the seeds was carried out using a semi-quantitative colorimetric method based on the Chen reagent [28]. Based on the availability of the free phosphorus inside the seeds, it was possible to classify the seeds into four categories: WT (wild type—Free P < 0.3 mg/g), W (weak—0.3 mg/g < Free P < 0.5 mg/g), I (intermediate—0.5 mg/g < Free P < 1.4 mg/g) and S (strong—Free P > 1.4 mg/g) (Figure 4a). This assay allowed us to confirm the HIP phenotype for 20 out the 41 seeds selected with the density assay, with six seeds belonging to the Strong and 14 to the Intermediate categories (Figure 2b).

2.4. Molecular Analysis of the F1 Plants of the Mutagenized Population

The lpa1-1 mutant was used as female parent in the initial cross we made to generate the mutagenized population. To identify and discard the contaminant seeds produced by the accidental self-fertilization of the lpa1-1 female parent, the 27 F1 plants, obtained from 180 F1 seeds sown in the open field (Figure 1), were genotyped by PCR analysis. The coding sequence of the *lpa1-1* allele is characterized, in comparison with the wild type allele, by the presence of a Single Nucleotide Polymorphism (SNP), C to T, in position 5759 with respect to the ATG on the genomic sequence. This allowed the design of two different forward primers, one specific for the wild type (ZmMRP430L) and the other specific for the *lpa1.1* allele (ZmMRP432L) [29,36]. We used the two specific forwards primers in combination with a reverse common primer (ZmMRP410R). Out of the 27 plants analyzed, 26 resulted in heterozygous *Lpa1/lpa1-1*; one plant was found to be a contaminant *lpa1-1/lpa1-1* and discarded (data not shown).

2.5. Screening and Selection of the Putative New lpa1 Mutant

Among the 26 F1 heterozygous plants (*Lpa1/lpa1-1*), we selected the 19 more vigorous that were self-fertilized (Figure 3). The four best F2 ears were tested for the HIP phenotype: 24 seeds were collected from each ear and the disruptive assay for free phosphate was performed. All the seeds tested were found to belong to the Strong or Intermediate categories, indicating an HIP phenotype for all the four selected ears and 50 seeds from each ear were sown in the field.

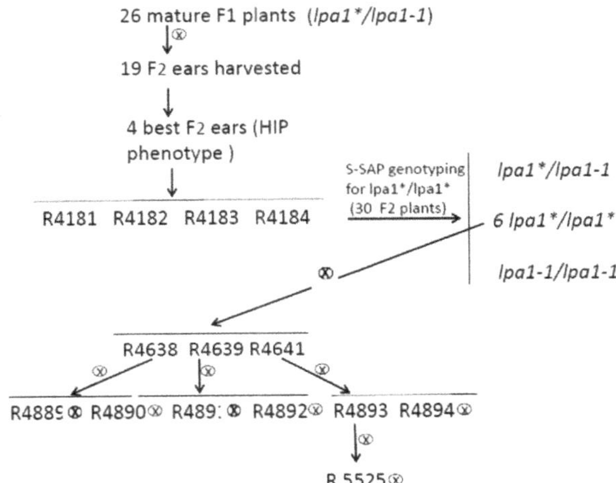

Figure 3. Scheme of the procedure used to obtain the NIL (Near Isogenic Line) lpa1-5525, homozygous for the new *lpa1* mutation.

Among the 90 plants that survived we selected the 30 more vigorous that were genotyped with the primers specific for *Lpa/lpa1* alleles, to identify the *Lpa1/Lpa1* F2 plants, i.e., the putative new lpa1 mutants, provisionally indicated as lpa1*.

The six F2 plants *lpa1*/lpa1* were selfed and F3 seeds were tested for the HIP phenotype. The best F3 ears were selected (R4638, R4639, R4641) and generated the following six F4 families: R4889, R4890, R4891, R4892, R4893 and R4894. The F4 seeds were tested for HIP phenotype (Table 2) and the F4 ear R4893, showing high amount of S-I seeds and the best agronomic performance, was selected, sown in the field and the F4 plants were selfed. The F5 seeds (R5525) representing the lpa1* mutant were further analyzed.

Table 2. Chen's assay performed on F5 progenies. The S+I/total ratio is reported. S: Strong; I: Intermediate; W: Weak; WT: Wild Type.

Code	N° of Seeds Tested	Phenotype		S/I %
		S/I	W/WT	
R4889 ⊗	24	13	11	54.17%
R4890 ⊗	36	15	21	41.67%
R4891 ⊗	36	8	28	22.22%
R4892 ⊗	42	41	1	97.62%
R4893 ⊗	18	17	1	94.44%
R4894 ⊗	36	29	7	80.56%
Total	192	123	69	64.06%

2.6. Quantitative Analysis of Phosphorus Content

The lpa1* mutant and wild type control seeds were tested for total P, free P and phytic acid content. No significant alterations in total P amount were observed between the putative new lpa1 mutant and the control: in fact, even if the new *lpa1* mutation caused approximately an eight-fold increase in the amount of free phosphate, this was balanced by a reduction of phytic acid that was nearly halved in comparison to the wild type content (1.28mg/g and 3mg/g, respectively) (Figure 4). In the lpa1 mutant, total phosphorus is 50% free P and 50% phytic P.

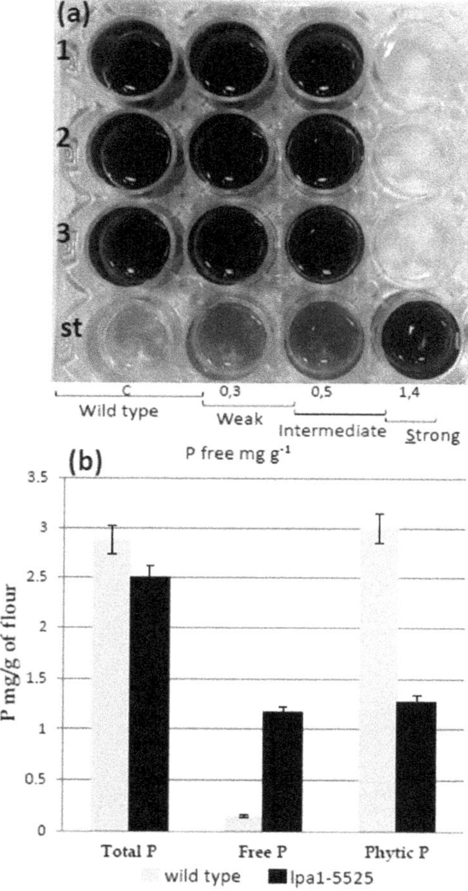

Figure 4. (**a**) Chen assay performed on F4 families mutant seeds. Microtiter rows 1, 2 and 3 show the results of the assay performed on nine mutant seeds, st row correspond to the standard used for the semiquantitative classification. (**b**) lpa1-5525 mutant and wild type control mature dry kernels were analyzed for total P, free inorganic P and phytic acid P amount. Values are expressed as milligrams of P (atomic weight 31) in 1 g of flour. Confidence intervals at 95% are shown.

2.7. Structure Analysis of the ZmMRP4 Locus

The lpa1* mutant was isolated through a transposon-mediated mutagenesis experiment with the *Ac* element. To check for the presence of the 4.6 kb sequence of the *Ac* transposon into the *ZmMRP4* locus, the sequence of *lpa1* gene, spanning from the 5′UTR (nucleotide -313) to the 3′UTR (nucleotide

6460) (Figure 5a) was amplified in the lpa1* mutant and in wild type. The gel electrophoresis of the amplicons obtained failed to reveal any difference in length between wild type and the lpa1* mutant (also in the promoter region till about -1500bp, data not shown), thus suggesting the absence of the *Ac* transposon inside the *ZmMRP4* locus (Figure 5b).

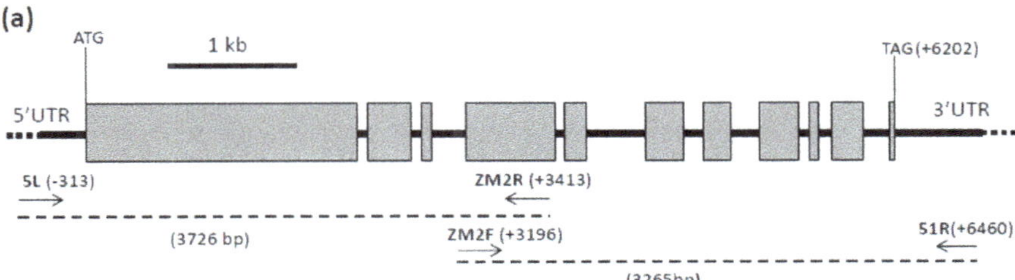

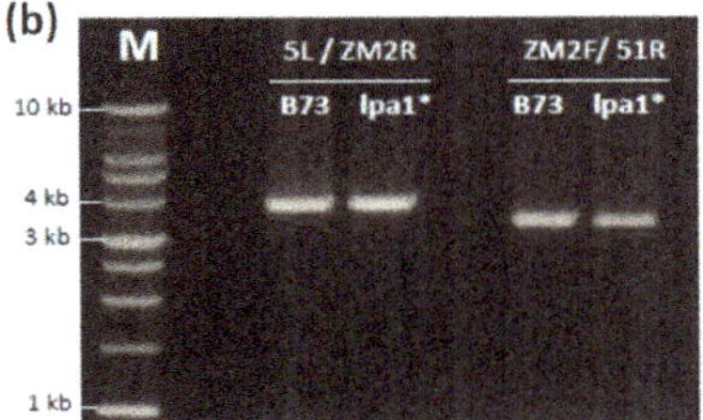

Figure 5. (**a**) Structure of the *ZmMRP4* gene. The primers used to amplify the coding sequence of *ZmMRP4* gene are indicated by the arrows. (**b**) Agarose gel showing the amplicons obtained amplifying wt (B73) and mutant (lpa1-5525) genomic DNA with primers 5L/ZM2r and ZM2F/51R.

3. Discussion

Phytic acid is an insoluble phosphate derivative, present in maize kernels, as in other seeds such as legumes. This element represents the main form of phosphorus reserve in the mature seed [3,37], degraded during germination by the activity of a group of enzymes called phytases [38]. Phytic acid has remarkable chelating properties and besides making phosphorus unavailable, it binds many minerals including calcium, iron, zinc, magnesium and manganese [10,39], which are not available for monogastric animals because of the lack of phytases in their digestive tract. According to the FAO data (www.FAO.org), maize is the leading cereal for production in the world, providing 1.04 billion tons of grain and occupying an area of 184.8 million hectares. For this reason, phytic acid is considered a strong worldwide food and feed antinutritional factor.

Furthermore, in recent years, phosphorus has become one of the main sustainability issues because it is a non-renewable resource. Being widely used in agriculture as an essential component of fertilizers and feed, the increasing production of food has increased the rate of mobilization of reserves and the price of the mineral is continuously increasing. It is estimated that, at the current consumption rates, the phosphorus reserves available with the current technology will run out in 90 years. To date, several strategies have been developed to improve the bioavailability of phosphate in animal feed based on seeds and to reduce its environmental impact [40–42]. Some strategies were aimed at increasing phytase activity in the feed by incubation of the feed in water before administration, or by addition of purified phytases, other projects were based on the use of GMO maize that accumulates fungal phytases in kernels [43] or on the development of pigs engineered for the presence of phytases in the saliva [44]. Classical breeding programs, furthermore, allowed the isolation of mutant plants whose seeds were characterized by a low phytic acid content, the low phytic acid (lpa) mutants, (i.e., mutants defective in phytine synthesis). The peculiarity of these mutants concerns the distribution of phosphorus fractions: despite the fact that a smaller quantity of phytin is produced, phosphorus is accumulated in the seed during its maturation, remaining in a free form, directly assimilable by the monogastrics [16,17]. Thus far, various studies have reported the isolation of low phytic acid mutations in different species by the screening of a mutagenized population through Chen's assay [12]. This disrupting assay is characterized by a high requirement of time and work [28]. In fact, F1 populations cannot be analyzed directly because of the disruption of the seeds analyzed, thus, we have to wait for F2 generation, and then for each F2 family, about 15–25 seeds have to be milled and the flour processed. For these reasons, to identify the putative mutagenized seeds in the F1 progeny, we developed and used a novel fast and cheap screening method based on the different density of lpa1 seeds in comparison with the wild type. In fact, in a previous paper [35], we showed that among the pleiotropic effects associated with the *lpa1* mutation there is also a significant reduction of seed density in comparison with the wild type. More recently, the role of the l*Lpa1* gene in influencing how densely starch granules are packed in the grain has also been confirmed in rice [45].

In this work, in order to isolate new mutations inside the *ZmMRP4/lpa1* locus, we generated a mutagenized population through a regional mutagenesis program based on the *Ac/Ds* transposons family. The initial crosses were performed between the *lpa1-1/lpa1-1* mutant line and a wild type line (*Lpa1/Lpa1*) carrying the *Ac* transposon [32] on the short arm of chromosome 1 (bin 1.02-03) where the *ZmMRP4* gene maps (Figure 1). This *Ac* line was chosen to increase the frequency of mutations at the locus *lpa1* because of the reported tendency of *Ac* transposons to reinsert in close linkage to the donor site [33]. The *lpa1-1* allele used in this work was characterized by a decrease in phytic acid synthesis (from 50% to 95%) and by a proportional increase of the level of free phosphorus inside the kernel. The *lpa-1-1* mutation leads to an alteration of the ABC transporter (ATP-Binding Cassette transporter) due to an SNP (Single Nucleotide Polymorphism) mutation, alanine to valine, on the amino acid 1432 [29]. This mutation impairs the transport and the subsequent impaired accumulation of phytic acid.

This density assay allowed the isolation of 271 low density putative lpa1 seeds among the 4787 F1 seeds tested (Figure 2 and Table 1).

The selection for low density also allowed the selection of moldy seeds and seeds with other kinds of mutations impairing seed (embryo and/or endosperm) development. For this reason, as confirmation of the validity of this test for isolating lpa mutants, we checked for the free phosphate content a sub set of the 271 low density seeds and we identified 50% of seeds (20 out 40 tested) showing the HIP phenotype (Figure 2).

Furthermore, the low density assay did not allow the identification of the out of type lpa1.1 seeds coming from self-pollination events. We performed this analysis by genotyping the 27 F1 mature plants obtained from the 180 low density seeds sown in the open field and we found 26 plants heterozygous *Lpa1/lpa1* and one plant was an out of type homozygous *lpa1/lpa1,* which was discarded (data not shown). The best four F2 ears were further analyzed by Chen's assay and sown in the open field, the six best F3 ears were analyzed by Chen's assay and we selected the plant coded R 4892 that showed almost 100% of HIP seeds (Figure 3 and Table 2). After two more cycles of self-fertilization, F5 seeds were tested for the HIP phenotype, allowing the confirmation of the isolation of a new lpa1* mutant that we named lpa1-5525. Performing analysis regarding the P repartition (Total P, Free P and Phytic P) in this new mutant we registered values similar to those reported for lpa1-1 [16,36] (Figure 4b).

Molecular analysis excluded the presence of the *Ac* element in the *ZmMRP4* coding sequence due to the fact that the length of the two amplicons obtained by PCR is the same compared to the control (Figure 5). Of course, this result leaves open the possibility that the lpa1 mutant phenotype is caused by an *Ac* insertion into the promoter or in other regulative sequences, which could be very distant from the *lpa1* locus. However the stability of the lpa1-5525 progeny for what concerns the Free P content (in Figure 4a, nine out of 50 seeds were shown to be assayed for HIP) led us to suppose that we have isolated a "solid spontaneous mutation" in *ZmMRP4* locus because usually the insertion of an active *Ac* element produces "unstable alleles" [46]. For the same reason we could exclude an epigenetic origin for the lpa1 mutant here reported [47].

In conclusion, we developed a cheap, rapid and easy method to isolate lpa mutants. This method, not only allowed us to save money (fewer samples to be analyzed) and time (no need to wait for the F2 generation) but it also has a great potential based on the possibility to screen populations obtained with any methods of mutagenesis chosen (chemical, physical, transpositional). Furthermore, it is open to being adapted to the screening of low density seeds of other species than maize. Finally, we reported data about the isolation of a new lpa1 mutant, vital and thus useful in future breeding programs aimed at improving the nutritional value and decreasing the environmental problems associated with the high phytic acid content in maize seeds. Future work will be necessary to characterize the molecular lesion responsible for the lpa1 phenotype we isolated.

4. Materials and Methods

4.1. Genetic Stocks and Sampling Material

The lpa1-1 mutant stock was kindly provided by Dr. Victor Raboy, USDA ARS, Aberdeen, ID, USA.

To generate the mutagenized population, we used a line (genetic background B73/W22) carrying the *Ac* transposon on short arm of chromosome 1 (bin 1.02/03) kindly provided by Dr. Thomas Brutnell, Shandong Agricultural University, China.

4.2. Density Assay

Previous analyses [35] indicated that the lpa1 seeds were characterized by a lower density in comparison to the corresponding wild type controls. Starting from these data, we developed a non-disrupting, fast, cheap and rapid test to screen the mutagenized populations looking for the putative mutants containing the *Ac* transposon, *lpa1-1/lpa1-Ac*.

The test was performed by putting the F1 seeds of the mutagenized populations in a concentrated sucrose solution, density 1.28 g/cm^3 or 1,122g/cm^3 at the temperature of 23 °C. This density assay allowed the isolation of the lpa1 putative mutants, which, because of their low density, were able to float while the wild type seeds stayed on the bottom of the beaker.

This assay is non-disrupting, i.e., the selected seeds can be recovered from the beaker, rinsed, dried and used for further analysis, stored or sown.

4.3. Assay for Free Phosphate Content in the Seeds

The Chen assay [28] was performed with some little modifications. The seeds were ground in a mortar with a steel pestle, and 100 mg of the flour obtained was extracted for 1 h at room temperature with 1 mL of 0.4 M HCl solution. After overnight incubation in a shaker at room temperature, 100 µL of extract was used for the free phosphate assay, adding 900 µL of Chen's reagent (6 N H$_2$SO$_4$, 2.5% ammonium molybdate, 10% ascorbic acid, H$_2$O [1:1:1:2, *v/v/v/v*]) in microtiter plates. After incubation of 1 h at room temperature the blue-colored phosphomolybdate complex was observed: the intensity of the blue color is directly proportional to the free phosphate content. The free phosphate content was quantified by using a spectrophotometer (λ 650 nm) and adopting a series of calibration standards obtained from a stock solution of KH$_2$PO$_4$.

4.4. Assay for Seed Phytate Content

In order to measure the content of phytic acid we used Megazyme's kit K-PHYT 11/15 (Astori-Tecnica). Flour samples were obtained using a ball mill (Retsch MM200, Retsch GmbH Germany), grinding the seeds for 1 minute at 21 oscillations s^{-1} frequency. For each sample, in a beaker we added 20 mL of hydrochloric acid (0.66 M) to 1 g of flour that was vigorously stirred overnight at room temperature. 1 mL of extract was transferred into a 1.5 mL microfuge tube and centrifuged at 13,000 rpm for 10 min; then 0.5 mL of supernatant was transferred in a fresh microfuge tube and neutralized by the addition of 0.5 mL of sodium hydroxide solution (0.75 M). The neutralized extracts were submitted to enzymatic dephosphorylation reaction using the solutions supplied by the kit and trichloroacetic acid (50% *w/v*). The reactions were done in duplicate, to determine free phosphorus and total phosphorus. For colorimetric determination of phosphorus, 1 mL of sample extract was transferred into a 1.5 mL microfuge tube with 0.50 mL of color reagent (Ascorbic acid: 10%, Sulfuric acid: 1 M, Ammonium molybdate: 5%). The samples were mixed by vortex and incubated in a water bath at 40 °C for 1 h.

The standard phosphorus solutions were prepared as described in manufacturers' instructions, with the only modification that after preparation the standards were not treated as samples, i.e., incubated at 40 °C, but were left at room temperature for 1 hour. The phytic acid content was quantified by using a spectrophotometer (λ 655 nm). The data obtained with the Megazyme software were expressed as g phytic acid/100 g of flour that we converted into mg phytic P/g of flour.

4.5. lpa1 Locus Molecular Genotyping

A molecular analysis was performed using *ZmMRP4* sequence-specific amplification polymorphism (S-SAP) markers that allowed the identification of the *lpa1-1* versus *Lpa1* allele. The allele-specific forward primers were designed on the single nucleotide substitution polymorphism in the *ZmMRP4* 10th exon [29,36]. The *Lpa1* wild type–specific forward primer was ZmMRP430L (5′-GTACTCGATGAGGCGACAGC-3′), whereas *lpa1-1* mutation-specific forward primer was ZmMRP432L (5′-GTACTCGATGAGGCGACAGTG-3′). The reverse primer ZmMRP410R (5′-CCTCTCTATATACAGCTCGAC-3′) was used to amplify both wild type and *lpa1-1* alleles.

The reaction mixture of the *Lpa1/lpa1-1* allele-specific amplifications contained an aliquot of genomic DNA, 1 × Green Go Taq buffer (Promega, Madison, WI), 2.5 μM MgCl2, 0.2 μM each of dATP, dCTP, dGTP, and dTTP, 0.3 μM of forward ZmMRP430L/ZmMRP432L-specific primer, 0.3 μM of reverse ZmMRP410R primer and 1.25 unit of Go Taq Flexy DNA polymerase (Promega), in a final volume of 25 μL.

The reaction mix underwent an initial denaturation step at 94 °C for 2 min, 37 cycles of denaturation at 94 °C for 45 s, annealing at 62 °C for 1 min, extension at 72 °C for 1.5 min. Extension at 72 °C for 5 min was performed to complete the reaction. The *Lpa1* and *lpa1-1* amplicons were 468 bp long. The amplicons were loaded on 1% (*w/v*) agarose gels and visualized by ethidium bromide staining under ultraviolet light.

4.6. Structural Analysis of ZmMRP4 Locus in Putative New lpa 1 Mutant

ZmMRP4 locus was amplified in wild type and in putative new lpa1 mutant using an high fidelity long range DNA polymerase (Platinum Super Fi DNA Polymerase, Invitrogen) with the primers 5L (5′-TGGTGAGGGGATCAGAGACG-3′) (forward primer, position -313) and ZM 2R (5′-GAACTTCCAAAGGCAAGGGACA-3′) (reverse primer, position +3413), ZM2F (5′-GGAAAAGTGAGCTCCAAAGTTTA-3′) (forward primer, position +3218) and 51R (5′-AAGCATCAGCTTCGGGTAATGT-3′) (reverse primer, position +6460). The primers positions are referred to the ATG on the genomic sequence.

The amplicons obtained were run on 1% agarose gel and visualized by ethidium bromide staining under UV light.

Author Contributions: Conceptualization, R.P.; methodology, R.P. and M.L.; data curation, G.B., C.R. and E.C.; writing—original draft preparation, G.B. and C.R.; writing—review and editing, R.P. and M.L.; funding acquisition, R.P.

Funding: This research was funded by Regione Lombardia, BIOGESTECA project, grant number 15083/RCC.

Acknowledgments: We thank V. Raboy (USDA ARS, Aberdeen, ID, USA) and T.P. Brutnell (Shandong Agricultural University, China) for the generous gifts of lpa1-1 and *Ac* line seeds, and Davide Reginelli for his hard work in the field.

Conflicts of Interest: The authors declare no conflict of interest. The funders had no role in the design of the study; in the collection, analyses, or interpretation of data; in the writing of the manuscript, or in the decision to publish the results.

References

1. O'Dell, B.L.; De Boland, A.R.; Koirtyohann, S.R. Distribution of phytate and nutritionally important elements among the morphological components of cereal grains. *J. Agric. Food Chem.* **1972**, *20*, 718–723. [CrossRef]
2. Raboy, V.; Dickinson, D.B.; Neuffer, M.G. A survey of maize kernel mutants for variation in phytic acid. *Maydica* **1990**, *35*, 383–390.
3. Raboy, V. Accumulation and Storage of Phosphate and Minerals. In *Cellular and Molecular Biology of Plant Seed Development*; Larkins, B.A., Vasil, I.K., Eds.; Springer: Dordrecht, The Netherlands, 1997; Volume 4, pp. 441–477.
4. Raboy, V. Progress in Breeding Low Phytate Crops. *J. Nutr.* **2002**, *132*, 503–505. [CrossRef] [PubMed]
5. Laboure, A.M.; Gagnon, J.; Lescure, A.M. Purification and characterization of a phytase (myo-inositol-hexakisphosphate phosphohydrolase) accumulated in maize (Zea mays) seedlings during germination. *Biochem. J.* **1993**, *295*, 413–419. [CrossRef] [PubMed]
6. Graf, E.; Eaton, J.W. Antioxidant functions of phytic acid. *Free Radic. Biol. Med.* **1990**, *8*, 61–69. [CrossRef]
7. Doria, E.; Galleschi, L.; Calucci, L.; Pinzino, C.; Pilu, R.; Cassani, E.; Nielsen, E. Phytic acid prevents oxidative stress in seeds: Evidence from a maize (Zea mays L.) low phytic acid mutant. *J. Exp. Bot.* **2009**, *60*, 967–978. [CrossRef] [PubMed]
8. Sharpley, A.; Meyer, M. Minimizing Agricultural Nonpoint-Source Impacts: A Symposium Overview. *J. Environ. Qual.* **1994**, *23*, 1–3. [CrossRef]
9. Mendoza, C.; Viteri, F.E.; Lönnerdal, B.; Young, K.A.; Raboy, V.; Brown, K.H. Effect of genetically modified, low-phytic acid maize on absorption of iron from tortillas. *Am. J. Clin. Nutr.* **1998**, *68*, 1123–1127. [CrossRef]
10. Hambidge, K.M.; Huffer, J.W.; Raboy, V.; Grunwald, G.K.; Westcott, J.L.; Sian, L.; Miller, L.V.; Dorsch, J.A.; Krebs, N.F. Zinc absorption from low-phytate hybrids of maize and their wild-type isohybrids. *Am. J. Clin. Nutr.* **2004**, *79*, 1053–1059. [CrossRef]
11. Hambidge, K.M.; Krebs, N.F.; Westcott, J.L.; Sian, L.; Miller, L.V.; Peterson, K.L.; Raboy, V. Absorption of calcium from tortilla meals prepared from low-phytate maize. *Am. J. Clin. Nutr.* **2005**, *82*, 84–87. [CrossRef]
12. Sparvoli, F.; Cominelli, E. Seed Biofortification and phytic acid reduction: A conflict of interest for the plant? *Plants* **2015**, *4*, 728–755. [CrossRef] [PubMed]
13. Panzeri, D.; Cassani, E.; Doria, E.; Tagliabue, G.; Forti, L.; Campion, B.; Bollini, R.; Brearley, C.A.; Pilu, R.; Nielsen, E.; et al. A defective ABC transporter of the MRP family, responsible for the bean lpa1 mutation, affects the regulation of the phytic acid pathway, reduces seed myo-inositol and alters ABA sensitivity. *New Phytol.* **2011**, *191*, 70–83. [CrossRef] [PubMed]
14. Naidoo, R.; Watson, G.M.F.; Derera, J.; Tongoona, P.; Laing, M.D. Marker-assisted selection for low phytic acid (lpa1-1) with single nucleotide polymorphism marker and amplified fragment length polymorphisms for background selection in a maize backcross breeding programme. *Mol. Breed.* **2012**, *30*, 1207–1217. [CrossRef]
15. Sureshkumar, S.; Tamilkumar, P.; Senthil, N.; Nagarajan, P.; Thangavelu, A.U.; Raveendran, M.; Vellaikumar, S.; Ganesan, K.N.; Balagopal, R.; Vijayalakshmi, G.; et al. Marker assisted selection of low phytic acid trait in maize (Zea mays L.). *Hereditas* **2014**, *151*, 20–27. [CrossRef] [PubMed]
16. Raboy, V.; Gerbasi, P.F.; Young, K.A.; Stoneberg, S.D.; Pickett, S.G.; Bauman, A.T.; Murthy, P.P.N.; Sheridan, W.F.; Ertl, D.S. Origin and seed phenotype of maize low phytic acid 1-1 and low phytic acid 2-1. *Plant Physiol.* **2000**, *124*, 355–368. [CrossRef] [PubMed]
17. Pilu, R.; Panzeri, D.; Gavazzi, G.; Rasmussen, S.K.; Consonni, G.; Nielsen, E. Phenotypic, genetic and molecular characterization of a maize low phytic acid mutant (lpa241). *Theor. Appl. Genet.* **2003**, *107*, 980–987. [CrossRef] [PubMed]
18. Shi, J.; Wang, H.; Hazebroek, K.; Ertl, D.S.; Harp, T. The maize low-phytic acid 3 encodes a myo-inositol kinase that plays a role in phytic acid biosynthesis in developing seeds. *Plant J.* **2005**, *42*, 708–719. [CrossRef]
19. Larson, S.R.; Young, K.A.; Cook, A.; Blake, T.K.; Raboy, V. Linkage mapping of two mutations that reduce phytic acid content of barley grain. *Theor. Appl. Genet.* **1998**, *97*, 141–146. [CrossRef]
20. Rasmussen, S.K.; Hatzack, F. Identification of two low-phytate barley (Hordeum vulgare L.) grain mutants by TLC and genetic analysis. *Hereditas* **1998**, *129*, 107–112. [CrossRef]
21. Bregitzer, P.; Raboy, V. Effects of four independent low-phytate mutations in barley (Hordeum vulgare L.) on seed phosphorus characteristics and malting quality. *Cereal Chem.* **2006**, *83*, 460–464. [CrossRef]

22. Guttieri, M.; Bowen, D.; Dorsch, J.A.; Raboy, V.; Souza, E. Identification and characterization of a low phytic acid wheat. *Crop Sci.* **2004**, *44*, 418–424. [CrossRef]

23. Liu, Q.L.; Xu, X.H.; Ren, X.L.; Fu, H.W.; Wu, D.X.; Shu, Q.Y. Generation and characterization of low phytic acid germplasm in rice (Oryza sativa L.). *Theor. Appl. Genet.* **2007**, *114*, 803–814. [CrossRef] [PubMed]

24. Wilcox, J.R.; Premachandra, G.S.; Young, K.A.; Raboy, V. Isolation of high seed inorganic P, low-phytate soybean mutants. *Crop Sci.* **2000**, *40*, 1601–1605. [CrossRef]

25. Hitz, W.D.; Carlson, T.J.; Kerr, P.S.; Sebastian, S.A. Biochemical and molecular characterization of a mutation that confers a decreased raffinosaccharide and phytic acid phenotype on soybean seeds. *Plant Physiol.* **2002**, *128*, 650–660. [CrossRef] [PubMed]

26. Yuan, F.J.; Zhao, H.J.; Ren, X.L.; Zhu, S.L.; Fu, X.J.; Shu, Q.Y. Generation and characterization of two novel low phytate mutations in soybean (Glycine max L. Merr.). *Theor. Appl. Genet.* **2007**, *115*, 945–957. [CrossRef] [PubMed]

27. Campion, B.; Sparvoli, F.; Doria, E.; Tagliabue, G.; Galasso, I.; Fileppi, M.; Bollini, R.; Nielsen, E. Isolation and characterisation of an lpa (low phytic acid) mutant in common bean (Phaseolus vulgaris L.). *Theor. Appl. Genet.* **2009**, *118*, 1211–1221. [CrossRef] [PubMed]

28. Chen, P.S.; Toribara, T.Y.; Warner, H. Microdetermination of phosphorus. *Anal. Chem.* **1956**, *28*, 1756–1758. [CrossRef]

29. Shi, J.; Wang, H.; Schellin, K.; Li, B.; Faller, M.; Stoop, J.M.; Meeley, R.B.; Ertl, D.E.; Ranch, J.P.; Glassman, K. Embryo-specific silencing of a transporter reduces phytic acid content of maize and soybean seeds. *Nat. Biotechnol.* **2007**, *25*, 930–937. [CrossRef]

30. Swarbreck, D.; Ripoll, P.J.; Brown, D.A.; Edwards, K.J.; Theodoulou, F. Isolation and characterisation of two multidrug resistance associated protein genes from maize. *Gene* **2003**, *315*, 153–164. [CrossRef]

31. Klein, M.; Burla, B.; Martinoia, E. The multidrug resistance-associated protein (MRP/ABCC) subfamily of ATP-binding cassette transporters in plants. *FEBS Lett.* **2006**, *580*, 1112–1122. [CrossRef]

32. Maes, T.; De Keukeleire, P.; Gerats, T. Plant tagnology. *Trends Plant Sci.* **1999**, *4*, 90–96. [CrossRef]

33. Vollbrecht, E.; Duvick, J.; Schares, J.P.; Ahern, K.R.; Deewatthanawong, P.; Xu, L.; Conrad, L.J.; Kikuchi, K.; Kubinec, T.A.; Hall, B.D.; et al. Genome-wide distribution of transposed Dissociation elements in maize. *Plant Cell* **2010**, *22*, 1667–1685. [CrossRef] [PubMed]

34. Dooner, H.K.; Belachew, A. Transposition pattern of the maize element Ac from the Bz-M2(ac) allele. *Genetics* **1989**, *122*, 447–457. [PubMed]

35. Landoni, M.; Badone, F.C.; Haman, N.; Schiraldi, A.; Fessas, F.; Cesari, V.; Toschi, I.; Cremona, R.; Delogu, C.; Villa, D.; et al. Low Phytic Acid 1 mutation in maize modifies density, starch properties, cations, and fiber contents in the seed. *J. Agric. Food Chem.* **2013**, *61*, 4622–4630. [CrossRef] [PubMed]

36. Cerino Badone, F.; Amelotti, M.; Cassani, E.; Pilu, R. Study of low phytic acid1-7 (lpa1-7), a new ZmMRP4 mutation in maize. *J. Hered.* **2012**, *103*, 598–605. [CrossRef] [PubMed]

37. Honke, J.; Kozłowska, H.; Vidal-Valverde, C.; Frias, J.; Górecki, R. Changes in quantities of inositol phosphates during maturation and germination of legume seeds. *Eur. Food Res. Technol.* **1998**, *206*, 279–283. [CrossRef]

38. Loewus, F.A.; Murthy, P.P.N. Myo-Inositol metabolism in plants. *Plant Sci.* **2000**, *150*, 1–19. [CrossRef]

39. Davidsson, L.; Almgren, A.; Juillerat, M.A.; Hurrell, R.F. Manganese absorption in humans: The effect of phytic acid and ascorbic acid in soy formula. *Am. J. Clin. Nutr.* **1995**, *62*, 984–987. [CrossRef]

40. Spencer, J.D.; Allee, G.L.; Sauber, T.E. Phosphorus bioavailability and digestibility of normal and genetically modified low-phytate corn for pigs. *J. Anim. Sci.* **2000**, *78*, 675–681. [CrossRef]

41. Veum, T.L.; Ledoux, D.R.; Raboy, V.; Ertl, D.S. Low-phytic acid corn improves nutrient utilization for growing pigs. *J. Anim. Sci.* **2001**, *79*, 2873–2880. [CrossRef]

42. Bohlke, R.A.; Thaler, R.C.; Stein, H.H. Calcium, phosphorus, and amino acid digestibility in low-phytate corn, normal corn, and soybean meal by growing pigs. *J. Anim. Sci.* **2005**, *83*, 2396–2403. [CrossRef] [PubMed]

43. Chen, R.; Xue, G.; Chen, P.; Yao, B.; Yang, W.; Ma, Q.; Fan, Y.; Zhao, Z.; Tarczynski, M.C.; Shi, J. Transgenic maize plants expressing a fungal phytase gene. *Transg. Res.* **2008**, *17*, 633–643. [CrossRef] [PubMed]

44. Golovan, S.P.; Meidinger, R.G.; Ajakaiye, A.; Cottrill, M.; Wiederkehr, M.Z.; Barney, D.J.; Plante, C.; Pollard, J.W.; Fan, M.Z.; Hayes, M.A.; et al. Pigs expressing salivary phytase produce low-phosphorus manure. *Nat. Biotechnol.* **2001**, *19*, 741–745. [CrossRef] [PubMed]

45. Edwards, J.D.; Jackson, A.K.; McClung, A.M. Genetic architecture of grain chalk in rice and interactions with a low phytic acid locus. *Field Crops Res.* **2017**, *205*, 116–123. [CrossRef]

46. Moreno, M.A.; Chen, J.; Greenblatt, I.; Dellaporta, S.L. Reconstitutional mutagenesis of the maize P gene by short-range Ac transpositions. *Genetics* **1992**, *131*, 939–956. [PubMed]
47. Pilu, R.; Panzeri, D.; Cassani, E.; Cerino Badone, F.; Landoni, M.; Nielsen, E. A paramutation phenomenon is involved in the genetics of maize low phytic acid1-241 (lpa1-241) trait. *Heredity* **2009**, *102*, 236–245. [CrossRef] [PubMed]

© 2019 by the authors. Licensee MDPI, Basel, Switzerland. This article is an open access article distributed under the terms and conditions of the Creative Commons Attribution (CC BY) license (http://creativecommons.org/licenses/by/4.0/).

Article

Mutation of *Inositol 1,3,4-trisphosphate 5/6-kinase6* Impairs Plant Growth and Phytic Acid Synthesis in Rice

Meng Jiang [1,2,3], **Yang Liu** [1], **Yanhua Liu** [1], **Yuanyuan Tan** [1], **Jianzhong Huang** [1,3] **and Qingyao Shu** [1,2,*]

[1] National Key Laboratory of Rice Biology, Institute of Crop Sciences, Zhejiang University, Hangzhou 310058, China; mengjiang@zju.edu.cn (M.J.); 21616041@zju.edu.cn (Y.L.); yanhua1624@163.com (Y.L.); tanyy@zju.edu.cn (Y.T.); jzhuang@zju.edu.cn (J.H.)
[2] Hubei Collaborative Innovation Center for Grain Industry, Yangtze University, Jingzhou 434025, China
[3] Institute of Nuclear Agricultural Sciences, Zhejiang University, Hangzhou 310058, China
* Correspondence: qyshu@zju.edu.cn; Tel.: +86-571-88982859

Received: 20 March 2019; Accepted: 24 April 2019; Published: 29 April 2019

Abstract: Inositol 1,3,4-trisphosphate 5/6-kinase (ITPK) is encoded by six genes in rice (*OsITPK1-6*). A previous study had shown that nucleotide substitutions of *OsITPK6* could significantly lower the phytic acid content in rice grains. In the present study, the possibility of establishing a genome editing-based method for breeding low-phytic acid cultivars in rice was explored, in conjunction with the functional determination of OsITPK6. Four *OsITPK6* mutant lines were generated by targeted mutagenesis of the gene's first exon using the CRISPR/Cas9 method, one (*ositpk6_1*) with a 6-bp in-frame deletion, and other three with frameshift mutations (*ositpk6_2, _3,* and *_4*). The frameshift mutations severely impaired plant growth and reproduction, while the effect of *ositpk6_1* was relatively limited. The mutant lines *ositpk6_1* and *_2* had significantly lower levels (−10.1% and −32.1%) of phytic acid and higher levels (4.12- and 5.18-fold) of inorganic phosphorus compared with the wild-type (WT) line. The line *ositpk6_1* also showed less tolerance to osmotic stresses. Our research demonstrates that mutations of *OsITPK6*, while effectively reducing phytic acid biosynthesis in rice grain, could significantly impair plant growth and reproduction.

Keywords: genome editing; growth; *ositpk6*; phytic acid; rice

1. Introduction

Myo-inositol-1,2,3,4,5,6-hexakisphosphate (IP_6), also known as phytic acid (PA), is the main storage form of phosphorous (P) (65–80%) in cereal and legume seeds, accounting for ~1.5% of the dry weigh [1]. In most cereal grains, PA exists as mixed salts (phytates) in protein storage bodies and can chelate several mineral cations, including Zn^{2+}, Fe^{2+}, Ca^{2+}, and Mg^{2+} [2]. During seed germination, endogenous grain phytase is activated to degrade phytate, releasing myo-inositol, phosphorus, and bound mineral cations [3], which are utilized by the developing seedlings. The PA biosynthetic pathway is still not well defined, but a number of genes involved in its biosynthesis or transport have already been cloned in several plants. Mutations of these genes could result in low-phytic-acid (*lpa*) grains in rice [4–14] and other plants, e.g., wheat [15] and maize [3,16,17]. In rice, 12 genes have been identified that catalyze the production of intermediate inositol polyphosphates in seeds [18].

Inositol 1,3,4-trisphosphate 5/6-kinase (ITPK) plays a pivotal role in phytic acid biosynthesis, whereby the inositol triphosphate (IP_3) molecule is further phosphorylated at the 5th or 6th position [19,20]. ITPK belongs to the ATP-grasp fold proteins group [21] and is conserved from plants to humans with diverse functions. ITPK has even been found in the anaerobic protozoan *Entamoeba histolytica* [22], where its transcription is slightly induced by heat shock, demonstrating its role in the cellular response to stress [21]. The first plant ITPK, AtITPK1, was identified in *Arabidopsis* [20]. AtITPK1 is involved in photomorphogenesis possibly by interacting with the constitutive photomorphogenic (COP) signalosome under red light [23]. The kinase activity of AtITPK1 is indispensable for maintaining inorganic phosphorus (Pi) homeostasis under Pi-replete conditions, and *itpk1* mutants exhibited decreased levels of IP_6 and diphosphoinositolpentakisphosphate (IP_7). Disruption of another ITPK family enzyme, ITPK4, also caused depletion of IP_6 and IP_7 but did not display similar Pi-related phenotypes as *itpk1* [24]. AtITPK4 is an outlier to its family and does not display inositol 3,4,5,6 tetrakisphosphate 1-kinase activity; rather, it displays inositol 1,4,5,6-tetrakisphosphate and inositol 1,3,4,5-tetrakisphosphate isomerase activity [21]. AtITPK2 was required for seed coat development and lipid polyester barrier formation [25], and ABA or phosphorus deficiency could induce *AtITPK2* expression. In maize (*Zea mays* L.), ZmITPK1 exhibits multiple inositol phosphate kinase activities and is involved in phytic acid biosynthesis in developing seeds [17]. In soybean (*Glycine max* L.), GmITPK1 is a potential candidate for developing low-phytate soybean [26], and GmITPK2 may play a role as a dehydration and salinity stress regulator [27].

In rice (*Oryza sativa* L.), the *OsITPK* genes can be divided into three sub-families [18]. *OsITPK1*, *OsITPK2*, and *OsITPK3* belong to subgroup I, each with 10 exons and 9 introns; *OsITPK4* and *OsITPK5* belong to subgroup II, which has no intron; and *OsITPK6* belongs to subgroup III, with 12 exons and 11 introns. OsITPK2 is a negative regulator of osmotic stress signaling [28], and its disruption could affect the expression of some of its homologous genes, *OsITPK1* and *OsITPK4* [29]. The expression of *OsITPK4*, but not of *OsITPK1, 2, 3*, and *5* can be strongly induced by cold and heat stresses [29]. The IP_3 level was not affected by the *ositpk2* mutation, probably owing to redundant functions of other homologs [29].The expression of *OsITPK6* could also be induced by heat [29], and mutations of *OsITPK6* were already demonstrated to result in significant reduction of IP_6 in rice grains [30], i.e., mutant lines with the amino acid substitution P522L had IP_6 content about half that of the wild-type (WT) line. Among the *ositpk6* mutants, one line with a P522L amino acid substitution had agronomic performance (seed weight, germination, and seedling growth) similar to that of its WT parent, suggesting *OsITPK6* could be a desirable target of mutagenesis for breeding yield-competitive *lpa* rice [30]. Since the binding site for nucleotide or ATP is between 200 and 500 amino acids in OsITPK6, the substitution mutation (P522L) is localized outside of this binding region. The effect of the P522L mutation in OsITPK6 on IP_6 biosynthesis could be related to the interaction of the enzyme with another substrate inositol polyphosphate. On the other hand, a splicing mutant of *OsITPK6* at the 9th intron showed a more severe *lpa* phenotype: lower phytic acid content with reduced seed set [30]. Hence, it would be worthwhile to examine the function of OsITPK6 by generating more and different mutants, particularly by disruption of the ATP-binding region.

The clustered regularly interspaced short palindromic repeats (CRISPR) and CRISPR-associated protein 9 (Cas9) system, CRISPR/Cas9, is an efficient and precise genome-editing technique and has the potential to be used for crop improvement [31–33], including rice [34–36]. In the present study, we explored the possibility of establishing a genome-editing-based method for the fast breeding of yield-competitive *lpa* rice by evaluating *OsITPK6* mutants generated by CRISPR/Cas9-mediated mutagenesis. Our results showed that mutation of *OsITPK6* not only significantly reduced the accumulation of IP_6 in rice grains but also impaired plant growth and tolerance to abiotic stress.

2. Results

2.1. Mutations of OsITPK6 and Development of Homozygous Transgene-Free Mutant Lines

In total, we obtained 23 hygromycin phosphotransferase (HPT)-positive T_0 plants transformed with the CRISPR/Cas9 vector pH-itpk6. Among them, seven plants were found mutated at the target region, which represents an editing efficiency of 30.4%. T_1 plants were tested for the presence of mutations and T-DNA, and transgene-free T_1 plants with four types of mutation were identified. The mutations included a single-nucleotide (nt) insertion and three types of deletion (Figure 1A). Seeds were harvested from T_1 plants with different mutations and developed into homozygous mutant lines, which were designated *ositpk6_1, _2, _3,* and *_4*.

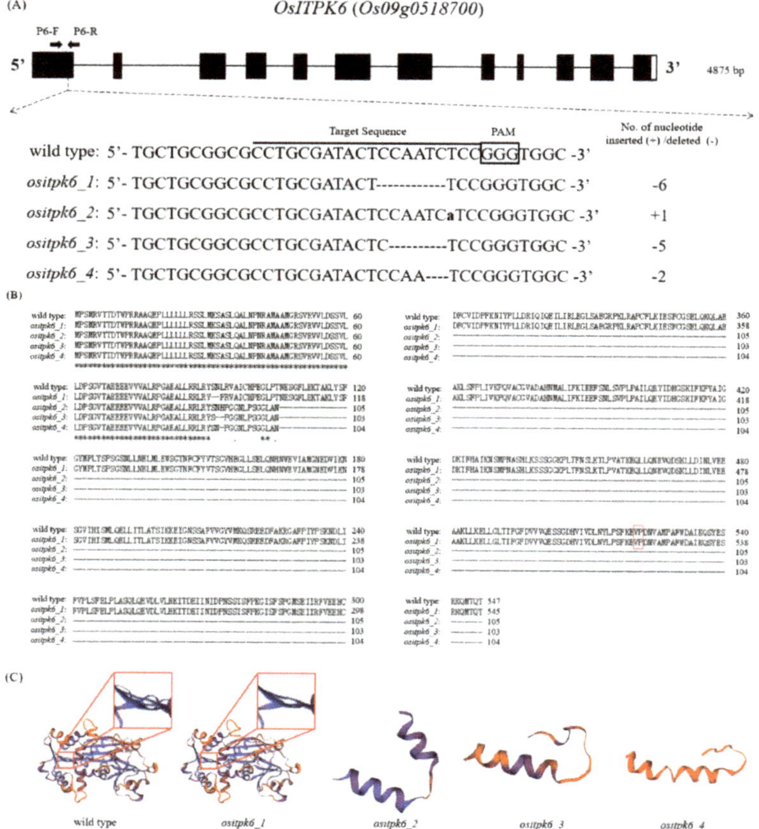

Figure 1. Schematic diagram of *OsITPK6* and sgRNA target site for CRISPR/Cas9-mediated mutagenesis of *OsITPK6*, and prediction of the related wild-type and mutant proteins. (**A**) Exons, introns, and UTRs are indicated by solid boxes, lines, and blank boxes, respectively. P6-F and P6-R are primers for mutation genotyping, and their positions are indicated by arrowheads. Mutation identified within the target site of *OsITPK6* generated through CRISPR/Cas9-mediated genome editing in rice. The PAM sequences (NGG) are boxed, and the 20-nt target sequences are underlined. Mutations are shown in lowercase letters (for insertions) or '–' (for deletions). (**B**) The amino acid sequences of mutant proteins were aligned to that of the wild-type protein using the Clustal Omega Multiple Sequence Alignment (https://www.ebi.ac.uk/Tools/msa/clustalo/). The numbers represent the total number of amino acids, and the amino acid V521 and P522 are highlighted in red box. (**C**) The three-dimensional structures of OsITPK6 and its mutants were analyzed on SWISS-MODEL (https://www.swissmodel.expasy.org/).

The *ositpk6_1* (a 6-nt in-frame deletion) mutation would result in the loss of two amino acids at positions 91 to 92 (Figure 1B). In contrast, all the other three mutations, i.e., *ositpk6_2* (a 1-nt insertion), *ositpk6_3* (a 5-nt deletion), and *ositpk6_4* (a 2-nt deletion), would generate a premature stop codon almost right after the mutation site and, hence, significantly shorten the encoded proteins (Figure 1B). Consequently, the *ositpk6_2, _3* and *_4* mutant alleles were predicted to produce proteins of only 105, 103, and 104 amino acids, respectively (Figure 1C). Analysis of ITPK6 proteins of six organisms indicated that the two amino acids missing in the *ositpk6_1* mutant were located in a highly conserved segment (Figure S1), suggesting the mutation of *ositpk6_1* could have a potential functional consequence.

2.2. Impact of ositpk6 Mutations on Plant Growth and Seed Germination

Plant growth of *ositpk6_2*, _3, and _4 was significantly impaired. First, their panicles were significantly shorter (−30.1%, −28.8%, and −29.1%, respectively) than that of the WT parental cultivar Xidao 1 (Figure 2A). Second, the mutant panicles had a high percentage of empty grains with darkened glumes (Figure 2B). Third, the height of the mutant plants was significantly reduced (−37.5%, −36.9%, and −39.1%, respectively) compared with that of their parental cultivar Xidao 1 (Figure 2C). The impact of *ositpk6_1* on plant growth and seed set was limited and not obvious (Figure 2A–C). No significant differences of tiller number per plant were observed between Xidao 1 and all four *ositpk6* mutant lines (Figure 2D). Compared with Xidao 1, the seed set and 1000-grain weight of *ositpk6_1* were also significantly decreased (−11.7% and −10.8%, respectively) (Figure 2E,F). Due to the extremely low seed set, we were not able to harvest enough seeds from *ositpk6_3* and *ositpk6_4* for the evaluation of 1000-grain weight and other characteristics. The germination of *ositpk6_1* was slower in the first three days (Figure 3A) but gradually caught up with that of the WT after five days and reached ~80% on the 7th day (Figure 3B). The germination rate of *ositpk6_2* was far lower than that of the WT (Figure 3A), being only ~20% on the 7th day (Figure 3B).

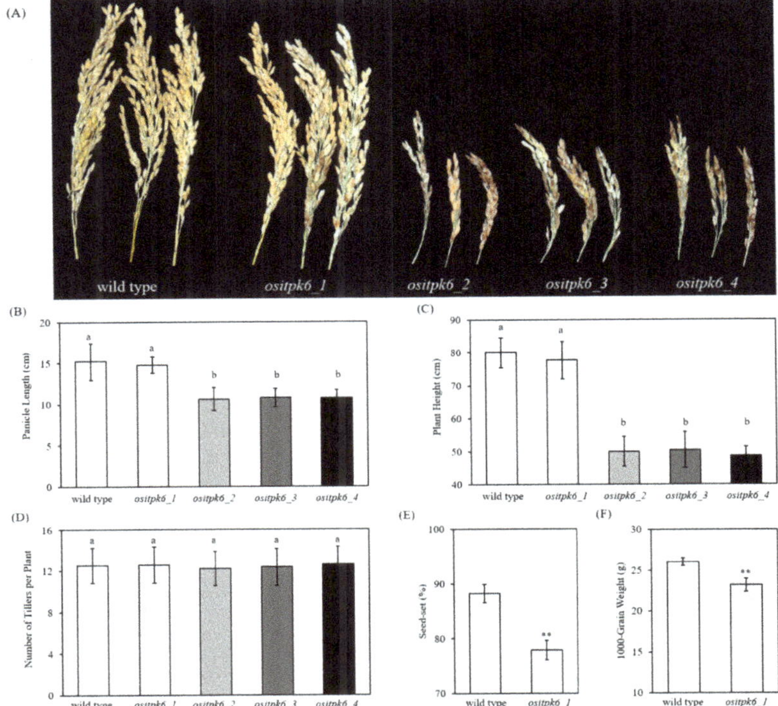

Figure 2. Agronomic traits of mutant *OsITPK6* line and wild-type plant. (**A**) Panicle phenotype of mutant *OsITPK6* and wild-type plants. (**B–D**) Twenty replicates were performed for four *OsITPK6* mutant lines and the wild-type plant. Error bars represent the standard error. The different letters show significant differences at a probability of $p < 0.05$. (**E,F**) Twenty replicates were performed for four *OsITPK6* mutant lines and the wild type. Error bars represent the standard error. Data with an asterisk(s) are significantly different from those of the wild type (* $p < 0.05$, ** $p < 0.01$).

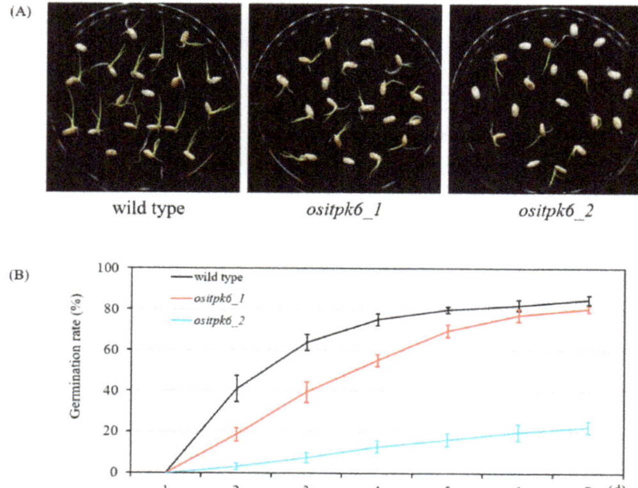

Figure 3. Germination rate of mutant seeds. (**A**) The pictures were taken on the 5th day after soaking. (**B**) The germination rate was recorded from 1 to 7 days after soaking, and three replicates were examined in each group.

2.3. Effect of ositpk6 Mutations on Inorganic Phosphorus (Pi), Phytic Acid Phosphorus (PA-P), and Total Phosphorus (TP) in Brown Rice

A colorimetric assay showed that *ositpk6_1* and *ositpk6_2* had significantly higher Pi levels than the control (Figure 4A). To quantify the mutational effect of *ositpk6_1* and *ositpk6_2*, the Pi, PA-P, and TP contents were assessed in seeds of these two mutant lines, together with Xidao 1 as the WT control. All mutant lines had significantly lower levels of PA-P and higher levels of Pi compared with the control, while TP was not significantly different from that of the control (Figure 4B–D). *ositpk6_1* and *ositpk6_2* had Pi levels of 1.13 mg g^{-1} and 1.43 mg g^{-1}, respectively, which were 4.12- and 5.18-fold higher than those of the control (0.28 mg g^{-1}), respectively (Figure 4B). Xidao 1 seeds had a PA-P content of 2.30 mg g^{-1}, which was significantly greater than those of the two mutant lines; the reduction of PA-P levels was 10.1% and 32.1% in *ositpk6_1* and *ositpk6_2*, respectively (Figure 4C). Xidao 1 seeds had a TP content of 3.9 mg g^{-1}, which was not significantly different from those of the two mutant lines (Figure 4D).

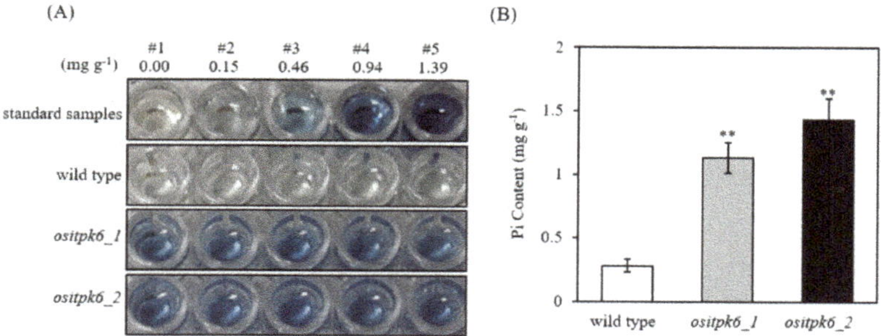

Figure 4. *Cont.*

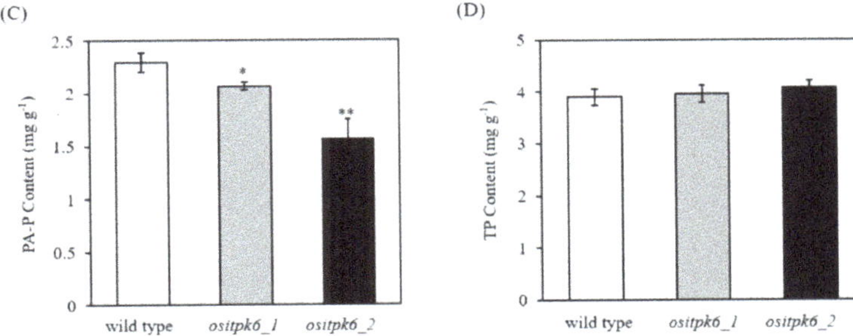

Figure 4. Inorganic P (Pi), phytic acid P (PA-P), and total phosphorus (TP) contents of the mutant *OsITPK6* lines and the wild type. (**A**) Qualitative assay of inorganic phosphorus (Pi) in mutant seeds. The concentration of the Pi standard samples is shown above. Five replicates were performed for two *OsITPK6* mutant lines and the wild type. (**B–D**) Six replicates were performed for two *OsITPK6* mutant lines and the wild type. Error bars represent the standard error. Data with an asterisk(s) are significantly different with respect to the wild-type data (* $p < 0.05$, ** $p < 0.01$).

2.4. Effect of ositpk6 Mutation on Stress Tolerance

To further test whether the mutation also had any impact on stress tolerance, we subjected the *ositpk6_1* and Xidao 1 plants to osmotic stress treatment (because of the limited number of seeds harvested and the low germination rate, *ositpk6_2*, was not further analyzed). The growth of *ositpk6_1* seedlings appeared to be inferior to that of the WT control grown either under normal or stressed conditions (Figure 5A). The shoot length of *ositpk6_1* was shorter than that of the WT with or without stress treatment (100 mM NaCl or 20 mM mannitol), while the root length of *ositpk6_1* was shorter than that of the WT only under stress (Figure 5A,B). There was no significant difference in the number of leaves and roots with or without stress treatment between *ositpk6_1* and WT (Figure 5C).

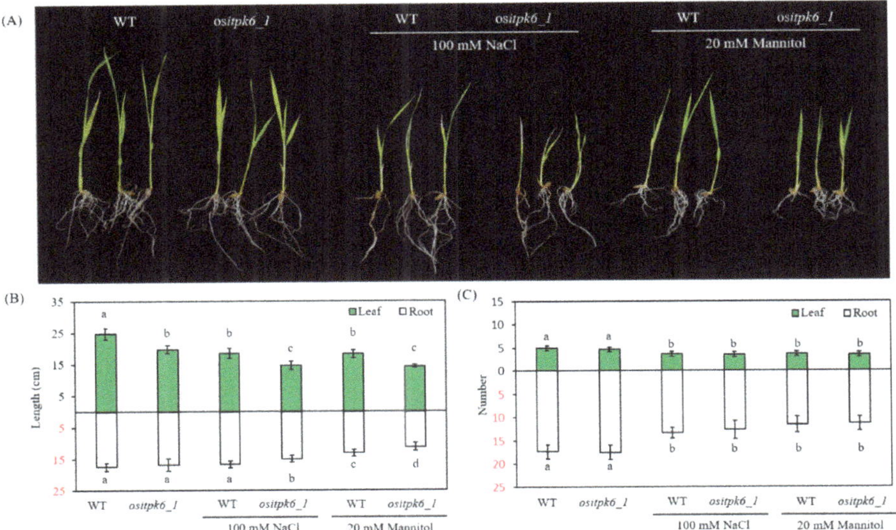

Figure 5. The phenotypes of the mutant *OsITPK6_1* and wild type under salt stress (100 mM NaCl) and drought stress (20 mM mannitol). (**A**) The picture was taken on the 7th day after treatment. (**B,C**) Six replicates were performed for the *OsITPK6* mutant line and the wild type. Error bars represent the standard error. The different letters show significant differences at a probability of $p < 0.05$.

3. Discussion

ITPK6 is a unique gene in the ITPK gene family, and knowledge of its function has so far been very limited. The identification and characterization in rice of two *itpk6* mutant lines were reported in a study, which is the only study on the function of ITPK6 in all organisms [30]. Our present study demonstrated that the knockout of *OsITPK6* could severely impair plant growth and reproduction, implying that ITPK6 may play important roles in plant growth and development, in addition to the biosynthesis of inositol polyphosphates.

First, we observed that the *OsIPTK6* knockout mutants (*ositpk6_2*, *_3*, and *_4*) generated in the present study grew poorly, e.g., their plant height was reduced to almost half of that of the WT, and their fertility was almost abolished (Figure 2). These results suggested that *ITPK6* plays an important role not only in the vegetative growth but also in the reproduction of rice. Because the reduction of phytic acid content in *ositpk6_1* grains (−32.1%) was less than in the P522L mutant line (−46%) [30], the reduction of phytic acid alone could not explain the inferior performance of *ositpk6_1*. Further studies are needed to uncover the biological basis leading to the discrepancy between our present study and that reported in reference [30] regarding the mutational effect on rice growth and reproduction.

In the present study, we identified an *ositpk6* mutant, i.e., *ositpk6_1*, with a 6-bp deletion. Though only two amino acids are expected to be removed from the derived protein, *ositpk6_1* did reduce IP$_6$ content. This may be because these two amino acids are located in a conserved region (Figure S1). We previously also observed a similar case of *lpa* rice, where a 6-bp deletion (and, hence, a deletion of two amino acids) in *OsSultr3;3* significantly reduced grain phytic acid content [14]. Although this mutation only reduced phytic acid content by less than 20%, it did exert a negative effect on seed set, grain weight, seed germination, and tolerance to abiotic stresses. This was somehow unexpected, because the P522L mutant line with a 46% reduction of phytic acid content still had normal plant growth as its WT parent [30]. Further studies are needed to fully evaluate the mutational effect by examining more *ositpk6* mutants.

The usefulness of CRISPR/Cas9-based mutagenesis for improving a particular trait is strictly dependent on the performance of the generated mutants. There are often trade-offs for mutating a gene for a specific purpose, and the overall performance of the mutated plant could be affected as a consequence of pleiotropic effects. Enlightened by the findings of Kim and Tai [30], we hoped to establish a fast and effective method for breeding yield-competitive, *lpa* rice cultivars by using genome-editing techniques. However, our results suggest that *OsITPK6* or its product plays an important role in multiple cellular processes, and simply knocking out *OsIPTK6* would impair rice growth and reproduction and, hence, would not work for our purpose.

Because [30] of the success in the production of *lpa* mutants without a significant negative impact on plant growth and seed development, it is still possible to generate *ositpk6* mutants without a significant effect on plant growth, if more appropriate vectors can be designed and more mutants are identified and assessed.

In summary, the present study demonstrates that the *OsITPK6* gene is essential for rice growth and reproduction.

4. Materials and Methods

4.1. CRISPR/Cas9 Vector Construction and Rice Transformation

To generate *OsITPK6* mutants, the 1st exon of *OsITPK6* (*Os09g0518700*) was chosen as a target (Figure 1A). The sgRNAs were designed by searching UniProt for precise positions (http://www.uniprot.org/), and CRISPR-P program (http://cbi.hzau.edu.cn/cgi-bin/CRISPR/) was used to minimize off-target effects [37]. Because of the homology of the target sequence among *OsITPK* genes, it is unlikely to cause mutations in the other five *OsITPK* homolog genes (Figure S2). DNA oligonucleotides were synthesized (Tsingke, Hangzhou, China) for the construction of a CRISPR/Cas9 vector, pH_itpk6, using the pHun4c12s as the starting plasmid, which harbors a *CYP81A6*-hpRNAi element [38] and was modified from pHun4c12 [39]. Correspondingly, the pH_itpk6 plasmid was transformed into *Agrobacterium tumefaciens* and used for rice transformation.

Rice calli were induced from mature seeds of the cultivar 'Xidao 1' (*O. sativa* L. *japonica*) and were transformed with the pH-itpk6 vector by *Agrobacterium*-mediated transformation according to reference [40]. Transgenic plantlets were regenerated from hygromycin-resistant calli and acclimatized inside a moist growth chamber (28 °C with a 12 h photoperiod) for one week before being transplanted to experimental facilities.

Ethics Approval and Consent to Participate: The experiments did not involve endangered or protected species. No specific permits were required for these locations/activities.

4.2. Mutation Detection in T_0 Plants

For detection of transgenes and mutations in regenerated T_0 plants, total genomic DNA was extracted from leaf tissues following a modified cetyltrimethylammonium bromide (CTAB) method [41]. The presence of the *HPT* gene was assessed by PCR using the primers HygR-F (5′-AGAAGAAGATGTTGGCGACCT-3′) and HygR-R (5′-GTCCTGCGGGTAAATAGCT-3′) [42]. Site-specific mutations were detected by PCR amplification using primer pairs flanking the designated target sites in *OsITPK6*, i.e., P6-F (5′-CTCGACCCATCCGGTGTTAC-3′) and P6-R (5′-AAATCGCAGGGGAGAGATCG-3′) (Figure 1A). The following generalized PCR program was used: 5 min at 94 °C, followed by 35 cycles of 30 s at 94 °C, 1 min at 60 °C, and 1 min at 72 °C, with a final extension of 10 min at 72 °C.

The PCR products were first subjected to HRM analysis for mutations according to reference [43], putative mutants were sequenced (TSINGKE, Hangzhou, China), and mutated sequences were decoded using the DSDecode program (http://skl.scau.edu.cn/dsdecode/) [44]. Mutations of a few selected plants were further confirmed by clone sequencing.

4.3. Development of Transgene-Free Mutant Lines

To obtain homozygous transgene-free mutants, T_1 seedlings were foliar-sprayed with 1000 mg/L bentazon (approximately 100 mL/m^2) at about the four-leaf stage according to a previous study [38]. At least five surviving T_1 plants from each independent T_0 plants were selected for further analysis of the presence of T-DNA and site-specific mutations. Eventually, four *OsITPK6* mutant lines (*ositpk6_1*, _2, _3, and _4) (Figure 1A) were identified; the T_2 and advanced-generation seeds were used for the experiments.

4.4. Agronomic Traits Assay

The transgene-free mutant lines and their wild-type parent control (Xidao 1) were grown at the Experimental Farm of Zhejiang Zhijiang Seed Tec. Ltd., Hangzhou, China, during the summer season, and their agronomic traits were evaluated either in fields or post-harvest. Plants were grown in randomized plots, each with 60 plants. Twenty inner plants of each plot were evaluated for each parameter.

4.5. Seed Germination Assay

A controlled germination test was performed to assess seed germination capability, each with 100 seeds and replicated for three times [45]. Seeds were soaked in water for 48 h at 30 °C and then germinated on filter paper soaked with distilled water at 30 °C in the dark for one week, and the germination percentage was recorded daily.

4.6. Seed Phosphorus Assay

Seed inorganic P (Pi) levels were assayed qualitatively according to the micro-determination method developed in reference [46] with modifications. The qualitative assay was used for the identification of the high-inorganic-P (HIP) phenotype. The seeds were transferred to 96-well plates and extracted in 0.4 M HCl solution (10 μL per mg sample) overnight at 4 °C. Aliquots of 10 μL of supernatant per each sample were used for Pi level determination, according to reference [4] with slight modifications [6]. The development of a blue color implied increased level of Pi (HIP), while colorless samples typified the WT levels of parent varieties (Figure S1).

Seed Pi levels were also quantitatively determined according to reference [47] in triplicates, as follows. Brown rice grains were ground into rice flour, and ~400 mg rice flour per sample were extracted in 12.5% (*w/v*) TCA (trichloroacetic acid containing 25 mM MgCl$_2$) by gentle shaking overnight at 4 °C. After centrifugation at 15,000× *g* for 10 min, the supernatants were used for Pi assay according to reference [2].

The PA content was determined for brown rice flour according to reference [48] by using a commercial assay kit (Megazyme, Irland) in triplicate. Briefly, ~1 g of brown rice grains was mixed with 20 mL of hydrochloric acid (0.66 M), stirred vigorously for 3 h at room temperature, and then centrifuged at 13,000 rpm for 10 min. Then, 0.5 mL of the supernatant was immediately neutralized by the addition of 0.5 mL of sodium hydroxide solution (0.75 M). The neutralized sample extract was subjected to an enzymatic dephosphorylation reaction procedure. The absorbances at 655 nm was determined using an ultraviolet spectrophotometer.

Total seed phosphorus and mineral elements were determined according to reference [49] in triplicate. Briefly, brown rice samples were digested in a microwave digestion system (Mars6, USA) at 160 °C for 40 min, using ~200 mg of sample in 7 mL of HNO$_3$. The digested solution was concentrated at 140 °C for 2 h until less than 1 mL of solution was left and then brought to 30 mL with ultrapure water [50]. Total seed phosphorus and mineral elements were analyzed by Inductively Coupled Plasma–Mass Spectrometry (ICP–MS) (PerkinElmer, MA, USA).

4.7. Stress Treatment

For stress treatment, 14-day-old seedlings were planted in 1× Murashige and Skoog (MS) liquid medium [51], supplemented with 100 mM NaCl or 20 mM mannitol and grown for 7 days before sampling at midday.

4.8. Statistical Analysis

All statistical analyses were performed using the student's *t*-test. The experimental data are presented with the mean standard errors (SE) based on three to six replications. The means were compared by ANOVA, and the significance of the differences between group means were calculated by the Bonferroni Post-tests.

Supplementary Materials: The following are available online at http://www.mdpi.com/2223-7747/8/5/114/s1. Figure S1: Multiple sequence alignment of ITPK6s. All the data are from RAP-DB (http://rapdb.dna.affrc.go.jp/) and Gramene (http://www.gramene.org/). Analysis of ITPK6 sequences from six organisms using BioEdit software. Identical amino acid residues are boxed in same color. Short lines indicate gaps introduced during alignment. The mutant site of OsITPK6 is marked by a redbox. Figure S2. Multiple sequence alignment of *OsITPKs*. All data are from RAP-DB (http://rapdb.dna.affrc.go.jp/) and Gramene (http://www.gramene.org/). Analysis of *OsITPK1*, *OsITPK2*, *OsITPK3*, *OsITPK4*, *OsITPK5*, and *OsITPK6* sequences using the BioEdit software. Identical DNA bases are in the same color. Dotted lines indicate gaps introduced during alignment. The target site of *OsITPK6* is marked by a red box.

Author Contributions: Q.S. and M.J. planned and designed the research. M.J., Y.L. (Yang Liu), Y.L. (Yanhua Liu), and Y.T. performed the laboratory experiments. M.J., Y.L. (Yang Liu), and Q.S. analyzed the data together. M.J. finished the first draft, which J.H. and Q.S. edited and converted into the final draft. All authors reviewed and approved the final manuscript.

Funding: This work was supported by the Zhejiang Provincial S&T Project on Breeding of Agricultural (Food) Crops (Grant No. 2016C02050-2).

Acknowledgments: We thank workers in the farms of Zhejiang Zhijiang Seed Tec. Ltd. for taking care of the paddy fields. We also appreciate the assistance in total seed phosphorus and mineral elements measurement of Wu Zhongchang and Xu Jiming from the College of Life Sciences, Zhejiang University, Hangzhou, China.

Conflicts of Interest: The authors declare no conflict of interest.

References

1. Lott, J.N.A.; Ockenden, I.; Raboy, V.; Batten, G.D. Phytic acid and phosphorus in crops seeds and fruits: A global estimate. *Seed Sci. Res.* **2000**, *10*, 11–33.
2. Raboy, V.; Young, K.A.; Dorsch, J.A.; Cook, A. Genetics and breeding of seed phosphorus and phytic acid. *J. Plant Physiol.* **2001**, *158*, 489–497. [CrossRef]
3. Raboy, V.; Gerbasi, P.F.; Young, K.A.; Stoneberg, S.D.; Pickett, S.G.; Bauman, A.T.; Murthy, P.P.N.; Sheridan, W.F.; Ertlet, D.S. Origin and seed phenotype of maize *low phytic acid 1-1* and *low phytic acid 2-1*. *Plant Physiol.* **2000**, *124*, 355–368. [CrossRef] [PubMed]
4. Larson, S.R.; Rutger, J.N.; Young, K.A.; Raboy, V. Isolation and genetic mapping of a non-lethal rice (*Oryza sativa* L.) *low phytic acid 1* mutation. *Crop Sci.* **2000**, *40*, 1397–1405. [CrossRef]
5. Kim, S.I.; Andaya, C.B.; Newman, J.W.; Goyal, S.S.; Tai, T.H. Isolation and characterization of a low phytic acid rice mutant reveals a mutation in the rice orthologue of maize MIK. *Theor. Appl. Genet.* **2008**, *117*, 1291–1301. [CrossRef] [PubMed]
6. Liu, Q.L.; Xu, X.H.; Ren, X.L.; Fu, H.W.; Wu, D.X.; Shu, Q.Y. Generation and characterization of low phytic acid germplasm in rice (*Oryza sativa* L.). *Theor. Appl. Genet.* **2007**, *114*, 803–814. [CrossRef]
7. Zhao, H.J.; Liu, Q.L.; Fu, H.W.; Xu, X.H.; Wu, D.X.; Shu, Q.Y. Effect of non-lethal low phytic acid mutations on grain yield and seed viability in rice. *Field Crop. Res.* **2008**, *108*, 206–211. [CrossRef]
8. Raboy, V. Approaches and challenges to engineering seed phytate and total phosphorus. *Plant Sci.* **2009**, *177*, 281–296. [CrossRef]
9. Xu, X.H.; Zhao, H.J.; Liu, Q.L.; Frank, T.; Engel, K.H.; An, G.; Shu, Q.Y. Mutations of the multi-drug resistance-associated protein ABC transporter gene 5 result in reduction of phytic acid in rice seeds. *Theor. Appl. Genet.* **2009**, *119*, 75–83. [CrossRef]
10. Ali, N.; Paul, S.; Gayen, D.; Sarkar, S.N.; Datta, K.; Datta, S.K. Development of low phytate rice by RNAi mediated seed specific seed specific silencing of inositol 1,3,4,5,6-pentakisphosphate 2-kinase gene (*IPK1*). *PLoS ONE* **2013**, *8*, e68161. [CrossRef]
11. Zhao, H.J.; Cui, H.R.; Xu, X.H.; Tan, Y.Y.; Fu, J.J.; Liu, G.Z.; Poirier, Y.; Shu, Q.Y. Characterization of *OsMIK* in a rice mutant with reduced phytate content reveals an insertion of a rearranged retrotransposon. *Theor. Appl. Genet.* **2013**, *126*, 3009–3020. [CrossRef]
12. Li, W.X.; Huang, J.Z.; Zhao, H.J.; Tan, Y.Y.; Cui, H.R.; Poirier, Y.; Shu, Q.Y. Production of low phytic acid rice by hairpin RNA- and artificial microRNA-mediated silencing of *OsMIK* in seeds. *Plant Cell Tiss. Org.* **2014**, *119*, 15–25. [CrossRef]
13. Li, W.X.; Zhao, H.J.; Pang, W.Q.; Cui, H.R.; Poirier, Y.; Shu, Q.Y. Seed-specific silencing of OsMRP5 reduces seed phytic acid and weight in rice. *Transgenic Res.* **2014**, *23*, 585–599. [CrossRef] [PubMed]
14. Zhao, H.J.; Frank, T.; Tan, Y.Y.; Zhou, C.G.; Jabnoune, M.; Arpat, A.B.; Cui, H.R.; Huang, J.Z.; He, Z.H.; Poirier, Y.; et al. Disruption of *OsSULTR3;3* reduces phytate and phosphorus concentrations and alters the metabolite profile in rice grains. *New Phytol.* **2016**, *211*, 926–939. [CrossRef] [PubMed]
15. Guttieri, M.; Bowen, D.; Dorsch, J.A.; Raboy, V.; Souza, E. Identification and characterization of a low phytic acid wheat. *Crop Sci.* **2014**, *44*, 418–424. [CrossRef]
16. Pilu, R.; Panzeri, D.; Gavazzi, G.; Rasmussen, S.K.; Consonni, G.; Nielsen, E. Phenotypic, genetic and molecular characterization of a maize low phytic acid mutant (*lpa 241*). *Theor. Appl. Genet.* **2003**, *107*, 980–987. [CrossRef] [PubMed]
17. Shi, J.; Wang, H.; Hazebroek, J.; Ertl, D.S.; Harp, T. The maize *low-phytic acid 3* encodes a myo-inositol kinase that plays a role in phytic acid biosynthesis in developing seeds. *Plant J.* **2005**, *42*, 408–419. [CrossRef] [PubMed]
18. Suzuki, M.; Tanaka, K.; Kuwano, M.; Yoshida, K.T. Expression pattern of inositol phosphate-related enzymes in rice (*Oryza sativa* L.): Implications for the phytic acid biosynthetic pathway. *Gene* **2007**, *405*, 55–64. [CrossRef] [PubMed]
19. Takazawa, K.; Perret, J.; Dumont, J.E.; Erneux, C. Molecular cloning and expression of a new putative inositol 1, 4, 5-trisphosphate 3-kinase isoenzyme. *Biochem. J.* **1991**, *278*, 883–886. [CrossRef]

20. Wilson, M.P.; Majerus, P.W. Characterization of a cDNA encoding *Arabidopsis thaliana inositol 1, 3, 4-trisphosphate 5/6-kinase*. *Biochem. Biophys. Res. Commun.* **1997**, *232*, 678–681. [CrossRef]

21. Sweetman, D.; Stavridou, I.; Johnson, S.; Green, P.; Caddick, S.E.; Brearley, C.A. *Arabidopsis thaliana inositol 1, 3, 4-trisphosphate 5/6-kinase 4 (AtITPK4)* is an outlier to a family of ATP-grasp fold proteins from *Arabidopsis*. *FEBS Lett.* **2007**, *581*, 4165–4171. [CrossRef]

22. Field, J.; Wilson, M.P.; Mai, Z.; Majerus, P.W.; Samuelson, J. An *Entamoeba histolytica* inositol 1, 3, 4-trisphosphate 5/6-kinase has a novel 3-kinase activity. *Mol. Biochem. Parasitol.* **2000**, *108*, 119–123. [CrossRef]

23. Qin, Z.X.; Chen, Q.J.; Tong, Z.; Wang, X.C. The *Arabidopsis* inositol 1, 3, 4-trisphosphate 5/6 kinase, AtItpk-1, is involved in plant photomorphogenesis under red light conditions, possibly via interaction with COP9 signalosome. *Plant Physiol. Biochem.* **2005**, *43*, 947–954. [CrossRef]

24. Kuo, H.F.; Hsu, Y.Y.; Lin, W.C.; Chen, K.Y.; Munnik, T.; Brearley, C.A.; Chiou, T.J. Arabidopsis inositol phosphate kinases ipk1 and itpk1 constitute a metabolic pathway in maintaining phosphate homeostasis. *Plant J.* **2018**, *95*. [CrossRef] [PubMed]

25. Tang, Y.; Tan, S.; Xue, H. *Arabidopsis inositol 1,3,4-trisphosphate 5/6 kinase 2* is required for seed coat development. *ActaBiochim. Biophys. Sin.* **2013**, *45*, 549–560. [CrossRef]

26. Krishnan, V.; Jain, P.; Vinutha, T.; Hada, A.; Manickavasagam, M.; Ganapathi, A.; Raj, D.R.; Archana, S. Molecular modeling and 'in-silico' characterization of '*Glycine max*' inositol (1, 3, 4) tris 5/6 kinase-1(*Gmitpk1*)—A potential candidate gene for developing low phytate transgenics. *Plant Omics*. **2015**, *8*, 381–391.

27. Marathe, A.; Krishnan, V.; Vinutha, T.; Dahuja, A.; Jolly, M.; Sachdev, A. Exploring the role of inositol 1, 3, 4-trisphosphate 5/6 kinase-2 (Gmitpk2) as a dehydration and salinity stress regulator in *Glycine max* (L.) merr through heterologous expression in *E.coli*. *Plant Physiol. Biochem.* **2017**, *8*, 50. [CrossRef] [PubMed]

28. Niu, X.; Chen, Q.; Wang, X. *Ositl1* gene encoding an inositol 1,3,4-trisphosphate 5/6-kinase is a negative regulator of osmotic stress signaling. *Biotechnol. Lett.* **2008**, *30*, 1687–1692. [CrossRef] [PubMed]

29. Du, H.; Liu, L.H.; You, L.; Yang, M.; He, Y.B.; Li, X.H.; Xiong, L.Z. Characterization of an inositol 1,3,4-trisphosphate 5/6-kinase gene that is essential for drought and salt stress responses in rice. *Plant Mol. Biol.* **2011**, *77*, 547–563. [CrossRef]

30. Kim, S.I.; Tai, T.H. Identification of novel rice low phytic acid mutations via TILLING by sequencing. *Mol. Breed.* **2014**, *34*, 1717–1729. [CrossRef]

31. Li, J.; Zhang, Y.; Chen, K.; Liang, Z. Targeted genome modification of crop plants using a CRISPR/CAS system. *Nat. Biotechnol.* **2013**, *31*, 686–688.

32. Cao, H.X.; Wang, W.; Le, H.T.T.; Vu, G.T.H. The power of CRISPR/Cas9-induced genome editing to speed up plant breeding. *Int. J. Genom.* **2016**, *10*, 5078796. [CrossRef] [PubMed]

33. Kumlehn, J.; Pietralla, J.; Hensel, G.; Pacher, M.; Puchta, H. The CRISPR/Cas revolution continues: From efficient gene editing for crop breeding to plant synthetic biology. *J. Integr. Plant Biol.* **2018**, *60*, 12. [CrossRef]

34. Jung, C.; Capistrano-Gossmann, G.; Braatz, J.; Sashidhar, N.; Melzer, S. Recent developments in genome editing and applications in plant breeding. *Plant Breed.* **2017**, *137*. [CrossRef]

35. Lu, H.P.; Luo, T.; Fu, H.W.; Wang, L.; Tan, Y.Y.; Huang, J.Z.; Wang, Q.; Ye, G.Y.; Gatehouse, A.M.R.; Lou, Y.G.; et al. Resistance of rice to insect pests mediated by suppression of serotonin biosynthesis. *Nat. Plants* **2018**, *4*, 338–344. [CrossRef]

36. Liu, S.M.; Jiang, J.; Liu, Y.; Meng, J.; Xu, S.L.; Tan, Y.Y.; Li, Y.F.; Shu, Q.Y.; Huang, J.Z. Characterization and Evaluation of *OsLCT1* and *OsNramp5* Mutants Generated Through CRISPR/Cas9-Mediated Mutagenesis for Breeding Low Cd Rice. *Rice Sci.* **2019**, *26*, 88–97.

37. Lei, Y.; Lu, L.; Liu, H.Y.; Li, S.; Xing, F.; Chen, L.L. CRISPR-P: A web tool for synthetic single-guide RNA design of CRISPR-system in plants. *Mol. Plant* **2014**, *7*, 1494–1496. [CrossRef] [PubMed]

38. Lu, H.P.; Liu, S.M.; Xu, S.L.; Chen, W.Y.; Zhou, X.; Tan, Y.Y.; Huang, J.Z.; Shu, Q.Y. CRISPR-S: An active interference element for a rapid and inexpensive selection of genome-edited, transgene-free rice plants. *Plant Biotechnol. J.* **2017**, *15*, 1371–1373. [CrossRef]

39. Xu, R.F.; Li, H.; Qin, R.Y.; Wang, L.; Li, L.; Wei, P.C.; Yang, J.B. Gene targeting using the Agrobacterium tumefaciens-mediated CRISPR-Cas system in rice. *Rice* **2014**, *7*, 5. [CrossRef] [PubMed]

40. Li, W.X.; Wu, S.L.; Liu, Y.H.; Jin, G.L.; Zhao, H.J.; Fan, L.J.; Shu, Q.Y. Genome-wide profiling of genetic variation in Agrobacterium-transformed rice plants. *J. Zhejiang Univ. Sci. B* **2016**, *17*, 992–996. [CrossRef] [PubMed]

41. Zhang, H.L.; Huang, J.Z.; Chen, X.Y.; Tan, Y.Y.; Shu, Q.Y. Competitive amplification of differentially melting amplicons facilitates efficient genotyping of photoperiod-and temperature-sensitive genic male sterility in rice. *Mol. Breed.* **2014**, *34*, 1765–1776. [CrossRef]

42. Li, W.L.; Xu, B.B.; Song, Q.J.; Liu, X.M.; Xu, J.M.; Brookes, P.C. The identification of 'hotspots' of heavy metal pollution in soil-rice systems at a regional scale in eastern China. *Sci. Total Environ.* **2014**, *472*, 407–420. [CrossRef]

43. Li, S.; Liu, S.M.; Liu, Y.H.; Lu, H.P.; Tan, Y.Y.; Huang, J.Z.; Wei, P.C.; Shu, Q.Y. HRM-facilitated rapid identification and genotyping of mutations induced by CRISPR/Cas9 mutagenesis in rice. *Crop Breed. Appl. Biotechnol.* **2018**, *18*, 184–191. [CrossRef]

44. Liu, W.Z.; Xie, X.R.; Ma, X.L.; Li, J.; Chen, J.H.; Liu, Y.G. DSDecode: Aweb-based tool for decoding of sequencing chromatograms for genotyping of targeted mutations. *Mol. Plant* **2015**, *8*, 1431–1433. [CrossRef]

45. Campion, B.; Sparvoli, F.; Doria, E.; Tagliabue, G.; Galasso, I.; Fileppi, M.; Bollini, R.; Nielsen, E. Isolation and characterization of an *lpa* (*low phytic acid*) mutant in common bean (*Phaseolus vulgaris* L.). *Theor. Appl. Genet.* **2009**, *118*, 1211–1221. [CrossRef]

46. Chen, P.S.; Toribara, T.Y.; Warner, H. Micro determination of phosphorous. *Anal. Chem.* **1956**, *28*, 1756–1758. [CrossRef]

47. Wilcox, J.R.; Premachandra, G.S.; Young, K.A.; Raboy, V. Isolation of high seed inorganic P, low-phytate soybean mutants. *Crop Sci.* **2000**, *40*, 1601–1605. [CrossRef]

48. McKie, V.A.; McCleary, B.V.A. Novel and rapid colorimetric method for measuring total phosphorus and phytic acid in foods and animal feeds. *J. AOAC Int.* **2016**, *99*, 738–743. [CrossRef]

49. Hu, L.F.; McBride, M.B.; Cheng, H.; Wu, J.J.; Shi, J.C.; Xu, J.M.; Wu, L.S. Root-induced changes to cadmium speciation in the rhizosphere of two rice (*Oryza sativa* L.) genotypes. *Environ. Res.* **2011**, *111*, 356–361. [CrossRef]

50. Meng, J.; Zhong, L.B.; Wang, L.; Liu, X.M.; Tang, C.X.; Chen, H.L.; Xu, J.M. Contrasting effects of alkaline amendments on the bioavailability and uptake of Cd in rice plants in a Cd-contaminated acid paddy soil. *Environ. Sci. Pollut. R.* **2018**, *25*, 8827–8835. [CrossRef]

51. Murashige, T.; Skoog, F. A revised medium for rapid growth and bio assays with tobacco tissue cultures. *Physiol. Plantarum* **1962**, *15*, 473–497. [CrossRef]

© 2019 by the authors. Licensee MDPI, Basel, Switzerland. This article is an open access article distributed under the terms and conditions of the Creative Commons Attribution (CC BY) license (http://creativecommons.org/licenses/by/4.0/).

Article

Genotypic Differences in the Effect of P Fertilization on Phytic Acid Content in Rice Grain

Ayaka Fukushima [1], Ishara Perera [2], Koki Hosoya [3], Tatsuki Akabane [3] and Naoki Hirotsu [1,3,*]

[1] Graduate School of Life Sciences, Toyo University, 1-1-1 Izumino, Itakura-machi, Oura-gun, Gunma 374-0193, Japan; s39101900193@toyo.jp

[2] Grain Legumes and Oil Crops Research and Development Centre, Department of Agriculture, Angunakolapelessa 82220, Sri Lanka; isharauip@gmail.com

[3] Faculty of Life Sciences, Toyo University, 1-1-1 Izumino, Itakura-machi, Oura-gun, Gunma 374-0193, Japan; s19101601054@toyo.jp (K.H.); s19101600929@toyo.jp (T.A.)

* Correspondence: hirotsu@toyo.jp; Tel.: +81-276-82-9027

Received: 11 December 2019; Accepted: 21 January 2020; Published: 23 January 2020

Abstract: Phytic acid (PA) prevents the absorption of minerals in the human intestine, and it is regarded as an antinutrient. Low PA rice is beneficial because of its higher Zn bioavailability and it is suggested that the gene expression level of *myo*-inositol 3-phosphate synthase 1 (*INO1*) in developing grain is a key factor to explain the genotypic difference in PA accumulation among natural variants of rice. P fertilization is also considered to affect the PA content, but it is not clear how it affects *INO1* gene expression and the PA content in different genotypes. Here, we investigated the effect of P fertilization on the PA content in two contrasting rice genotypes, with low and high PA accumulation, respectively. Based on the results of the analysis of the PA content, inorganic P content, *INO1* gene expression, and xylem sap inorganic P content, we concluded that the effect of P fertilization on PA accumulation in grain differed with the genotype, and it was regulated by multiple mechanisms.

Keywords: inorganic P; P fertilizer; phytic acid; rice

---

1. Introduction

Many people living in developing countries have a higher risk of malnutrition due to Zn deficiency, as they mainly take micronutrients from cereals, such as legumes, wheat, and rice. Atmospheric carbon dioxide (CO_2) concentration is increasing and expected to reach 700 ppm by the end of this century [1]. Elevated CO_2 (e[CO_2]) causes a reduction in the mineral content in staple crops [2]. In the future, Zn deficiency is expected to expand globally, especially in developing countries, due to e[CO_2]. Phytic acid (*myo*-inositol-1,2,3,4,5,6-hexakisphate; PA) is the storage form of P in cereal grains; it accounts for 75% of total P in grains [3]. Phytic acid is also known to chelate with minerals, such as Fe, Mg, Ca, K, and Zn. It prevents the absorption of minerals in the human intestine and is regarded as an antinutrient. Research conducted using a suckling rat pup model showed a negative correlation between dietary PA and Zn absorption from the grains of maize, rice, and barley [4]. In addition, PA inhibits enzymes needed for protein degradation and disturbs proteolysis in the stomach and small intestine [5]. Ruminants, such as cows, secrete phytase, an enzyme digesting PA, but humans lack phytase and therefore cannot digest PA. Thus, reducing the PA content in cereals is essential to overcome Zn deficiency.

The regulation of PA content has to take into account the P status in plants, P absorption from the roots, and remobilization from plant organs. P transporters are known to be involved in the uptake of inorganic P (Pi) from the root and transport to plant organs. So far, 13 Pi transporters belonging to the PHT1 family have been identified in rice [6,7]. OsPT8 known as a high affinity Pi transporter is essential for Pi translocation from vegetative organs into rice gain [8]. SULTR-like phosphorus distribution transporter (SPDT) controls the allocation of P to the grains [9]. These findings suggest that it might be possible to control the PA content by manipulating P transporter.

To identify the biosynthetic pathway of PA and reduce the PA content in the grain, low phytic acid (*lpa*) mutants of wheat [10], maize [11–13], soybean [14,15], barley [16–18], and rice [19] have been used. These mutants have disrupted PA biosynthesis genes and exhibit low PA accumulation in the grains; however, in most cases, these mutants showed a significant reduction in germination and yield [20]. On the contrary, mutants that repress *INO1* with an 18-kDa oleosin promoter showed an approximately 70% reduction in the PA content, with no negative effects on plant growth [21]. Similarly, no undesirable yield reduction was observed in *lpa* mutants of barley [22] and soybean [23]. Developing stable *lpa* mutants without yield loss is indispensable to overcome malnutrition.

P is one of the most essential elements for plant growth. P fertilizer is indispensable in practical agriculture and therefore P application cannot be reduced [24]. The applied P is absorbed from the root and remobilized to the shoot, and then PA is synthesized in developing seeds using the transported P [25]. It has been reported that the PA content is affected by the amount of supplied P in various crops [26–28], including rice [29]. In a low-phytate soybean line, which derived from a cross of the normal-phytate Japanese cv. Tanbakuro and the low-phytate line CX1834, no negative effects were reported on plant growth and yield, leaf photosynthesis, and nitrogen fixation by different levels of P fertilizer application [30]. These results suggest that the *lpa* mutant can exhibit a low PA phenotype without any loss of growth-related performances at various levels of P fertilizer application. Thus, information on the effect of P fertilizer on seed PA content and yield in the *lpa* mutant is accumulating; however, it is not clear how P fertilizer affects the PA content in natural genotypic variants.

Previously, we examined variations in the PA content in 69 accessions of the World Rice Core Collection (WRC) and identified WRC 5 and WRC 6 as cultivars with the lowest and highest PA content in the collection [31]. We then compared WRC 5 and WRC 6 to identify the molecular determinant of the natural variation in the PA content in rice. The results suggested the gene expression level of *myo*-inositol 3-phosphate synthase 1 (*INO1*) was the genetic basis explaining the natural variation in PA accumulation in rice [32]. Interestingly, DNA sequences of the coding and promoter region (1000 bp) of the *INO1* gene were identical between WRC 5 and WRC 6. This suggests there are different regulation mechanisms of PA content besides DNA mutation of the biosynthesis gene. To elucidate these regulation mechanisms, WRC 5 and WRC 6 will be useful cultivars. In this study, we evaluated the effect of P fertilization on the PA content in WRC 5 and WRC 6 to clarify how the PA content is regulated under a different P status in natural variants of rice.

2. Results

2.1. Differences in Yield-Related Traits

2.1.1. Effect of P Fertilization in Low and High PA Rice

Yield-related traits of WRC 5 and WRC 6 are shown in Table 1. There were no significant differences in the panicle number and panicle weight between low and high PA plants. The panicle length of WRC 6 was significantly shorter than that of WRC 5. The panicle weight of WRC 5 was slightly higher than that of WRC 6. No significant difference was observed in the total yield per plant.

Table 1. Yield-related traits in WRC 5 and WRC 6. Data are shown as the mean ± SD of 5–10 replicates. Data with asterisks indicate a significant difference between WRC 5 and WRC 6 (Student's *t*-test, ***$p < 0.001$).

WRC No.	Panicle Number	Panicle Length (cm)	Panicle Weight (g)	1000 Seed Weight (g)	Total Yield per Plant (g)
WRC 5	16.6 ± 1.1	24.7 ± 2.0	2.8 ± 0.5	20.2 ± 0.7	35.3 ± 4.3
WRC 6	15.2 ± 1.1	20.3 ± 0.7	2.5 ± 0.3	23.4 ± 0.6	36.8 ± 2.2
	ns	***	ns	***	ns

2.1.2. Effect of the Amount of P Fertilizer on Initial Growth and Yield

To investigate the effect of P fertilizer on yield-related traits, the panicle weight and number (Table 2) and initial growth at the seedling stage (Table 3) were measured in WRC 5 and WRC 6. No significant differences were observed in WRC 5 and WRC 6.

Table 2. Panicle number and total yield per plant of WRC 5 and WRC 6. Data are shown as the mean ± SD of four replicates. Different letters indicate statistical differences (Tukey's HSD test, $p < 0.05$).

Cultivar	Treatment	Panicle Number	Total Yield per Plant (g)
WRC 5	Control	8.8 ± 0.5 [ab]	19.9 ± 3.6 [ab]
	P1	6.3 ± 1.0 [b]	13.3 ± 1.7 [b]
	P4	10.0 ± 1.4 [a]	20.5 ± 4.5 [a]
WRC 6	Control	9.0 ± 2.7 [ab]	14.6 ± 2.6 [ab]
	P1	9.0 ± 0.8 [ab]	16.1 ± 2.5 [ab]
	P4	10.0 ± 1.4 [a]	17.9 ± 3.4 [ab]

Table 3. Initial growth of WRC 5 and WRC 6 at the seedling stage. Data are shown as the mean ± SD of four to five replicates. ns indicates that there was no significant difference between WRC 5 and WRC 6 (Student's *t*-test).

	Shoot		Root	
	Weight (g)	Length (mm)	Weight (g)	Length (mm)
WRC 5	0.76 ± 0.087	333 ± 39	0.52 ± 0.15	151 ± 23
WRC 6	0.73 ± 0.033	306 ± 24	0.62 ± 0.037	171 ± 17
	ns	ns	ns	ns

2.2. Effects of the Time of P Fertilizer Application

WRC 5 showed a lower PA content than WRC 6 under the control condition (Figure 1). When P fertilizer was applied at the seedling stage, WRC 5 presented increased PA content compared with that of the control, whereas WRC 6 presented no increase. When P fertilizer was applied in the heading stage, there was no significant difference in the PA content compared with that when P fertilizer was applied in the seedling stage.

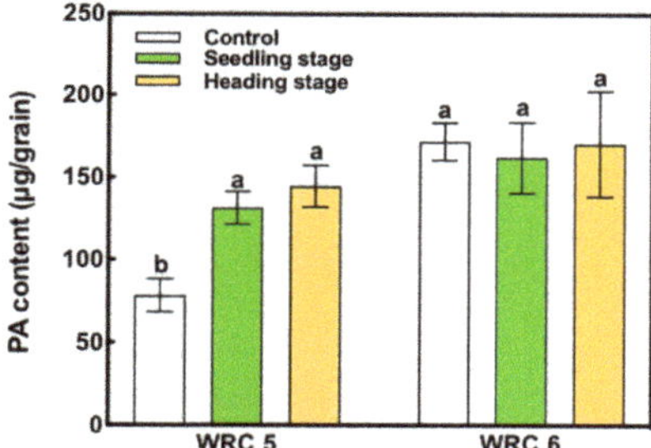

Figure 1. Effect of the time of P fertilizer application on the PA content in WRC 5 and WRC 6. White, green, and orange indicate control, P applied in the seedling stage, and P applied in the heading stage, respectively. Each value represents the mean ± SD of four replicates. Different letters indicate statistical differences (Tukey's HSD test, $p < 0.05$).

2.3. Variation in the PA Content with Change in the P Concentration in the Soil

To examine the effect of the amount of P fertilizer on PA accumulation in the grain, we applied different amounts of P fertilizer at the seedling stage and measured the PA concentration and content in harvested grains (Figure 2). In the control, WRC 6 showed a higher PA concentration and content than WRC 5 as expected (Figure 2a,b). We observed that the PA concentration and content in both cultivars increased under P4 treatment.

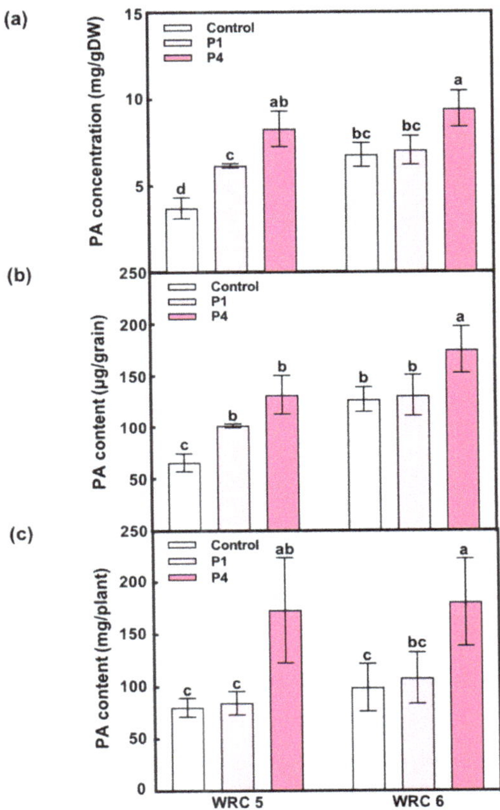

Figure 2. Effect of the amount of P fertilizer on PA accumulation in WRC 5 and WRC 6. (**a**) PA concentration per gram dry weight (mg/gDW), (**b**) PA content per grain (µg/grain), (**c**) PA content per plant (mg/plant). White, pale pink, and pink indicate control, P1 treatment, and P4 treatment, respectively. Each value represents the mean ± SD of four replicates. Different letters indicate statistical differences (Tukey's HSD test, $p < 0.05$).

2.4. PA-P and Pi Content

The Pi content and PA-P/Pi ratio at 10 days after flowering (DAF) and at harvest are shown in Figure 3. We found that the Pi content showed no response to P fertilizer treatment (Figure 3a). WRC 6 showed a significantly higher Pi content at 10 DAF than WRC 5. The PA-P/Pi ratio presented no significant difference between P treatments or genotypes at 10 DAF (Figure 3b). The PA-P/Pi ratio at harvest was increased under P4 treatment and was associated with the PA content per grain (Figures 2b and 3b).

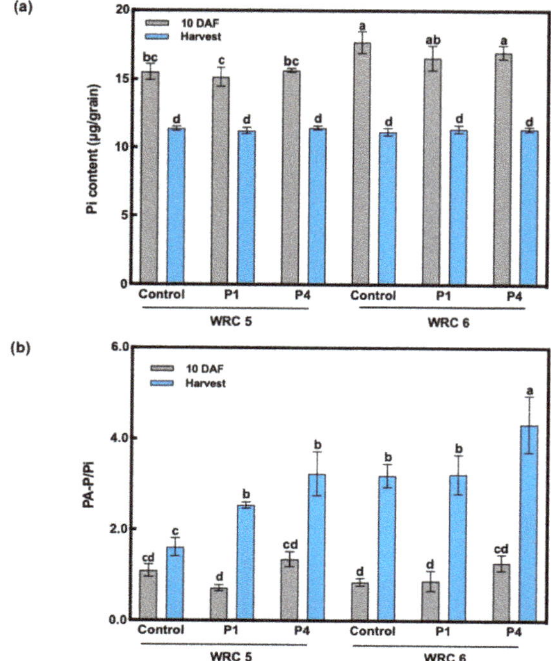

Figure 3. Effect of the amount of P fertilizer on Pi and PA-P/Pi in WRC 5 and WRC 6. (**a**) Pi content per gram dry weight (µg/grain) and (**b**) PA-P/Pi ratio at 10 DAF and harvest. Gray and blue represent the value at 10 DAF and harvest, respectively. Each value represents the mean ± SD of four replicates. Different letters indicate statistical differences (Tukey's HSD test, $p < 0.05$).

2.5. INO1 Expression and PA Content at 10 DAF

The expression of *INO1* at 10 DAF was significantly increased by P fertilizer treatment in WRC 6, whereas WRC 5 showed no response to P fertilizer treatment (Figure 4). We also investigated the expression levels of eight PA biosynthesis genes in Nipponbare (Figure 5). Among these eight genes, no significant differences in the expression level were observed except for the multidrug resistance-associated protein 13 (*MRP13*) gene, which showed a significant decrease under the P4 treatment.

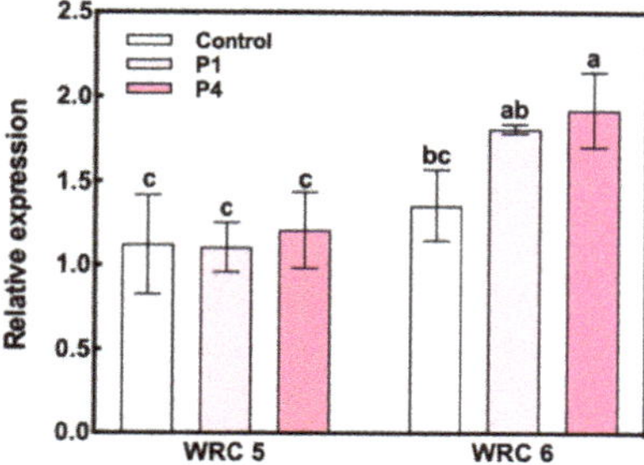

Figure 4. Relative gene expression of *INO1* at 10 DAF in WRC 5 and WRC 6. White and pink indicate the control and P treatment, respectively. Each value represents the mean ± SD of three replicates. Different letters indicate statistical differences (Tukey's HSD test, $p < 0.05$).

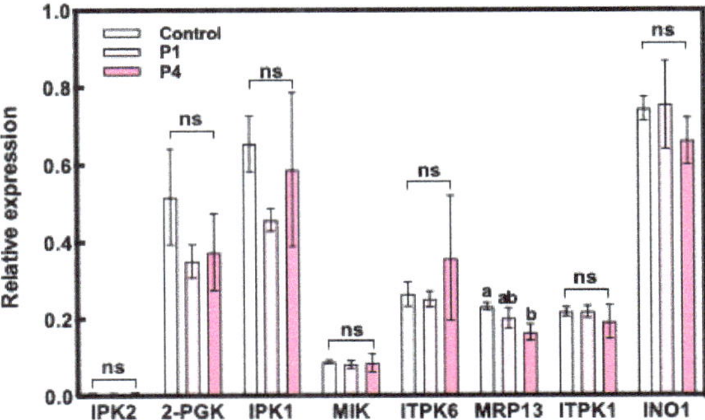

Figure 5. Expression level of PA biosynthesis genes in Nipponbare. White, pale pink, and pink indicate the control, P1 treatment, and P4 treatment, respectively. Each value represents the mean ± SD of three replicates. Different letters indicate statistical differences (Tukey's HSD test, $p < 0.05$).

2.6. Short-Term Response of Pi Uptake in WRC 5 and WRC 6

To evaluate Pi uptake on WRC 5 and WRC 6 in the initial growth stage, we measured the Pi content in xylem sap using the molybdenum blue method (Figure 6). When additional P fertilizer was applied, the Pi content in xylem sap per root dry weight (μg/gDW/2 h) was significantly increased, whereas there was no significant difference between WRC 5 and WRC 6.

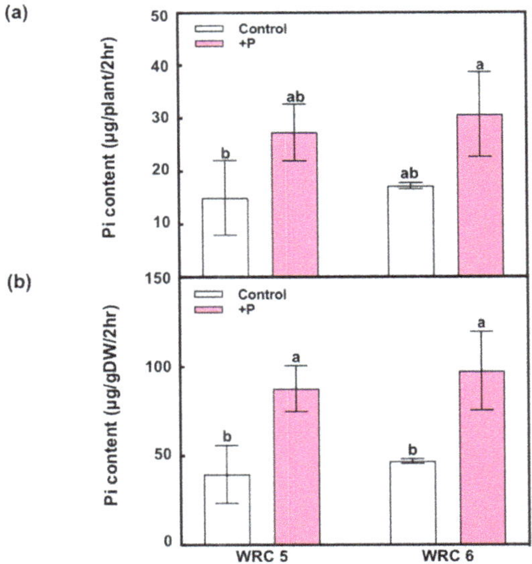

Figure 6. Short-term response of Pi uptake using the xylem sap analysis in WRC 5 and WRC 6. (**a**) Pi content in xylem sap oozed out from one seedling for 2 h (μg/plant/2 h) and (**b**) Pi content per root DW (μg/gDW/2 h). White and pink indicate the control and P treatment, respectively. Each value represents the mean ± SD of three replicates. Different letters indicate statistical differences (Tukey's HSD test, $p < 0.05$).

3. Discussion

3.1. Effect of P Fertilizer on the PA Content

To develop *lpa* mutants, DNA mutations in the PA biosynthesis genes have been used. These *lpa* mutants helped reveal the mechanism of PA biosynthesis; however, *lpa* mutant rice often exhibited low plant biomass and yield compared with those of wild type [20]. Therefore, the use of *lpa* rice mutants in commercial agriculture is not very competitive. In our study, two contrasting rice cultivars selected in terms of PA content from natural variation in the rice germplasm did not show any significant difference in the initial growth (Table 3) and yield-related traits (Table 1). P fertilization did not affect the panicle weight and yield in both WRC 5 and WRC 6 (Table 2). It suggested that the sensitivity of P application in terms of plant growth and yield was similar between WRC 5 and WRC 6. Thus, these contrasting lines are effective in analyzing the mechanism of P fertilizer response of PA synthesis without consideration of the plant yield and growth differences.

In rice, P is accumulated in the leaves until the heading stage, and then transported to the stem in the heading stage. Finally, P is stored in the grain at maturity [33]. According this P translocation system, we expect that rice would show a different response of PA content with the time of P fertilizer application. Our study showed that the PA content increased with P fertilizer in WRC 5, irrespective of the time of P fertilizer application (Figure 1). Moreover, the PA content did not change in WRC 6 with P fertilizer. These results indicate that the effect of P fertilizer on the assimilation and translocation of P is not stage dependent but differs with the genotype.

When a higher amount of P was applied, the PA content increased in both cultivars (Figure 2a). Because the grain yield was not affected by higher P (P4) application, the PA content per grain (Figure 2b) or per plant (Figure 2c) also increased, indicating that PA biosynthesis was upregulated by P application. On the contrary, the PA content under the control treatment differed with the genotype; WRC 5 showed a significantly lower PA content than WRC 6. This result suggests that WRC 5 and WRC 6 have a different responsiveness toward soil P, and WRC 6 might be sensitive to PA accumulation under lower soil P conditions. The Pi content in WRC 6 at 10 DAF was higher than that in WRC 5 (Figure 3a), and this might be due to the higher PA content in WRC 6. Interestingly, the Pi content at 10 DAF did not respond to P fertilizer application, and Pi influx in the early development stage of grain might determine the genotypic differences in the PA content.

3.2. Effect of P Fertilizer on PA Biosynthesis

While the PA content varied in response to P application, the Pi content remained constant and was not affected by the P application and genotype (Figure 3a). The Pi content in the grain was found to be regulated at a constant level. While the Pi content was constant, the ratio of PA-P to Pi responded to P fertilizer, and it was associated with the PA content (Figure 3b). This suggests that the biosynthesized PA using influxed Pi determines the PA response to P application. To assess the PA biosynthesis level, we analyzed *INO1* gene expression, which we previously reported as the key determinant that explains genotypic differences between WRC 5 and WRC 6 [32]. The expression of *INO1* differed with genotype; WRC 6 responded to P fertilizer, whereas WRC 5 did not show any response to P fertilizer (Figure 4). This differential response to P fertilizer might be one of the key factors explaining the genotypic difference in the PA content and the difference in P responsiveness. Simultaneously, while *INO1* showed no response to P fertilizer, the PA content was stimulated by P fertilizer in WRC 5 (Figure 2b), suggesting there might be other factor(s), besides *INO1* gene expression, that regulate the PA biosynthesis response to P fertilizer.

3.3. Other Factors Regulating Grain PA Content

The biosynthesis of PA continues until 25 DAF [32,34]; we investigated the expression level of other genes related to PA synthesis identified from the Nipponbare genome database [35]. Seven genes, namely, inositol 1, 3, 4-trisphosphate 5/6-kinase 2 (*ITPK2*), 2-phospho-glycerate kinase (*2-PGK*), inositol-pentakisphosphate 2-kinase 1 (*IPK1*), myo-inositol kinase (*MIK*), inositol 1, 3, 4-trisphoshate 5/6-kinase 2 (*ITPK6*), inositol 1, 3, 4-triskisphosphate 5/6-kinase 1 (*ITPK1*), and *INO1* showed no response to P fertilizer treatment in Nipponbare (Figure 5). We could not compare these gene expression levels in WRC 5 and WRC 6 directly because WRC 5 and WRC 6 might have different DNA sequences in the priming site of the polymerase chain reaction (PCR) and this interferes with the accurate comparison of the gene expression level. Although we should further analyze gene expression using WRC 5 and WRC 6, the results indicated that the PA content response to P fertilizer might not be regulated by PA biosynthesis in Nipponbare.

There are three potential steps that determine the grain PA content, namely, uptake of soil-Pi, translocation of Pi within the plant body, and biosynthesis and accumulation of PA in the grain. To analyze Pi uptake by the root system and Pi translocation capacity from the root to shoot, we analyzed the xylem sap in seedlings grown by hydroponic culture (Figure 6). The Pi content in xylem sap increased with P fertilizer application, but there was no difference between WRC 5 and WRC 6. These results indicate that the Pi uptake and translocation in the seedling stage are stimulated by P fertilizer; this additional Pi might be used for additional PA biosynthesis. Contrarily, we could not find any genotypic difference in the Pi content in xylem sap in the seedling stage. Further investigation of xylem sap collected at various growth stages will be required to asses Pi uptake and Pi translocation capacities. Recently, SPDT is reported to regulate Pi distribution between the leaves and grains [9]. A comprehensive analysis throughout Pi uptake, translocation, and grain accumulation will be important to further elucidate the effect of P fertilizer on PA accumulation in grains.

4. Materials and Methods

4.1. Plant Materials

We obtained WRC accessions from the National Agriculture and Food Research Organization (NARO) Genebank in Tsukuba, Ibaraki, Japan. The selected WRC 5 (NABA) and WRC 6 (PULUIK ARANG) lines were grown in an outside paddy field or in 1/5000 Wagner pots filled with soil (Bon-sol #2; Sumitomo Chemical, Tokyo, Japan; containing 1.5 g each N, P, and K) in Itakura, Gunma, Japan (36°13′23″N 139°36′37″E) in 2018 to 2019. First, to compare the effect of the time of application of P fertilizer, 1 g of P fertilizer (as $Ca(H_2PO_4)_2$) containing 0.26 g of P was applied into the pots at the seedling stage or heading stage in 2018. Second, we applied different amounts of P fertilizer, 1 g (P1) and 4 g (P4), containing respectively 0.26 and 1.1 g P at the seedling stage in 2019. Developing (10 DAF) and matured (30 DAF) grains were used for further analysis. In the pot experiments, 2 plants were planted in a pot and we used 2 pots to generate 4 replicates. Rice plants grown in the outside paddy field were used for the analysis of yield-related traits. Panicle number, 1000 seed weight, and total yield per plant were analyzed in 5 replicates and panicle length and weight were analyzed in 10 replicates.

4.2. Determination of the PA and Pi Content

The Pi and PA content were determined using the enzymatic method [36]. Pi and PA were extracted for 17 h in 0.66 M HCl from one grain of brown rice, and then the Pi and PA content was determined colorimetrically using the Phytic Acid (Phytate)/Total Phosphorus kit (K-PHYT; Megazyme International, Wicklow, Ireland) according to the manufacturer's instruction after neutralization. This kit measures Pi released from the extracted grain sample after treatment with phytase and alkaline phosphatase. The free Pi content was estimated from samples not treated with phytase.

4.3. Analysis of Gene Expression

Developing ovaries collected at 10 DAF were used for gene expression analysis. RNA was extracted using the cetyl trimethyl ammonium bromide method from approximately 6–7 frozen ovaries. First strand cDNA was synthesized from 5 ng of total RNA using the Primer Script RT Reagent kit (PR037A; Takara Bio Inc., Shiga, Japan) according to the manufacturer's instructions. Quantitative

RT-PCR was performed with the LightCycler system (Light Cycler 480, Roche Diagnostics, Basel, Switzerland) using Universal Probe Library (UPL, Roche Diagnostics, Basel, Switzerland). *GAPDH* was used as the reference gene, and the relative expression levels were calculated using the $2^{-\Delta\Delta CT}$ method [37]. The primers and probes are listed in Table 4.

Table 4. List of primer and probes used in the gene expression analysis.

Gene	Forward (5′–3′)	Reverse (5′–3′)	Probe#
INO1	CAGGGTCGGGAGCTACAA	AAGGTCATCAGGGTTCACCA	#52
ITPK1	AAGGTGAAGAGCTTCCTCCAG	TTCAGAGAGAGGACGAGTCTCA	#72
MRP13	GCACTAGCAAGCAAGACCGTA	TGATATGACCATCCTTAAGAACCA	#146
ITPK6	GGCAAACCTCTTACATTCAACAGT	TTGCACCTCGTTTTGCAG	#68
MIK	CGATGAAGGAGTTTGTGTCACT	GAAGATCCTTGGCACTTAGCC	#120
IPK1	GCTCTTCTAATTTCTGACCACACA	GCCTTTATCTCCACTGCTATGC	#38
2-PGK	TGAAGAAAGAGATATGCACGAGA	CAGATCGTCGTATTCCTCGTC	#22
ITPK2	GTACGCCCTCACCAAGAAGA	CAATTGCTACAAGATTAATTCCCTTC	#41
GAPDH	GCTGCTGCTCACTTGAAGG	AAACATCGGAGCATCTTTGC	#142

4.4. Analysis of xylem sap

Rice was grown for 4 weeks hydroponically in a growth chamber as described by Nagai et al. [38]. The shoots were cut 3–4 cm above the shoot meristem; the section was covered with silicon tube containing glass wool, and then xylem sap was collected for 2 h (10:00–12:00 h). The glass wool was transferred into an ultrafiltration filter tube (Nanosep, ODGHPC34; Pall Corporation, NY, USA) and centrifuged at 17 800 g for 3 min. Xylem saps were stored at −80 °C. Xylem sap diluted 10 times with distilled water was used to determine the Pi content. Diluted saps (50 µL) were treated with 25 µL of color reagent containing 10% (w/v) ascorbic acid, 97% (w/w) H_2SO_4, and 5% (w/v) ammonium molybdate. After incubation at 40 °C for 1 h, the absorbance of the sample at 655 nm was measured using a microplate spectrophotometer (xMark, Bio-Rad, Hercules, CA, USA). A P calibration curve was prepared using P standard solution.

4.5. Data Analysis

We conducted statistical analyses using JMP (SAS Institute, Cary, NC, USA) software. Tukey's HSD test and student's *t*-test were used to determine significant differences between means at $p < 0.05$.

5. Conclusions

In this study, we investigated how P fertilization affects the PA content in rice. WRC 5 showed a lower PA content than WRC 6, and the Pi content at 10 DAF might be related to this genotypic difference in the PA content. In both genotypes, the PA content increased with P fertilizer treatment; this might be due to the stimulation of Pi uptake and translocation. On the contrary, the *INO1* gene expression responded to P fertilizer only in WRC 6, and the response of PA biosynthesis differed with the genotype. It can be concluded that the PA content is regulated by multiple mechanisms, and the response of plants to P fertilizer differed with the genotype. Further comprehensive investigation is required to understand the regulation mechanisms of PA accumulation by different P fertilizer treatments.

Author Contributions: A.F., I.P., and N.H. conceptualized the research project and designed the experiment. A.F. carried out all experiments. K.H. and T.A. performed the plant growing and phytic acid analyses. A.F., I.P., and N.H. wrote the manuscript and all authors were involved in revising the manuscript. All authors have read and agree to the published version of the manuscript.

Funding: This research received no external funding.

Acknowledgments: WRC seed materials used in the study were provided by National Agriculture and Food Research Organization (NARO) gene bank, Tsukuba, Japan.

Conflicts of Interest: The authors declare no conflict of interest.

References

1. IPCC. *Climate Change 2013: The Physical Science Basis. Contribution of Working Group I to the Fifth Assessment Report of the Intergovernmental Panel on Climate Change*; Cambridge University Press: Cambridge, UK, 2013.
2. Myers, S.S.; Zanobetti, A.; Kloog, I.; Huybers, P.; Leakey, A.D.; Bloom, A.J.; Carlisle, E.; Dietterich, L.H.; Fitzgerald, G.; Hasegawa, T.; et al. Increasing CO_2 threatens human nutrition. *Nature* **2014**, *510*, 139–142. [CrossRef] [PubMed]
3. Roboy, V. Approaches and challenges to engineering seed phytate and total phosphorus. *Plant Sci.* **2009**, *177*, 281–296. [CrossRef]
4. Lönnerdal, B.; Mendoza, C.; Brown, K.H.; Rutger, J.N.; Raboy, V. Zinc absorption from *low phytic acid* genotypes of Maize (*Zea mays* L.), Barley (*Hordeum vulgare* L.), and Rice (*Oryza sativa* L.) assessed in a suckling rat pup model. *J. Agric. Food Chem.* **2011**, *59*, 4755–4762. [CrossRef] [PubMed]
5. Kies, A.K.; De Jonge, L.H.; Kemme, P.A.; Jongbloed, A.W. Interaction between protein, phytate, and microbial Phytase. In vitro studies. *J. Agric. Food Chem.* **2006**, *54*, 1753–1758. [CrossRef] [PubMed]
6. Mlodzinska, E.; Zboinska, M. Phosphate uptake and allocation—A closer look at *Arabidopsis thaliana* L. and *Oryza sativa* L. *Front. Plant Sci.* **2016**, *7*, 1198. [CrossRef] [PubMed]
7. Liu, F.; Chang, X.J.; Ye, Y.; Xie, W.B.; Wu, P.; Lian, X.M. Comprehensive sequence and whole-life-cycle expression profile analysis of the phosphate transporter gene family in rice. *Mol. Plant.* **2011**, *4*, 1105–1122. [CrossRef] [PubMed]
8. Li, Y.; Zhang, J.; Zhang, X.; Fan, H.; Gu, M.; Qu, H.; Xu, G. Phosphate transporter OsPht1; 8 in rice plays an important role in phosphorus redistribution from source to sink organs and allocation between embryo and endosperm of seeds. *Plant Sci.* **2014**, *230*, 23–32. [CrossRef]
9. Yamaji, N.; Takemoto, Y.; Miyaji, T.; Mitani-Ueno, N.; Yoshida, K.T.; Ma, J.F. Reducing phosphorus accumulation in rice grains with an impaired transporter in the node. *Nature* **2017**, *541*, 92–95. [CrossRef]
10. Guttieri, M.; Bowen, D.; Dorsch, J.A.; Raboy, V.; Souza, E. Identification and characterization of a low phytic acid wheat. *Crop Sci.* **2004**, *44*, 418–424. [CrossRef]
11. Raboy, V.; Gerbasi, P.F.; Young, K.A.; Stoneberg, S.D.; Pickett, S.G.; Bauman, A.T.; Murthy, P.P.N.; Sheridan, W.F.; Ertl, D.S. Origin and seed phenotype of maize *low phytic acid 1-1* and *low phytic acid 2-1*. *Plant Physiol.* **2000**, *124*, 355–368. [CrossRef]
12. Shi, J.; Wang, H.; Hazebroek, J.; Ertl, D.S.; Harp, T. The maize *low-phytic acid* 3 encodes a *myo*-inositol kinase that plays a role in phytic acid biosynthesis in developing seeds. *Plant J.* **2005**, *42*, 708–719. [CrossRef] [PubMed]
13. Pilu, R.; Panzeri, D.; Gavazzi, G.; Rasmussen, S.K.; Consonni, G.; Nielsen, E. Phenotypic, genetic and molecular characterization of a maize low phytic acid mutant (lpa241). *Theor. Appl. Genet.* **2003**, *107*, 980–987. [CrossRef] [PubMed]
14. Wilcox, J.R.; Premachandra, G.S.; Young, K.A.; Raboy, V. Isolation of high seed inorganic P, low-phytate soybean mutants. *Crop Sci.* **2000**, *40*, 1601–1605. [CrossRef]
15. Frank, T.; Habernegg, R.; Yuan, F.J.; Shu, Q.Y.; Engel, K.H. Assessment of the contents of phytic acid and divalent cations in low phytic acid (*lpa*) mutants of rice and soybean. *J. Food Compos. Anal.* **2009**, *22*, 278–284. [CrossRef]
16. Dorsch, J.A.; Cook, A.; Young, K.A.; Anderson, J.M.; Bauman, A.T.; Volkman, C.J.; Murthy, P.P.; Raboy, V. Seed phosphorus and inositol phosphate phenotype of barley low phytic acid genotypes. *Phytochemistry* **2003**, *62*, 691–706. [CrossRef]
17. Moreau, R.A.; Bregitzer, P.; Liu, K.; Hicks, K.B. Compositional equivalence of barleys differing only in low- and normal-phytate levels. *J. Agric. Food Chem.* **2012**, *60*, 6493–6498. [CrossRef]
18. Ye, H.; Zhang, X.Q.; Broughton, S.; Westcott, S.; Wu, D.; Lance, R.; Li, C. A nonsense mutation in a putative sulphate transporter gene results in low phytic acid in barley. *Funct. Integr. Genom.* **2011**, *11*, 103–110. [CrossRef]
19. Frank, T.; Meuleye, B.S.; Miller, A.; Shu, Q.Y.; Engel, K.H. Metabolite profiling of two low phytic acid (*lpa*) rice mutants. *J. Agric. Food Chem.* **2007**, *55*, 11011–11019. [CrossRef]
20. Zhao, H.J.; Liu, Q.L.; Fu, H.W.; Xu, X.H.; Wu, D.X.; Shu, Q.Y. Effect of non-lethal low phytic acid mutations on grain yield and seed viability in rice. *Field Crops Res.* **2008**, *108*, 206–211. [CrossRef]

21. Kuwano, M.; Mimura, T.; Takaiwa, F.; Yoshida, K.T. Generation of stable 'low phytic acid' transgenic rice thorough antisense repression of the 1D-*myo*-inositol 3-phosphate synthase gene (*RINO1*) using the 18-kDa oleosin promoter. *Plant Biotechnol. J.* **2009**, *7*, 96–105. [CrossRef]

22. Raboy, V.; Peterson, K.; Jackson, C.; Marshall, J.M.; Hu, G.; Saneoka, H.; Bregitzer, P. A substantial fraction of barley (*Hordeum vulgare* L.) low phytic acid mutations have little or no effect on yield across diverse production environments. *Plants* **2015**, *4*, 225–239. [CrossRef] [PubMed]

23. Yuan, F.J.; Zhao, H.J.; Ren, X.L.; Zhu, S.L.; Fu, X.J.; Shu, Q.Y. Generation and characterization of two novel low phytate mutations in soybean (*Glycine max* L. Merr.). *Theor. Appl. Genet.* **2007**, *115*, 945–957. [CrossRef] [PubMed]

24. Srilatha, M.; Sharma, S.H.K. Influence of long term use of fertilizers and manures on available nutrient status and inorganic "Phosphorus" fractions in soil under continuous rice—Rice cropping system. *IJAR* **2015**, *3*, 960–964.

25. Shen, J.; Yuan, L.; Zhang, J.; Li, H.; Bai, Z.; Chen, X.; Zhang, W.; Zhang, F. Phosphorus dynamics: From soil to plant. *Plant Physiol.* **2011**, *156*, 997–1005. [CrossRef]

26. Buerkert, A.; Haake, C.; Ruckwied, M.; Marschner, H. Phosphorus application affects the nutritional quality of millet grain in the Sahel. *Field Crops Res.* **1998**, *57*, 223–235. [CrossRef]

27. Coelho, C.M.M.; Santos, J.C.P.; Tsai, S.M.; Vitorello, V.A. Seed phytate content and phosphorus uptake and distribution in dry bean genotypes. *Braz. J. Plant Physiol.* **2002**, *14*, 51–58. [CrossRef]

28. Saneoka, H.; Koba, T. Plant growth and phytic acid accumulation in grain as affected by phosphorus application in maize (*Zea maize* L.). *Grassl. Sci.* **2003**, *48*, 485–489.

29. Su, D.; Zhou, L.; Zhao, Q.; Pan, G.; Cheng, F. Different phosphorus supplies altered the accumulations and quantitative distributions of phytic acid, zinc, and iron in rice (*Oryza sativa* L.) grains. *J. Agric. Food Chem.* **2018**, *66*, 1601–1611. [CrossRef]

30. Taliman, N.A.; Dong, Q.; Echigo, K.; Raboy, V.; Saneoka, H. Effect of phosphorus fertilization on the growth, photosynthesis, nitrogen fixation, mineral accumulation, seed yield, and seed quality of a soybean low-phytate line. *Plants* **2019**, *8*, 119. [CrossRef]

31. Perera, I.; Fukushima, A.; Arai, M.; Yamada, K.; Nagasaka, S.; Seneweera, S.; Hirotsu, N. Identification of low phytic acid and high Zn bioavailable rice (*Oryza sativa* L.) from 69 accessions of the world rice core collection. *J. Cereal Sci.* **2019**, *85*, 206–213. [CrossRef]

32. Perera, I.; Fukushima, A.; Akabane, T.; Genki, H.; Seneweera, S.; Hirotsu, N. Expression regulation of *myo*-inositol 3-phosphate synthase 1 (INO1) in determination of phytic acid accumulation in rice grain. *Sci. Rep.* **2019**, *9*, 14866. [CrossRef] [PubMed]

33. Delin, L.; Zhaomin, Z. Effect of available phosphorus in paddy soils on phosphorus uptake of rice. *J. Radioanal. Nucl. Chem.* **1996**, *205*, 235–243. [CrossRef]

34. Iwai, T.; Takahashi, M.; Oda, K.; Terada, Y.; Yoshida, K.T. Dynamic changes in the distribution of minerals in relation to phytic acid accumulation during rice seed development. *Plant Physiol.* **2012**, *160*, 2007–2014. [CrossRef] [PubMed]

35. Perera, I.; Seneweera, S.; Hirotsu, N. Manipulating the phytic acid content of rice grain toward improving micronutrient bioavailability. *Rice* **2018**, *11*, 4. [CrossRef]

36. Alkarawi, H.H.; Zotz, G. Phytic acid in green leaves of herbaceous plants–temporal variation in situ and response to different nitrogen/phosphorus fertilizing regimes. *AoB Plants* **2014**, *6*. [CrossRef]

37. Livak, K.J.; Schmittgen, T.D. Analysis of relative gene expression data using real-time quantitative PCR and the $2^{-\Delta\Delta CT}$ method. *Methods* **2001**, *25*, 402–408. [CrossRef]

38. Nagai, Y.; Matsumoto, K.; Kakinuma, Y.; Ujiie, K.; Ishimaru, K.; Gamage, D.M.; Thompson, M.; Milham, P.J.; Seneweera, S.; Hirotsu, N. The chromosome regions for increasing early growth in rice: Role of source biosynthesis and NH_4^+ uptake. *Euphytica* **2016**, *211*, 343–352. [CrossRef]

© 2020 by the authors. Licensee MDPI, Basel, Switzerland. This article is an open access article distributed under the terms and conditions of the Creative Commons Attribution (CC BY) license (http://creativecommons.org/licenses/by/4.0/).

Article

Effect of Phosphorus Fertilization on the Growth, Photosynthesis, Nitrogen Fixation, Mineral Accumulation, Seed Yield, and Seed Quality of a Soybean Low-Phytate Line

Nisar Ahmad Taliman [1], Qin Dong [1], Kohei Echigo [1], Victor Raboy [2,†] and Hirofumi Saneoka [1,*]

[1] Graduate School of Biosphere Science, Hiroshima University, 1-4-4 Kagamiyama, Higashi-Hiroshima 739-8528, Japan; nisarahmad_kandahar@yahoo.com (N.A.T.); d170372@hiroshima-u.ac.jp (Q.D.); echigo-kohei@hiroshima-u.ac.jp (K.E.)

[2] USDA-ARS, Small Grains and Potato Research Unit, 1600 South 2700 West, Aberdeen, ID 83210, USA; vraboy@pdx.edu

* Correspondence: saneoka@hiroshima-u.ac.jp; Tel.: +81-82-424-7917

† Current Address: Department of Biology, Portland State University, 1719 SW 10th Avenue, Portland, OR 97201, USA.

Received: 8 March 2019; Accepted: 5 May 2019; Published: 8 May 2019

Abstract: Crop seed phosphorus (P) is primarily stored in the form of phytate, which is generally indigestible by monogastric animals. Low-phytate soybean lines have been developed to solve various problems related to seed phytate. There is little information available on the effects of P fertilization on productivity, physiological characteristics, and seed yield and quality in low-phytate soybeans. To address this knowledge gap, studies were conducted with a low-phytate line and two normal-phytate cultivars from western Japan when grown under high- and low-P fertilization. The whole plant dry weight, leaf photosynthesis, dinitrogen fixation, and nodule dry weight at the flowering stage were higher in the higher P application level, but were not different between the low-phytate line and normal-phytate cultivars. As expected, seed yield was higher in the higher level of P application for all lines. Notably, it was higher in the low-phytate line as compared with the normal-phytate cultivars at both levels of fertilizer P. The total P concentration in the seeds of the low-phytate line was the same as that of the normal-phytate cultivars, but the phytate P concentration in the low-phytate line was about 50% less than that of the normal-phytate cultivars. As a result the molar ratio of phytic acid to Zn, Fe, Mn, and Cu in seed were also significantly lower in the low-phytate line. From these results, it can be concluded that growth after germination, leaf photosynthesis, nitrogen fixation, yield and seed quality were not less in the low-phytate soybean line as compared with two unrelated normal-phytate cultivars currently grown in Japan, and that low-phytate soybeans may improve the bioavailability of microelements.

Keywords: low-phytate soybean; microelement; phosphorus; phytic acid; seed yield; seed quality

---

1. Introduction

Phosphorus (P) is one of the most essential macro elements, along with nitrogen, required by plants to grow. Adequate P fertilization is essential for effective crop production to attain optimum yields. However, excessive P fertilization increases the risk of P losses to surface and ground waters, with detrimental effects on aquatic ecosystems through eutrophication [1,2]. Phosphorus absorbed by plants is translocated from the roots and leaves to the seeds during seed development, where phytic acid (phytate; myo-inositol 1, 2, 3, 4, 5, 6-hexakisphosphate) is synthesized [3,4]. Phytate accounts for up to 80% of the total P concentration in seeds and its breakdown during germination supplies P and other cations needed for early seedling growth. However, phytate is poorly digested by non-ruminants such as pigs, chickens, and humans, which don't express phytase, an enzyme that degrades phytate [2,5]. To provide the nutritional requirements of P for animals, and thereby ensure optimal productivity, feeds are supplemented with inorganic phosphorus (inorganic P). Therefore, poultry and chicken manure contain a large amount of phytate-derived P. When these manures are applied to fields, inorganic P is released by the phytase expressed by microorganisms in the soil. If this P is not absorbed by plants, a large amount of P flows into rivers and lakes through rainwater. In addition, dietary phytate chelates divalent cations, including iron (Fe), zinc (Zn), copper (Cu), manganese (Mn), magnesium (Mg), and calcium (Ca), and this reduces the bioavailability and utilization of these essential nutrients [6–8].

One approach to these phytate-related problems that increases the bioavailability of P and minerals in seeds is to breed low-phytate crops. Low-phytate lines have been isolated from various agronomic plants such as major cereal crops such as barley (*Hordeum vulgare* L.) [9,10], rice (*Oryza saviva* L.) [11], maize (*Zea mays* L.) [12], wheat (*Triticum aestivum* L.) [13], soybean (*Glycine max* (L.) Merr.) [14], and common bean (*Phaseolus vulgaris* L.) [15,16]. Compared with normal-phytate lines, with one known exception [10] these low-phytate lines have a total P concentration that is no different from normal-phytate lines. They have a 40 to 80% reduction in phytate P, resulting in up to an 80% increase in inorganic P (available P) for animals. From animal feeding experiments using low-phytate lines, it has been confirmed that the P-utilization rates of animals is improved, and the amount of P excreted is decreased [1].

The phytate concentration of seed depends both on the crop variety or its genetics and environmental factors, especially P fertilization. It has been reported that the phytate concentration of seed gradually increases and is positively correlated with applied P levels in soybean [17], oat [18], and maize [19]. However, there have been relatively few reports of analyses of the productivity and seed quality of low-phytate lines grown under varying P fertilization levels [10,20]. Therefore in this study we compared the yield, mineral and phytate concentration of the seeds, and physiological factors such as nitrogen fixation, between a soybean low-phytate line (hereafter referred to as LP) descended (F_{10} and F_{11}) from a cross of the normal-phytate Japanese cv. Tanbakuro and the low-phytate line CX1834 [14,21], and two normal-phytate soybean cultivars Enrei and Akimaro, when grown under two different levels of nutrient P fertilization. We asked if phytate accumulation in the seeds of the low-phytate soybean line in response to P fertilization differs from that observed in normal-phytate cultivars and whether a low P fertilizer condition differentially affects seed yield and quality, as well as plant growth and physiological factors such as photosynthesis and nitrogen fixation.

2. Results

2.1. Biomass Production at the Flowering Stage

In Experiment 1 we compared the LP-F_{10} progeny with the cultivar Enrei (Figures 1–3 and Table 1). The whole plant weight of both the LP-F_{10} line and Enrei was significantly affected by the higher P application at the flowering stage (Figure 1). The whole plant weight in both the LP line and normal-phytate cultivar in the P150 treatment was 1.5 to 1.6 times higher than that in the P50 treatment, respectively. However, there was no difference in the whole plant weight between the LP line and normal-phytate cv. Enrei within the same P treatment.

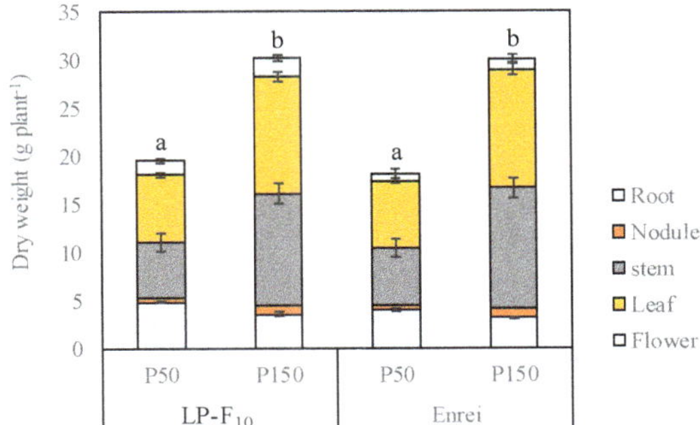

Figure 1. Effect of phosphorus fertilization on whole plant dry weight of the LP-F_{10} progeny and the normal-phytate cv. Enrei (Experiment 1). The same letter indicates no significant difference ($p \le 0.05$).

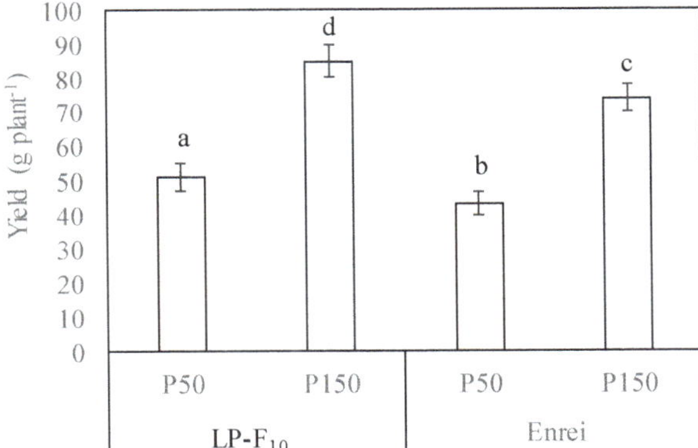

Figure 2. Effect of phosphorus fertilization on seed yield in LP-F_{10} and the normal-phytate cv. Enrei in Experiment 1. The same letter indicates no significant difference ($p \le 0.05$).

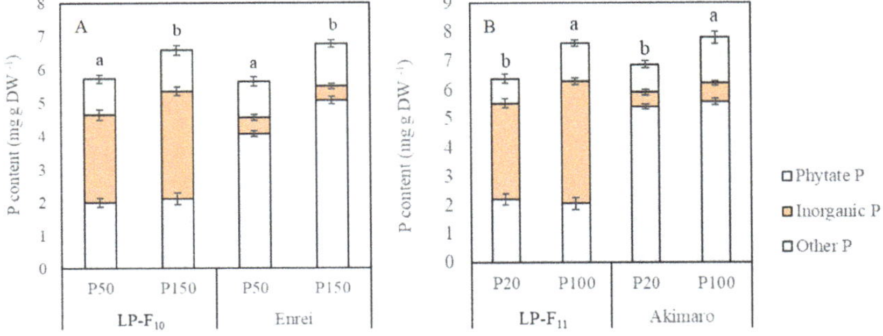

Figure 3. Effect of phosphorus fertilization on seed total P, phytate P and inorganic P concentrations in LP lines and normal-phytate soybean cultivars of soybean in Experiment 1 (**A**) and Experiment 2 (**B**). The same letter indicates no significant difference ($p \le 0.05$).

2.2. Photosynthesis and Dinitrogen Fixation Activity at the Flowering Stage

In Experiment 1 the photosynthetic rate and dinitrogen fixation activity in LP-F_{10} and cv. Enrei were 1.3 and 1.5 times higher (respectively) in the P150 treatment than in the P50 treatment, but there was no substantial differences between LP-F_{10} and Enrei within a given P treatment (Table 1). The nodule number was higher in the higher P application level for both LP-F_{10} and Enrei however the nodule number of LP-F_{10} was higher than that of Enrei when grown under the P50 treatment. The specific nodule activity in both LP-F_{10} and Enrei, obtained by dividing the nitrogen fixation rate by the root nodule weight, was greater in the P50 treatment than in the P150 treatment. No significant differences were observed in specific nodule activity of LP-F_{10} as compared with Enrei within a given P treatment.

Table 1. Effect of phosphorus fertilization on photosynthesis and nitrogen fixation in LP-F_{10} and the normal-phytate cv. Enrei (Experiment 1). The same letter indicates no significant difference ($p \leq 0.05$).

Line or Cultivar	Treatment	Photosynthetic Rate (μmol CO_2 m^{-2} s^{-1})	Nitrogen Fixation (μmol C_2H_4 $plant^{-1}$ h^{-1})	Nodule Number (number $plant^{-1}$)	Specific Nodule Activity (μmol g^{-1} nodule weight h^{-1})
LP-F_{10}	P50	$14^b \pm 1$	$158^b \pm 9$	$119^b \pm 6$	$310^a \pm 12$
	P150	$18^a \pm 1$	$227^a \pm 8$	$146^a \pm 5$	$236^b \pm 15$
Enrei	P50	$13^b \pm 2$	$151^b \pm 6$	$102^c \pm 3$	$336^a \pm 14$
	P150	$18^a \pm 2$	$203^a \pm 10$	$138^a \pm 4$	$246^b \pm 16$

2.3. Seed Yield and Yield Components

In Experiment 1, seed yield of both LP-F_{10} and cv. Enrei was 1.7 times higher in the P150 treatment as compared with the P20 treatment (Figure 2). In Experiment 2 seed yield was 2.4 times higher in LP-F_{11} and 2.2 times higher in Akimaro in the P100 treatment as compared to the P50 treatment (Table 2). Importantly, in both experiments, in a given P treatment level the seed yield of the LP line was significantly ($p \leq 0.05$) higher than the normal-phytate line, by 16% to 34%. Pod number data was not collected in Experiment 1. In Experiment 2, the number of pods per plant was significantly higher ($p \leq 0.05$) in the P100 treatment as compared with the P20 treatment in both the LP-F_{11} line (2.1 times higher) and Akimaro (2.3 times higher, Table 2). Of importance here is that the pod number per plant of the LP-F_{11} line was higher than that of Akimaro in both P treatments; 1.41 and 1.28 times higher in the P20 and P100 treatments, respectively, but only significantly different, $p \leq 0.05$ in the P100 treatment. P treatment did not greatly affect 100 seed weight in either line or cultivar (Table 2). The 100 seed weight was about 1.6 times higher in LP-F_{11} than in Akimaro in both P treatments, but only significantly greater ($p = \leq 0.05$) in the P20 treatment.

Table 2. Effect of phosphorus fertilization on yield and yield attributes in LP-F_{11} and the normal-phytate cv. Akimaro in Experiment 2. The same letter indicates no significant difference ($p \leq 0.05$).

Line or Cultivar	Treatment	Yield (g $plant^{-1}$)	Pod Number (number $plant^{-1}$)	100 Seed Weight (g)
LP-F_{11}	P20	$42^c \pm 4$	$75^c \pm 2$	$24^a \pm 1$
	P100	$99^a \pm 4$	$156^a \pm 3$	$26^a \pm 1$
Akimaro	P20	$33^d \pm 4.$	$53^c \pm 2$	$15^b \pm <1$
	P100	$74^b \pm 4.$	$122^b \pm 5$	$16^b \pm 1$

2.4. Seed Total P, Phytate P and Inorganic P Concentration

In Experiment 1, the seed total P concentration in both LP-F_{10} and cv. Enrei was ~5–6 mg g^{-1} DW in the P50 treatment and ~6–7 mg/g DW in the P 150 treatment but did not differ between LP-F_{10} and Enrei (Figure 3A). The percentage of total P represented by phytate P in LP-F_{10} (~32–35%) was about half that of Enrei (~72–75%), and the percentage of total P represented by inorganic P in LP-F_{10} (~46–49%) was ~5–6 times greater than that of Enrei (~6–9%). Seed phytate P concentration within in a given line or cultivar was not consistently impacted by P treatment, whereas inorganic P was 15% higher in LP lines grown with P150 as compared with P50. Overall, similar results were observed in Experiment 2 (Figure 3).

2.5. Seed Crude Protein, Lipid and Mineral Concentrations

Data for seed crude protein, lipid and mineral concentrations was only collected in Experiment 2 (Table 3). Seed crude protein and lipid concentrations in both LP-F_{11} and cv. Akimaro were about 10% higher in the P100 versus the P20 P treatments (Table 3). The seed crude protein concentration in both P treatments was 3% to 12% higher in LP-F_{11} than in the Akimaro; however, the lipid concentration was 5% to 12% higher in Akimaro that in LP-F_{11}.

The K and Ca concentrations in the seeds of both LP-F_{11} and cv. Akimaro were 13% to 37% higher in the P100 treatment as compared with the P20 treatment (Table 3). While the seed Mg concentration in LP-F_{11} was not different between the P20 and P100 treatments, the seed Mg concentration in Akimaro was 15% higher in the higher P application level. Seed K concentration in the P20 treatment was 17% higher in LP-F_{11} than in Akimaro, but no differences between lines were observed in the P100 treatment. Statistically significant differences in seed Ca and Mg concentrations were not observed between LP-F_{11} and Akimaro at a given level of P application. In contrast, the seed Zn, Mn, and Cu concentrations in both LP-F_{11} and Akimaro were 20% to 30% lower in the P100 fertilization level as compared with the P20 level, probably due to a dilution effect resulting from the higher seed yields at the higher P fertilization rate. With the exception of seed Fe levels in the P100 treatment and Cu at the P100 level, these element's concentrations were 11% to 43% higher in LP-F_{11} seed than in Akimaro in both P application levels. For both LP-F_{11} and Akimaro there was little difference in the molar ratio of phytic acid to Zn, Fe or Mn between the two levels of P fertilization. However for Cu, the phytic acid:Cu molar ratio observed for the P100 treatment was 1.68 and 1.73 times as great in the P100 versus P20 fertilization levels for LP-F_{11} and Akimaro, respectively. Of importance for this study, due to Akimaro's 2.5 to 2.7-times higher level of phytic acid as compared with LP-F_{11}, phytic acid molar ratios to Zn, Fe, Mn, and Cu were 2.5 to 3.5-times higher in Akimaro seed as compared with LP-F_{11} seed across both P fertilization treatments (Table 3).

Table 3. Effect of phosphorus fertilization on seed quality in the low-phytate LP-F_{11} as compared with the normal-phytate cv. Akimaro soybean in Experiment 2. The same letter indicates no significant difference ($p \leq 0.05$).

Seed Quality	LP-F_{11}		Akimaro	
	P20	P100	P20	P100
Protein (%)	$33^b \pm 2$	$38.9^a \pm 1.9$	$31.9^b \pm 1.3$	$34.8^b \pm 1.5$
Lipid (%)	$19^c \pm <1$	$23^a \pm <1$	$22^b \pm <1$	$24^a \pm <1$
K (mg g^{-1})	$203^b \pm 16$	$229^a \pm 17$	$174^c \pm 15$	$238^a \pm 14$
Ca (mg g^{-1})	$0.97^b \pm 0.1$	$1.35^a \pm 0.2$	$1.13^b \pm 0.1$	$1.49^a \pm 0.1$
Mg (mg g^{-1})	$2.2^a \pm 0.2$	$2.2^a \pm 0.3$	$2.5^a \pm 0.3$	$2.9^a \pm 0.3$
Zn (µg g^{-1})	$48^a \pm 3$	$39^b \pm 3$	$39^b \pm 3$	$35^c \pm 3$
Fe (µg g^{-1})	$51^a \pm 7$	$49^a \pm 6$	$41^b \pm 6$	$49^a \pm 7$
Mn (µg g^{-1})	$63^a \pm 6$	$50^b \pm 4$	$44^{bc} \pm 4$	$38^c \pm 4$
Cu (µg g^{-1})	$7.3^a \pm 0.7$	$4.1^b \pm 0.4$	$6.4^a \pm 0.6$	$3.8^b \pm 0.4$
Phytic acid (mg g^{-1})	$7.8^b \pm 04$	$7.4^b \pm 0.2$	$19.2^a \pm 1$	$19.8^a \pm 1$
Phytic acid:Zn (Molar ratio)	$18.4^c \pm 0.7$	$18.0^c \pm 0.9$	$45.3^b \pm 4.2$	$56.3^a \pm 3.4$
Phytic acid:Fe (Molar ratio)	$12.9^b \pm 0.6$	$12.8^b \pm 0.8$	$39.7^a \pm 4.2$	$34.4^a \pm 3.3$
Phytic acid:Mn (Molar ratio)	$10.3^c \pm 0.8$	$12.4^c \pm 0.6$	$36.1^b \pm 1.6$	$43.7^a \pm 2.5$
Phytic acid:Cu (Molar ratio)	$103.9^d \pm 6.8$	$174.5^c \pm 8.7$	$291.0^b \pm 20.5$	$503.6^a \pm 18$

3. Discussion

Many studies on the influence of P application on the growth, photosynthesis, nitrogen fixation, and yield of soybean and other crop species have been reported [17,22–26]. Most of these studies were carried out with crop cultivars that have "standard" (non-mutant) levels of seed phytic acid. In contrast relatively few such studies have been conducted with low-phytate variants of crops species [10,20] including one with low-phytate soybean lines [20] which will be discussed below. The proportion of phytate P to total P concentration in the seed of normal soybean cultivars is about 70–80%, but that of the low-phytate lines that we bred is about 30–35%. It has been reported that many low-phytate mutants impact plant growth and productivity and have decreased yield as compared with sibling normal-phytate/wild-type lines [27–29]. However in the case of soybean lines developed from crosses using the same low-phytate parent used here, CX1834 [14], previous studies demonstrated that while yields of low-phytate progeny were 85% to 90% of sibling normal-phytate lines, the yield loss was largely due to negative impacts on germination and emergence that led to reduced stand establishment, rather than impacts on plant growth and productivity subsequent to emergence and germination [30,31]. Therefore, in this study, we further investigated whether in the low-phytate lines that we bred, there was an effect on the growth, photosynthesis, nitrogen fixation, seed yield, and seed quality subsequent to germination and emergence. In addition, it has been reported that the phytate concentration of the seed of normal-phytate cultivars is increased by P application [17,19], and that there is a highly positive correlation between total P and phytate in seeds [17]. Therefore, we investigated whether the phytate concentration of seed increases with a higher P application level even in a low-phytate line.

Highly significant effects of the P application level on the plant growth, leaf photosynthesis, biological nitrogen fixation of nodules, and yield of seed were observed in this study. With higher application of P, the whole plant dry weight was markedly higher in Experiment 1 (Figure 1). However, while an increase in whole plant dry weight in the P 150 treatment compared to the P 50 treatment was observed, there were no differences in whole plant dry weight between the low-phytate line and normal-phytate cultivar in either P treatment. These results suggest that the decrease of phytate in the seeds of the low-phytate line did not substantially affect the growth of plants subsequent to germination at any P application level.

At the low P level, the photosynthetic rate and dinitrogen fixation activity of both the low-phytate line and normal phytate cultivar were less than those of the higher P application level (Table 1). There are many reports that photosynthesis increases with P application. Wang et al. [26] reported that the activity of Rubisco, which is a key enzyme of photosynthesis, and the protein concentration of leaves were higher in a high P application than in a low P application. Furthermore, Singh and Reddy [32] also reported that Rubisco activity, RUBP regeneration, and maximum quantum yield related to the photochemical system were decreased under a P-deficient condition. Dinitrogen fixation activity was also increased by P application. It is assumed that the increase in dinitrogen fixing activity by P application is due to an increase in nodule weight (Table 1). In any case, photosynthesis and dinitrogen fixation were lower in the low P level, but the reduction was not significantly different between the low-phytate line and normal-phytate cultivar. These results indicate that both seed phytate concentration and low-phytate genotype, at least those low-phytate lines derived from crosses using the CX-1834 source used here, have little impact on photosynthesis and dinitrogen fixation in soybeans at the growth stages following germination.

In this study, yield and yield components of both the low-phytate line and normal-phytate cultivar were significantly affected by P application. As expected, the yield of both the low-phytate line and normal-phytate cultivar was higher in the high P level than in the low P level. However, the low-phytate line had significantly higher seed yields than the normal-phytate cultivar in all the treatments (Figure 2 and Table 1). These differences were mainly attributed to the number of productive pods and 100 seed weight (Table 1).

Since comparisons here were made between low-phytate soybean lines obtained from the cross of cv. Tanbakuro and the low-phytate line CX1834 [14], and the unrelated normal-phytate cultivars

Enrei and Akimaro, and not between the low-phytate line and a sibling normal-phytate/wild-type line derived from the same Tanbakuro by CX1834 cross, we cannot attribute any observed differences in yield or any trait to allelic differences in genes conditioning the low-phytate trait. We can only conclude that on a single-plant basis and not taking into account any effect on germination and emergence, yield and other traits of these specific low-phytate soybean progeny was observed to be equal to or better than either of these unrelated normal-phytate cultivars currently grown in Japan.

A previous study of several low-phytate soybean lines grown under varying levels of P supply in hydroponics reported that two such lines were more productive in terms of seed yield at given level of nutrient P than were the non-mutant, normal-phytic acid parental line, even when these lines were grown under a very growth-limiting low-nutrient P level [20]. However, one other low-phytate soybean line evaluated in this earlier study displayed reduced seed yield when grown under limiting nutrient P. Therefore one can conclude from the present and earlier studies that in soybean the impact of the low-phytate trait on growth and yield when grown under varying levels of nutrient P is variable, depending on genotype and level of nutrient P, but that some low-phytate soybean lines appear to be as productive if not more productive than normal-phytate lines.

In the two closely related food legume species soybean and common bean, there are two copies of the gene perturbed in the mutant studied here, a member of the MRP gene family that encodes an ABC transporter specific to phytic acid transport [16,33,34]. The low-phytate trait in the original low-phytate soybean parental line CX1834 line [14] that was used as the trait donor in breeding the low-phytate line studied here [21], was found to be conditioned by mutations in both copies of this particular MRP gene in the soybean genome [33]. In contrast, in the common bean low-phytate mutants that are conditioned by mutations in its MRP ortholog, the low-phytate trait is the result of a mutation in only one of the two MRP duplications, and there is little or no negative impact on germination, emergence or subsequent plant growth, performance or yield [15,16]. It was hypothesized that this lack of negative impact or pleiotropic effect is due to a buffering effect resulting from functional complementation provided by the second wild-type duplication of the gene [16,34]. Therefore it is possible that when transferring the trait from CX1834 into the parental background cv. Tanbakuro used to breed the of the low-phytate line studied here, a similar buffering effect was obtained, either provided by variant alleles of one of the two MRP orthologs, or perhaps by other components of the Tanbakuro genome. However the most likely explanation for the good growth and performance of the low-phytate soybean lines studied here is that problems with performance of this specific genotype are probably due to negative impacts on germination and emergence, rather than to negative impacts on subsequent growth and performance [29–31].

Highly significant effects of higher P application on the total P and phytate P concentrations of the seeds were found in both Experiment 1 (Figure 3A) and Experiment 2 (Figure 3B). While total P concentration was higher in the higher level of P application there was little difference in total P between the low-phytate line and normal-phytate cultivar. However the substantial effects of P fertilization on phytate P and inorganic P concentrations differed between low-phytate lines and normal-phytate cultivars. For example, in Experiment 1 the phytate P concentration in the normal-phytate cultivar was 4.09 mg g^{-1} at the 50 P application level and 5.26 mg g^{-1} at the 150 P level. However, in the low-phytate line, phytate P concentrations were similar at both P fertilization levels, about 2.0 mg g^{-1} dry weight. Similar results were obtained in Experiment 2 (Figure 3B). In the low-phytate line, inorganic P increases occurred instead of increasing phytate P in response to increasing nutrient P supply. As an outcome of this, the proportion of phytic acid P to total P concentration was reduced with increased P fertilization in the low-phytate lines but not greatly affected by the level of P application in the normal phytate lines (Figure 3). These results support the finding of previous studies that indicate that the accumulation and synthesis of phytic acid in seed is highly dependent on the soil P level [35]. A highly significant positive relationship between P supply and phytic acid P in seeds has been observed in soybean [17] and maize [19]. The results of this study agree with that of Oltmans et al. [36] who reported that the mean total P of a similar low-phytate line and its sibling normal-phytate lines was not significantly different, but the mean phytate P, inorganic P, and other P concentrations were significantly different.

Lipids and protein are important factors in seed quality. There are some reports that the lipid and protein concentrations of seed were increased or decreased by P fertilization, and the influence an lipids and proteins by P application was different depending on the study. For example, in soybean, Abbasi et al. [37] reported that P application increased the oil (lipid) and protein concentrations, but Yi et al. [38] reported that the protein concentration was increased, and oil concentration was decreased with increased P application. Krueger et al. [39] reported that oil (lipid) and protein concentrations were not affected by P fertilization. Bethlenfalvay et al. [40] also found that soybean seed lipid and protein concentrations were not significantly correlated, and there was a highly significant negative correlation between seed P and lipid concentration. In this study, the lipid and protein concentration of seed in both the low-phytate line and normal-phytate cultivar were significantly higher in the higher P application level (Table 3). From this study and previous reports, the influence of P application on the protein and lipid concentration of soybean seed may be different depending on the P application level, soil moisture condition, and characteristics of the cultivar used in the study. In any case, from this study, even when the low-phytate line and normal-phytate cultivar were cultivated under the same P application condition, there was no difference in the lipid and protein concentration of the seed.

The differences in seed K, Ca, and Mg concentrations observed here could be attributed to the differences in seed phytate concentration that result either from differences in P fertilization or low-phytate genotype (Table 3). A similar result was found by Toda et al. [41]: since Ca and Mg ions form salts with phytic acid and as a result diminish its buffering effect, unbound phosphate groups of phytic acid would result from the low concentration of Ca and Mg. In contrast to K, Ca, and Mg, microelements, such as Zn, Fe, Mn, and Cu in the both low-phytate line and normal-phytate cultivar were less in the higher P application level. It is well known that the concentration of these microelements is affected by P fertilization. Naeem et al. [23] reported that the Zn concentration of seed in wheat was decreased by P application and Sánchez-Rodríguez et al. [42] reported the same in wheat and barley. Zhang et al. (2012) [43] also reported that P application significantly decreased Zn and Cu, but had no effect on the Fe and Mn concentrations of wheat seed. Zhang et al. (2017) [44] also reported that total P to Zn, Cu, Fe, and Mn molar ratios in maize seed were increased, which indicates that P application in maize may affect the bioavailability of these minerals. In this study, Zn, Fe, and Mn levels and the phytic acid:mineral molar rations for Zn, Fe and Mn were not greatly affected by differing P fertilization in either the low-phytate line or normal-phytate cultivar (Table 3). However the higher level of phytic acid in the normal-phytate cultivar as compared with the low-phytate line when grown on low or high fertilizer P resulted in much higher (2.5 to 3.5-fold) phytic acid:mineral molar ratios. From this result, it can be suggested that the low-phytate line maintained high bioavailability of these microelements in comparison to the normal-phytate cultivar regardless of P application, and will likely have positive effects on the nutrition of monogastric animals.

4. Materials and Methods

4.1. Plant Material and Growing Condition

The low-phytate soybean line CX1834, isolated from the mutant originally referred to as M153 [14], was obtained from USDA-ARS, National Small Seeds Germplasm Research Unit, Aberdeen, Idaho. This low-phytate line was used as the pollen parent in a cross with the Japanese commercial cultivar "Tanbakuro" [21]. These two parents were crossed in August 2004 under greenhouse conditions, and ten progeny populations were developed from the F1 seeds. The F1 generation and the parents were grown in a field of the Graduate School of Biosphere Science, Hiroshima University, Higashi-Hiroshima, Japan, during the summer of 2005, and low-phytate lines were selected by measuring the phytate and inorganic and total phosphorus concentrations in seeds after harvesting. In this study, we conducted two experiments using one selected low-phytate line each (F10 line for Experiment 1 and F11 line for Experiment 2) and one normal-phytate cultivar each (Enrei for Experiment 1 and Akimaro for Experiment 2), that are cultivated in western Japan. Both the low-phytate and normal-phytate soybeans were sown in a seed bed containing a soil mixture of granite regosol soil, perlite, peat moss, and nursery soil with compost (2:1:1:1 per volume base). Inoculation of root nodule bacteria (*Bradyrhizobium japonicum*) was performed by mixing a small amount of soil from the previous year's cultivation into the seedbed. Both experiments were conducted in the greenhouse for prevention of rainfall.

Experiment 1: Both soybean F_{10} line and cv. Enrei were planted in seed beds containing a mixture of soil as above on 3 June, 2016. After 13 days of germination, uniform seedlings were transplanted to a wooden container (30 cm in width, 10.5 m in length, and 18 cm in depth) filled with a mixture of soil. The plant-to-plant space was 20 cm in containers separated by 100 cm distance from each other. Each treatment was separated by inserting a water proof plastic sheet in order to prevent moisture and nutrient movement between the treatments in the container. Before transplanting, the soil mixture was fertilized with a basal fertilization of 100 kg ha^{-1} of K_2O as potassium sulfate. The soil pH (H_2O) was adjusted to about 6.0 with dolomitic calcium carbonate. P treatment provided the following two fertilization levels: 50 and 150 kg P_2O_5 ha^{-1} as single super phosphate in a randomized complete block design with four replications. At the flowering stage in mid-July, the leaf photosynthetic rate and nitrogen fixation of the nodule were measured. At the full-maturity stage in mid-October, plants were harvested and the stem was separated. The seeds and stem were dried at 80 °C in an air-forced oven for more than 3 days, and weighed.

Experiment 2: Both soybean F11 line and cv. Akimaro were planted in seed beds containing a mixture of soil on 11 June, 2017. After 14 days of germination, one seedling each was transplanted to 9 liters pots (25 cm in diameter, and 23 cm in depth) filled with soil mixture. The pot-to-pot distance was 30 cm with no interaction between the plant roots. Before transplanting, as in Experiment 1, the soil mixture was fertilized with a basal fertilization of 100 kg ha^{-1} of K_2O as potassium sulfate. The soil pH (H_2O) was adjusted to about 6.0 with dolomitic calcium carbonate. P treatment provided the following two fertilization levels: 20 and 100 kg P_2O_5 ha^{-1} as single super phosphate in a randomized complete block design with four replications. The plants were harvested at full-maturity stage in mid-October. After pod and seed numbers were counted and the seeds were weighed, the seeds were oven dried at 70 °C for 72 h for analysis of the mineral and oil concentration.

4.2. Determination of Leaf Photosynthesis and Dinitrogen Fixation

The dinitrogen fixation activity and leaf photosynthetic rate were determined at flowering stage in Experiment 1. The photosynthetic rate was measured on the upper most fully-expanded leaf using portable infrared gas analyzer (Model LI-6400, Licor Co. Ltd., NE, USA) between 10:30 a.m. and 12:30 p.m. While taking the measurements, the photosynthetic active radiation was adjusted at 1200 µmol m^{-2} S^{-1}, the humidity was 65%, the leaf temperature was 28 °C, and the ambient CO_2 concentration was 350 mol L^{-1}. Dinitrogen fixation was analyzed by measuring the reduction of acetylene to ethylene at flowering time as follows. Intact root systems were excised and gently separated from the soil. The root system and attached nodules were quickly placed in a 1000 mL glass bottle. The bottle was sealed with a rubber stopper and 100 mL of air in the bottle was replaced with acetylene. Samples were incubated at ambient laboratory temperatures. During the incubation periods, 0.3 mL of gas samples were extracted at 10 min and 1 hour and injected into a gas chromatograph (Shimazu GC-14B, Shimazu Co., Kyoto, Japan) fitted with a flame ionization detector to determine the ethylene concentration. Specific nitrogenase activity was calculated by dividing the nitrogenase activity of each sample by the dry weight of the nodules.

4.3. Determination of Seed Mineral Concentration

Samples of dried seeds were finely ground using a vibrating sample mill (TI-100, Heiko, Japan) and the mineral concentrations of the seed were measured. Finely ground samples were digested by Sulfuric Acid-Hydrogen Peroxide (H_2SO_4–H_2O_2), and the K concentration was measured using a flame photometer (ANA 135, Tokyo Photoelectric, Tokyo, Japan). Ca and Mg were measured using an atomic absorption flame emission spectrophotometer (AA-6200, Shimadzu, Japan). The total P was determined using a UV-Spectrophotometer (U-3310, Hitachi Co. Ltd. Tokyo, Japan) following the molybdenum reaction solution method [45]. The total nitrogen was measured using the Kjeldahl method after sample digestion with concentrated H_2SO_4 and H_2O_2. The crude protein concentration was calculated based on converting the seed nitrogen to protein percentage by multiplying by a conversion factor of 6.25 [46].

4.4. Determination of Phytate P and Inorganic P Concentration

Seed phytate P was measured according to the method by Raboy et al. [47], where aliquots of flour were extracted in extraction media (0.2 M HCl: 10% Na_2SO_4) overnight at 4 °C with shaking. Extracts were centrifuged, and phytate was obtained as a ferric precipitate and assayed for P calorimetrically using ammonium molybdate reaction reagent. Inorganic phosphorus was extracted in trichloroacetic acid (12.5%) + $MgCl_2$ (2 mmol L^{-1}) while stirring overnight, and Pi was measured using colorimetric methods [45]. The phytic acid concentration was calculated by multiplying phytic acid P values by 3.55 as described by Raboy and Dickinson [48].

4.5. Determination of Zn, Fe and Mn Concentration

Determination the Zn, Fe, and Mn concentration, the seeds were digested by HNO_3 and H_2O_2 (4:1) and measured using an inductively coupled argon atomic emission spectrometer (iCAP 6000, Thermo Fisher Scientific Inc. USA).

4.6. Determination of Lipid Concentration

The lipid concentration was measured by chloroform–methanol 2:1 (v/v) according to the method by Folch et al. [49]. For each sample, a mixed solution of chloroform–methanol (2:1) was added to the dried fine seed samples in a vented conical Erlenmeyer flask, and heated at 65°C for 30 min and boiled for 1 hour. After extraction, the samples were cooled and filtered in an eggplant-type flask using a glass filter with additional mixed solution of chloroform-methanol. The chloroform–methanol solution was evaporated from the sample and were petroleum ether and sodium sulfate were added, the solution was shaken. The samples were centrifuged at 3000 rpm, and the supernatant was transferred to a weighing glass tube and dried at 105°C. After evaporation of the ether, the glass tube was weighed and the lipid concentration calculated.

4.7. Statistical Analysis

All the collected data were subjected to analysis of variance using SPSS statistics package, Student Version 19, and means (n = 4) were separated using the Duncan Multiple Range Test at $p = 0.05$.

5. Conclusions

It is evident from previous literature that phosphorus fertilizer will significantly increase the whole plant growth, leaf photosynthesis, dinitrogen fixation of nodules, and yield of seed. In this study, we further found no significant differences in these physiological parameters between the low-phytate line and normal-phytate cultivar under any fertilization conditions. While seed crude protein and lipid concentrations were found to be higher in the higher levels of P application in both the low-phytate line and normal-phytate cultivar, seed crude protein concentration was higher in the low-phytate line than in the normal-phytate cultivar. These results suggest that reducing phytic acid in seeds does not affect growth after leaf photosynthesis, dinitrogen fixation of nodules, or final yield of plants following germination and stand establishment. P application had little effect on seed Zn, Fe, and Mn, but increasing P fertilization increased seed Cu by 1.7-fold regardless of genotype or cultivar background. The lower molar ratios of phytic acid to Zn, Fe, Cu, and Mn in seed of the low-phytate soybean lines as compared with the normal-phytate soybean cultivars should enhance the bioavailability of these microelements.

Author Contributions: Conceived and designed the experiments, N.A.T., H.S.; breeded and selected the low phytate soybean lines, H.S.; performed experiments, N.A.T., Q.D. and K.E.; wrote the paper, N.A.T. and H.S.; data analysis, writing and editing, V.R.

Funding: This research was funded by Grans-in-Aid for Scientific Research, KAKEN, from Japan Society for the Promotion of Science: Grant Number 15K14678 and 18K05948 to Saneoka H.

Conflicts of Interest: The authors declare no conflict of interest.

References

1. Raboy, V. Seeds for a better future: 'low phytate' seeds help to overcome malnutrition and reduce pollution. *Trends Plant Sci.* **2001**, *6*, 458–462. [CrossRef]
2. Vats, P.; Bhattacharyya, M.; Banerjee, U. Use of phytases (myo-inositol hexakisphosphstate phosphohydrolases) for combatting environmental pollution: A biological approach. *Environ. Sci. Technol.* **2005**, *35*, 469–486. [CrossRef]
3. Lott, J.N.A.; Ockenden, I.; Raboy, V.; Batten, G.D. Phytic acid and phosphorus in crop and fruits: A global estimate. *Seed Sci. Res.* **2000**, *10*, 11–33.
4. Urbano, G.; López-Jurado, M.; Aranda, P.; Vidal-Valverde, C.; Porres, T.J. The role of phytic acid in legumes: Antinutrient or beneficial function? *J. Physiol. Biochem.* **2000**, *56*, 283–294. [CrossRef]
5. Raboy, V. Approaches and challenges to engineering seed phytate and total phosphorus. *Plant Sci.* **2009**, *177*, 281–296. [CrossRef]
6. Erdman, J. Bioavailability of trace minerals from cereals and legumes. *Cereal Chem.* **1981**, *58*, 21–26.
7. Persson, H.; Türk, M.; Nyman, M.; Sandberg, A.S. Binding of Cu^{2+}, Zn^{2+}, and Cd^{2+} to inositol tri-, tera-, penta-, and hexaphosphates. *J. Agric. Food Chem.* **1998**, *46*, 3194–3200. [CrossRef]
8. Holm, P.B.; Kristiansen, K.N.; Pedersen, H.B. Transgenic approaches in commonly consumed cereals to improve iron and zinc concentration and bioavailability. *J. Nutr.* **2002**, *132*, 514S–516S. [CrossRef] [PubMed]
9. Larson, S.R.; Young, K.A.; Cook, A.; Blake, T.K.; Raboy, V. Linkage mapping of two mutations that reduce phytic acid concentration of barley seed. *Theor. Appl. Genet.* **1998**, *97*, 141–146. [CrossRef]
10. Raboy, V.; Cichy, K.; Peterson, K.; Reichman, S.; Sompong, U.; Srinives, P.; Saneoka, H. Barley (*Hordeum vulgare* L.) *Low phytic acid 1-1*: An endosperm-specific, filial determinant of seed total phosphorus. *J. Hered.* **2014**, *105*, 656–665. [CrossRef] [PubMed]
11. Larson, S.R.; Rutger, J.N.; Young, K.A.; Raboy, V. Isolation and genetic mapping of a non-lethal rice (*Oryza sativa* L.) low phytic acid 1 mutation. *Crop. Sci.* **2000**, *40*, 1397–1405. [CrossRef]
12. Raboy, V.; Gerbasi, P.F.; Young, K.A.; Stoneberg, S.D.; Pickett, S.G.; Bauman, A.T.; Murthy, P.P.; Sheridan, W.F.; Ertl, D.S. Origin and seed phenotype of maize *low phytic acid 1-1* and *low phytic acid 2-1*. *Plant Physiol.* **2000**, *124*, 355–368. [CrossRef]
13. Guttieri, M.; Bowen, D.; Dorsch, J.A.; Raboy, V.; Souza, E. Identification and characterization of low phytic acid wheat. *Crop. Sci.* **2004**, *44*, 418–424. [CrossRef]
14. Wilcox, J.R.; Premachandra, G.S.; Young, K.A.; Raboy, V. Isolation of high seed inorganic P, low-phytate soybean mutants. *Crop. Sci.* **2000**, *40*, 1601–1605. [CrossRef]
15. Campion, B.; Sparvoli, F.; Doria, E.; Tagliabue, G.; Galasso, I.; Fileppi, M.; Bollini, R.; Nielsen, E. Isolation and characterisation of an lpa (low phytic acid) mutant in common bean (*Phaseolus vulgaris* L.). *Theor. Appl. Genet.* **2009**, *118*, 1211–1221. [CrossRef] [PubMed]
16. Cominelli, E.; Confalonieri, M.; Carlessi, M.; Cortinovis, G.; Daminati, M.G.; Porch, T.G.; Losa, A.; Sparvoli, F. Phytic acid transport in Phaseolus vulgaris: A new low phytic acid mutant in the PvMRP1 gene and study of the PvMRPs promoters in two different plant systems. *Plant Sci.* **2018**, *270*, 1–12. [CrossRef]
17. Raboy, V.; Dickinson, D.B. Phytic acid levels in seeds of *Glycine max* and *G. soja* as influenced by phosphorus status. *Crop. Sci.* **1993**, *33*, 1300–1305. [CrossRef]
18. Miller, G.A.; Youngs, V.L.; Oplinger, E.S. Effect of available soil-phosphorus and environment on the phytic acid concentration in oats. *Cereal Chem.* **1980**, *57*, 192–194.

19. Saneoka, H.; Koba, T. Plant growth and phytic acid accumulation in seed as affected by phosphorus application in maize (*Zea maize* L.). *Grassl. Sci.* **2003**, *48*, 485–489.

20. Kumar, V.; Singh, T.R.; Hada, A.; Jolly, M.; Ganapathi, A.; Sachdev, A. Probing phosphorus efficient low phytic acid concentration soybean genotypes with phosphorus starvation in hydroponics growth system. *Appl. Biochem. Biotech.* **2015**, *177*, 689–699. [CrossRef]

21. Fukuda, Y.; Tatsukawa, E.; Saneoka, H.; Hoshina, T.; Uefuji, M.; Raboy, V. Growth characteristics, phytate concentrations, and coagulation properties of soymilk from a low-phytate Japanese soybean (*Glycine max* (L.) Merr.) line. *Soil Sci. Plant Nutr.* **2011**, *57*, 674–680. [CrossRef]

22. Sun, J.S.; Simpson, R.J.; Sands, R. Nitrogenase activity of two genotypes of *Acacia mangium* as affected by phosphorus nutrition. *Plant Soil* **1992**, *144*, 51–58. [CrossRef]

23. Naeem, A.; Aslam, M.; Lodhi, A. Improved potassium nutrition retrieves phosphorus-induced decrease in zinc uptake and seed zinc concentration of wheat. *J. Sci. Food Agric.* **2018**, *98*, 4351–4356. [CrossRef]

24. Ao, X.; Guo, X.H.; Zhu, Q.; Zhang, H.J.; Wang, H.Y.; Ma, Z.H.; Han, X.R.; Zhao, M.H.; Xie, F.T. Effect of phosphorus fertilization to P uptake and dry matter accumulation in soybean with different P efficiencies. *J. Integr. Agric.* **2014**, *13*, 326–334. [CrossRef]

25. Singh, S.K.; Reddy, V.R.; Fleisher, D.H.; Timlin, D.J. Phosphorus nutrition affects temperature response of soybean growth and canopy photosynthesis. *Front. Plant Sci.* **2018**, *9*, 116. [CrossRef]

26. Wang, J.; Chen, Y.; Wang, P.; Li, Y.S.; Khan, A. Leaf gas exchange, phosphorus uptake, growth and yield responses of cotton cultivars to different phosphorus rate. *Photosynthetica* **2018**, *56*, 1414–1421. [CrossRef]

27. Ertl, D.; Young, K.A.; Raboy, V. Plant genetic approaches to phosphorus management in agricultural production. *J. Environ. Qual.* **1998**, *27*, 299–304. [CrossRef]

28. Raboy, V.; Young, K.A.; Dorsch, J.A.; Cook, A. Genetics and breeding of seed phosphorus and phytic acid. *J. Plant Physiol.* **2001**, *158*, 489–497. [CrossRef]

29. Meis, S.J.; Fehr, W.R.; Schnebly, S.R. Seed source effect on field emergence of soybean lines with reduced phytate and raffinose saccharides. *Crop. Sci.* **2003**, *43*, 1336–1339. [CrossRef]

30. Oltmans, S.E.; Fehr, W.R.; Welke, G.A.; Raboy, V.; Peterson, K.L. Agronomic and seed traits of soybean lines with low–phytate phosphorus. *Crop. Sci.* **2005**, *45*, 593–598. [CrossRef]

31. Wiggins, S.J.; Smallwood, C.J.; West, D.R.; Kopsell, D.A.; Sams, C.E.; Pantalone, V.R. Agronomic Performance and Seed Inorganic Phosphorus Stability of Low-Phytate Soybean Line TN09-239. *J. Am. Oil Chem. Soc.* **2018**, *95*, 787–796. [CrossRef]

32. Singh, S.K.; Reddy, V.R. Methods of mesophyll conductance estimation: Its impact on key biochemical parameters and photosynthetic limitations in phosphorus-stressed soybean across CO_2. *Physiol. Plant* **2016**, *157*, 234–254. [CrossRef]

33. Gillman, J.D.; Pantalone, V.R.; Bilyeu, K. The low phytic acid phenotype in soybean line CX1834 is due to mutations in two homologs of the maize low phytic acid gene. *Plant Genome* **2009**, *2*, 179–190. [CrossRef]

34. Panzeri, D.; Cassani, E.; Doria, E.; Tagliabue, G.; Forti, L.; Campion, B.; Bollini, R.; Brearley, C.A.; Pilu, R.; Nielsen, E.; et al. A defective ABC transporter of the MRP family, responsible for the bean lpa1 mutation, affects the regulation of the phytic acid pathway, reduces seed myo-inositol and alters ABA sensitivity. *New Phytol.* **2011**, *191*, 70–83. [CrossRef] [PubMed]

35. Raboy, V.; Dickinson, D.B. The timing and rate of phytic acid accumulation in developing soybean seeds. *Plant Physiol.* **1987**, *85*, 841–844. [CrossRef] [PubMed]

36. Oltmans, S.E.; Fehr, W.R.; Welke, G.A.; Raboy, V.; Peterson, K.L.; Peeler, H.T. Biological availability of nutrients in feeds: Availability of major mineral ions. *J. Anim. Sci.* **1972**, *35*, 695–712.

37. Abbasi, M.K.; Tahir, M.M.; Abbas, W.A.; Rahim, N. Soybean yield and chemical composition in response to phosphorus-potassium nutrition in Kashmir. *Agron. J.* **2012**, *104*, 1476–1484. [CrossRef]

38. Yi, X.; Bellaloui, N.; MaClure, A.M.; Tyler, D.D.; Mengistu, A. Phosphorus fertilization differentially influences fatty acids, protein, and oil in soybean. *Am. J. Plant Sci.* **2016**, *7*, 1975–1992.

39. Krueger, K.; Goggi, A.S.; Mallarino, A.P.; Mullen, R.E. Phosphorus and potassium fertilization effects on soybean seed quality and composition. *Crop. Sci.* **2013**, *53*, 602–610. [CrossRef]

40. Bethlenfalvay, G.J.; Schreiner, R.P.; Mihara, K.L. Mycorrhizal fungi effects on nutrient composition and yield of soybean seeds. *J. Plant Nutr.* **1997**, *20*, 581–591. [CrossRef]

41. Toda, K.; Takahashi, K.; Ono, T.; Kitamura, K.; Nakamura, Y. Variation in the phytic acid concentration of soybeans and its effect on consistency of tofu made from soybean cultivars with high protein concentration. *J. Sci. Food Agric.* **2006**, *86*, 212–219. [CrossRef]

42. Sánchez-Rodríguez, A.R.; del Campillo, M.C.; Torrent, J. Phosphorus reduces the zinc concentration in cereals pot-grown on calcareous vertisols from southern Spain. *J. Sci. Food Agric.* **2017**, *97*, 3427–3432. [CrossRef]

43. Zhang, Y.Q.; Deng, Y.; Chen, R.Y.; Cui, Z.L.; Chen, X.P.; Yost, R.; Zhang, F.S.; Zou, C.Q. The reduction in zinc concentration of wheat seed upon increased phosphorus-fertilization and its mitigation by foliar zinc application. *Plant Soil* **2012**, *361*, 143–152. [CrossRef]

44. Zhang, W.; Liu, D.Y.; Li, C.; Chen, X.P.; Zou, C.Q. Accumulation, partitioning, and bioavailability of micronutrients in summer maize as affected by phosphorus supply. *Eur. J. Agron.* **2017**, *86*, 48–59. [CrossRef]

45. Chen, P.S.J.; Toribara, T.Y.; Warner, H. Microdetermination of phosphorus. *Anal. Chem.* **1956**, *28*, 1756–1758. [CrossRef]

46. Mariotti, F.; Tomé, D.; Mirand, P.P. Converting nitrogen into protein-beyond 6.25 and Jones' factors. *Crit. Rev. Food Sci. Nutr.* **2008**, *48*, 177–184. [CrossRef]

47. Raboy, V.; Dickinson, D.B.; Below, F.E. Variation in seed total phosphorus, phytic acid, zinc, calcium, magnesium, and protein among lines of *Glycine max* and *G. soja* L. *Crop. Sci.* **1984**, *24*, 431–434. [CrossRef]

48. Raboy, V.; Dickinson, D.B. Effect of phosphorus and zinc nutrition on soybean seed phytic acid and zinc. *Plant Physiol.* **1984**, *75*, 1094–1098. [CrossRef]

49. Folch, J.; Lees, M.; Sloane Stanley, G.H. A simple method for the isolation and purification of total lipids from animal tissues. *J. Biol. Chem.* **1957**, *226*, 497–509.

© 2019 by the authors. Licensee MDPI, Basel, Switzerland. This article is an open access article distributed under the terms and conditions of the Creative Commons Attribution (CC BY) license (http://creativecommons.org/licenses/by/4.0/).

Review

Can Inositol Pyrophosphates Inform Strategies for Developing Low Phytate Crops?

Catherine Freed, Olusegun Adepoju and Glenda Gillaspy *

Department of Biochemistry, Virginia Tech, Blacksburg, VA 24061, USA; freedc@vt.edu (C.F.); ade1704@vt.edu (O.A.)
* Correspondence: gillaspy@vt.edu; Tel.: +1-540-231-3062

Received: 24 December 2019; Accepted: 15 January 2020; Published: 17 January 2020

Abstract: Inositol pyrophosphates (PP-InsPs) are an emerging class of "high-energy" intracellular signaling molecules, containing one or two diphosphate groups attached to an inositol ring, that are connected with phosphate sensing, jasmonate signaling, and inositol hexakisphosphate (InsP$_6$) storage in plants. While information regarding this new class of signaling molecules in plants is scarce, the enzymes responsible for their synthesis have recently been elucidated. This review focuses on InsP$_6$ synthesis and its conversion into PP-InsPs, containing seven and eight phosphate groups (InsP$_7$ and InsP$_8$). These steps involve two types of enzymes: the ITPKs that phosphorylate InsP$_6$ to InsP$_7$, and the PPIP5Ks that phosphorylate InsP$_7$ to InsP$_8$. This review also considers the potential roles of PP-InsPs in plant hormone and inorganic phosphate (Pi) signaling, along with an emerging role in bioenergetic homeostasis. PP-InsP synthesis and signaling are important for plant breeders to consider when developing strategies that reduce InsP$_6$ in plants, as this will likely also reduce PP-InsPs. Thus, this review is primarily intended to bridge the gap between the basic science aspects of PP-InsP synthesis/signaling and breeding/engineering strategies to fortify foods by reducing InsP$_6$.

Keywords: inositol; inositol phosphate; inositol pyrophosphate; inositol phosphate signaling; inositol phosphate kinases; PPIP5K; ITPK; phytate

---

1. Introduction

The inositol phosphate (InsP) signaling pathway has been implicated in a diverse array of cellular processes across eukaryotic organisms (for reviews, see [1–3]). Inositol phosphates are intracellular signaling molecules built around a simple 6-carbon *myo*-inositol (inositol) scaffold used in signal transduction (Figure 1). Differing phosphorylation patterns on the inositol ring convey specific cellular information, and several combinations of phosphorylation events are possible. Inositol hexakisphosphate (InsP$_6$), also called phytate when complexed with cations, is a molecule where all hydroxyl (OH) groups on the inositol ring are phosphorylated, that serves as both a storage pool of phosphate as well as a signaling molecule in plants [4–6].

InsP$_6$ is the major phosphorus storage sink within the plant seed, comprising up to ~1% of an Arabidopsis seed's dry weight, and roughly 65–85% of total seed phosphorus in cereal crops [6]. InsP$_6$ is one of the most highly electronegative molecules present in the cell, resulting in a greatly limited ability to cross cell membranes. It can be transported into the vacuole by MRP5, an ABC-transporter localized in the vacuolar membrane that has been shown to specifically transport InsP$_6$ [4]. During seed development, the electronegative properties of InsP$_6$ result in the chelation of positively charged metal ions such as Mg^{2+}, Fe^{2+}, Zn^{2+}, Mn^{2+}, Ca^{2+}, and these InsP$_6$ metal complexes accumulate within the protein storage vacuole (PSV) (Figure 2) [7,8].

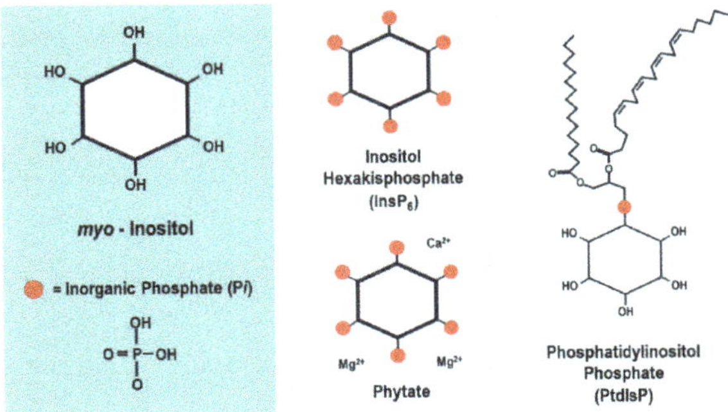

Figure 1. Simplified structures of *myo*-inositol, InsP$_6$, phytate, and PtdInsP. Inorganic phosphate (P*i*) groups, covalently bound to the *myo*-inositol ring by an oxygen molecule, are represented in red. For simplification purposes, this figure does not take into account the axial and equatorial positions of the moieties.

While InsP$_6$ is important for plant survivorship and yield, it is also, unfortunately, an anti-nutrient, as it can prevent useful cations and minerals from being absorbed by the animal gut [9–11]. Additionally, non-ruminant animals cannot digest InsP$_6$, and that which is excreted from livestock is a major environmental concern as it leads to phosphorus pollution, eutrophication, and toxified watersheds [12,13]. Given this, plant breeders and biotechnologists have sought to limit the production of InsP$_6$ in plants, resulting in the so-called low-phytate crops [14]. However, reducing InsP$_6$ synthesis and accumulation in plants can come with consequences, such as a significant decline in vital plant signaling molecules derived from InsP$_6$.

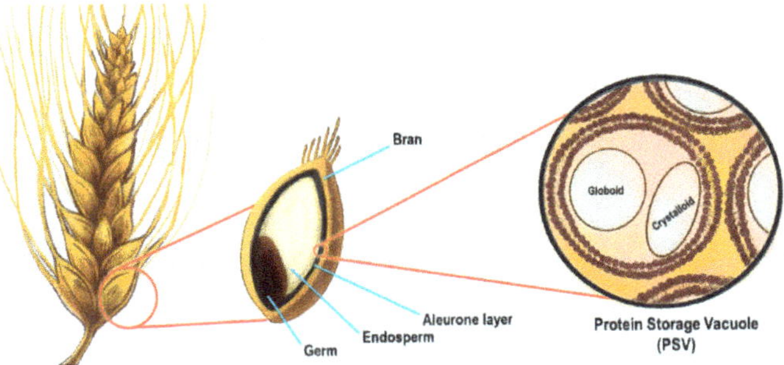

Figure 2. Diagram illustrating phytate storage in wheat granules. Phytate and other cations are stored in the protein storage vacuole (PSV) within small, crystalline storage bodies, known as globoids [9,15]. While the location of globoids varies across species, a majority of these are found within the aleurone layer of cereal crops, such as wheat and barley, as well as oilseed crops including peanuts, hemp, sunflower, and cotton [6,9,16]. Maize globoids are concentrated within the germ, whereas soybean phytate is distributed throughout the entire seed [6,17,18]. PSVs also contain crystalloid bodies, containing stored proteins and lipids [9].

Our primary expertise and focus are on an emerging class of InsPs derived from InsP$_6$ that contain one or two diphosphate or pyrophosphate groups (PP) attached to the inositol ring (Figure 1) [5,19]. These inositol pyrophosphates (PP-InsPs) are gaining significant attention due to their newly discovered roles in energetic metabolism as well as hormone signaling and P*i* sensing (for a review, see [20]). The presence of pyrophosphate bonds in PP-InsPs has resulted in their consideration as "energetic signaling" molecules, and we note that PP-InsPs are similar in structure to ADP or ATP. PP-InsP biosynthesis is well described in non-plant eukaryotes, such as yeast and mammals, and many physiological roles have been linked to these molecules [1–3]. Elucidating the route of PP-InsP biosynthesis in plants has recently been accomplished and is fundamentally critical to our understanding of these molecules, which is described in the following sections. We will start with an overview of how plants synthesize the precursor to PP-InsPs, $InsP_6$, as several of the enzymes in this pathway are key to understanding the PP-InsP pathway.

2. $InsP_6$ Synthesis: The Lipid-Dependent Pathway

$InsP_6$ can be synthesized by two interconnected pathways in plants. The pathways are named for their starting material: the Lipid-Dependent pathway and the Lipid-Independent pathway (Figure 3). The Lipid-Dependent pathway is present in all eukaryotic organisms [21–25]. The Lipid-Independent pathway for synthesizing $InsP_6$ was originally discovered in *Dictyostelium* in a landmark paper by Stephens and Irvine in 1990, and followed up on in *Spirodela polyrhiza* [26,27]. This pathway was thought be unique to these organisms, along with land plants. However, a very recent publication by Desfougères et al. shows that the Lipid-Independent pathway is also present in mammals [28]. This work is the first to report evidence of the Lipid-Independent pathway in mammals and will be crucial for exploring the evolution of enzymes across organisms.

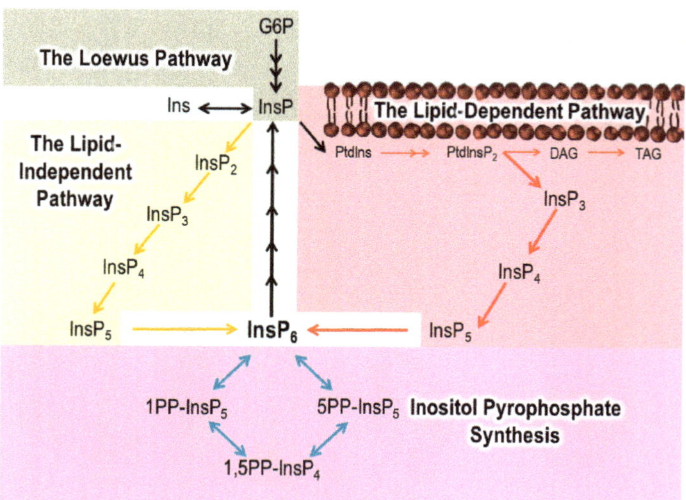

Figure 3. A simplified view of the InsP synthesis and degradation pathway. InsP synthesis starts in the Loewus Pathway (tan), where InsP is synthesized from Glucose-6-Phosphate (G6P). InsPs are synthesized through the Lipid-Dependent (pink) or Lipid-Independent (yellow) pathways. PP-InsPs in plants are synthesized from $InsP_6$ (purple). The enzymes involved in the pathway are discussed throughout the review.

The lipid component in the Lipid-Dependent pathway is phosphatidylinositol phosphate (PtdInsP), a molecule containing inositol as the head group (Figure 1). While plants synthesize a myriad of lipid-soluble PtdInsPs, phosphatidylinositol (4,5) bisphosphate ($PtdIns(4,5)P_2$), is important for the Lipid-Dependent pathway as it is acted on by the enzyme phospholipase C (PLC) [29]. Phospholipases, by definition, hydrolyze phospholipids. The hydrolysis of $PtdIns(4,5)P_2$ by PLC produces $Ins(1,4,5)P_3$ and diacylglycerol (DAG), which essentially converts a phosphorylated lipid-signaling molecule ($PtdIns(4,5)P_2$) into a water-soluble, InsP-signaling molecule ($Ins(1,4,5)P_3$) (Figure 3) [29].

Ins(1,4,5)P$_3$ can be subsequently phosphorylated into InsP$_4$, and then InsP$_5$, by a dual specific inositol polyphosphate multikinase (IPMK). The IPMK enzyme is encoded by two genes in Arabidopsis, *AtIPK2α* and *AtIPK2β* (Table 1) [30]. Both genes encode enzymes with a 6/3-kinase activity, catalyzing the conversion of Ins(1,4,5)P$_3$ to Ins(1,4,5,6)P$_4$ and to a final Ins(1,3,4,5,6)P$_5$ product [31]. 5-kinase activity towards Ins(1,3,4,6)P$_4$ and Ins(1,2,3,4,6)P$_5$ was also reported by Stevenson-Paulik et al. Notably, these genetic studies show that the *AtIPK2β* gene can complement a yeast *ipk2* mutant, and that Arabidopsis T-DNA loss-of-function *atipk2β* mutants have a 35% reduction in seed InsP$_6$ (Table 1) [30]. *atipk2α* mutants are not easily studied as those that are recovered are lethal. This, along with the generally ubiquitous expression of *AtIPK2α*, supports the idea that *AtIPK2α* supplies the major IPK2 or IPMK activity in the plant cell.

Table 1. Loss-of-function Arabidopsis mutants and impacts on InsP$_6$ and PP-InsP levels. Arabidopsis is a simple model system that can be used to gauge the impacts of genetic changes on InsPs. The table shows the impact on InsP$_6$ and PP-InsPs in Arabidopsis mutants for enzymes involved in InsP synthesis. Mutants for enzymes important in both the Lipid-Dependent and Lipid-Independent pathways are indicated (*).

Pathway	Mutant	Impact on InsP$_6$	Impact on PP-InsPs
Lipid-Dependent Pathway	*plc*	Nine genes; characterized single mutants have no change in InsP$_6$ [32]	No Change [32]
	ipk2α *	Lethal Knock-Out [30]	Unknown
	ipk2β *	35% reduction in mass seed InsP$_6$; no change in seedling tissue as measured by radiolabeling [30]	Unknown
	ipk1 *	83% reduction in mass seed InsP$_6$; 93% reduction in seedlings as measured by radiolabeling [30]	InsP$_7$ and InsP$_8$ are reduced [33,34]
Lipid-Independent Pathway	*mik*	62–66% reduction in mass seed InsP$_6$ [35]	Unknown
	lpa1	47–57% reduction in mass seed InsP$_6$ [35]	Unknown
	itpk1–4	46% reduction in mass seed InsP$_6$ in *Atitpk1* mutants; no changes in *Atitpk2* or *Atitpk3*; 40–51% reduction in *Atitpk4* [35]	*Atitpk1* and *Atitpk4* have reduced InsP$_7$ and InsP$_8$ [33,36]
Phytate Storage	*mrp5*	73–80% reduction in mass seed InsP$_6$ [35]; decreased InsP$_6$ in seeds and vegetative tissue as measured by radiolabeling [4]	Elevated InsP$_7$ and InsP$_8$ [5]
PP-InsP Synthetic Pathway	*vip1*	No change reported [34,37]	Increased InsP$_7$ and Decreased InsP$_8$ [34,37]
	vip1/vip2	No change reported [37]	Increased InsP$_7$ and Decreased InsP$_8$ [37,38]

The last step in the Lipid-Dependent pathway of InsP$_6$ biosynthesis is the phosphorylation of Ins(1,3,4,5,6)P$_5$ to InsP$_6$. This step is catalyzed by only one type of enzyme in nature, the inositol pentakisphosphate 2-kinase (IPK1), so named because all known IPK1 enzymes phosphorylate the 2-position of InsP$_5$ [30]. While there are seven genes in Arabidopsis that are predicted to encode IPK1 enzymes, the *At5g42810* gene (*AtIPK1*) is the only one actively expressed in plants [30]. Complementation assays reveal that *AtIPK1* is able to complement a yeast *ipk1* mutant, restore InsP$_6$ levels, and rescue the mutant's temperature-sensitive growth phenotype [39]. As loss of IPK1 function results in an 83% reduction in InsP$_6$ in seeds (Table 1), this shows that IPK1 plays a major role in maintaining seed InsP$_6$ levels [30].

3. InsP$_6$ Synthesis: The Lipid-Independent Pathway

Given the importance of *Pi* storage in plants, it is not surprising that plants evolved a separate way to synthesize InsP$_6$, apart from the PtdInsP pathway. In the Lipid-Independent pathway, "free" *myo*-inositol is acted on by a series of inositol kinases. The first is the *myo*-inositol kinase (MIK), which was first identified in maize as a product of the *Low Phytic Acid* gene [40]. Loss-of-function maize *lpa3* mutants have reduced InsP$_6$ and elevated inositol levels in the seeds [40]. Arabidopsis *atmik* mutants have a large reduction in total seed mass InsP$_6$ levels (Table 1) [35]. The second step in the Lipid-Independent pathway is likely catalyzed by a gene/protein named LPA1 in rice [41]. This gene product was originally categorized as a potential 2-phosphoglycerate kinase that impacts InsP$_6$ accumulation [41]. However, recent structural modeling indicates that InsP$_3$ can be accommodated in the active site of this kinase, supporting a role for it as an InsP kinase [42].

The next step in this pathway should involve an inositol kinase capable of phosphorylating an InsP$_2$ substrate. As no such kinase has been identified, this prompted speculation that InsP$_2$'s conversion to InsP$_3$ is catalyzed by a moonlighting function of another inositol phosphate kinase, which has yet to be identified. The last novel component of the Lipid-Independent pathway is the inositol triphosphate kinase (ITPK), which can phosphorylate specific InsP$_3$ and InsP$_4$ molecules [39,43]. A gene encoding ITPK1 was also identified in a maize mutant named *lpa2* [44]. There are four ITPK enzymes (AtITPK1–4) in Arabidopsis [45]. It is interesting to note that only Arabidopsis *atitpk1* and *atitpk4* mutants have reduced seed InsP$_6$ levels (Table 1) [33,35]. Our own work with AtITPK1 and AtITPK2 enzymes shows that these proteins are also efficient at converting InsP$_6$ to InsP$_7$ [38]. Laha et al. additionally used NMR to show that 5PP-InsP$_5$ is the isoform synthesized by AtITPK1 and AtITPK2 [36]. These are key findings that highlight the catalytic flexibility of the ITPKs, as well as indicating that the Lipid-Independent pathway may have an important relationship with, and impact on, PP-InsP synthesis.

The ITPKs are thought to act in concert with the IPK2 enzymes in producing Ins(1,3,4,5,6)P$_5$ in the Lipid-Independent pathway [27,31,43]. The final step is the conversion of Ins(1,3,4,5,6)P$_5$ to InsP$_6$. Both pathways utilize IPK1 to synthesize InsP$_6$ and converge at this last step in InsP synthesis. This further highlights the importance of IPK1 in InsP$_6$ synthesis. A very recent publication on non-plant organisms suggests that there might be some conserved functions of the Lipid-Independent pathway in other eukaryotes. Humans, for example, contain ITPK genes, and the expression of these enzymes appears to complement mutants defective in PLC, which cannot synthesize InsP$_6$ [28,46]. Although more work needs to be done, this suggests that animals may also utilize ITPKs in a Lipid-Independent InsP$_6$ pathway. In the same work, these authors also found that the plant ITPK1 could complement the yeast PLC mutant, suggesting that the plant ITPK may also have a very flexible substrate preference, and may act at several different steps in the Lipid-Independent pathway [28].

4. PP-InsP Synthetic Pathway

While a great majority of InsP$_6$ is stored as phytate, a small pool can be further phosphorylated to form PP-InsPs. Our group and others have examined a class of enzymes involved in PP-InsP synthesis named the diphosphoinositol-pentakisphosphate kinases (PPIP5Ks), known as VIP or VIH in plants, and Vip1 in *Chlamydomonas reinhardtii* (algae) [5,34,47]. Mammalian and yeast PPIP5K enzymes phosphorylate the 1-position on InsP$_6$ and 5PP-InsP$_5$ to generate an InsP$_7$ molecule, 1PP-InsP$_5$, and an InsP$_8$ molecule, 1,5(PP)$_2$-InsP$_4$, respectively [48–52]. Two Arabidopsis genes, *AtVIP1* and *AtVIP2*, are orthologous to the mammalian *PPIP5K* genes [5,34]. AtVIP1 (also referred to as AtVIH2) and AtVIP2 (AtVIH1) are dual-domain enzymes, consisting of an ATP-grasp N-terminal kinase domain (KD) and a C-terminal histidine phosphatase domain (PD) [5]. We recently found that the KD of both AtVIP enzymes can phosphorylate 5PP-InsP$_5$ in vitro [38]. Additionally, the Hothorn group used NMR and showed that the product of the AtVIP enzymes is indeed 1,5-PP-InsP$_4$ [37].

A second class of enzymes, known as the inositol hexakisphosphate kinases (IP6Ks), function in non-plant organisms by phosphorylating the 5-position of InsP$_5$ and InsP$_6$ to generate InsP$_7$ [53]. While the genes coding for IP6Ks are present in humans and yeast, there is no identifiable *IP6K* gene in the plant genome [5]. This prompted us and other groups to speculate the AtVIPs might be bifunctional enzymes, phosphorylating both InsP$_6$ and InsP$_7$. While our biochemical analyses of the AtVIP KDs did not rule out the possibility that these enzymes can phosphorylate InsP$_6$, they suggested that other enzymes likely had to exist in plants to drive this reaction [38]. As the ITPKs phosphorylate the 5-position of lower InsPs, we decided to target this class of enzymes, and found that both AtITPK1 and AtITPK2 are able to phosphorylate InsP$_6$ in vitro [36,38]. We also demonstrated that the AtITPK1 product could be further phosphorylated by the AtVIP1-KD, resulting in InsP$_8$ [38]. Based on our findings, as well as recent work by Laha et al., we conclude that the AtITPKs are the missing enzyme in the pathway [36,38].

5. How Do PP-InsPs Function in Plants?

Williams et al. suggested, in a review published in 2015, that a major function of PP-InsPs in plants was as a "glue" to bring together various protein binding partners [20]. At the time of the review, it was known that InsP$_6$ could bind to the transport inhibitor response 1 (TIR1) auxin receptor [54]. Additionally, InsP$_7$ was hypothesized to bind to the jasmonate (JA) receptor based on structural modeling experiments [34]. Exciting data, of importance to crop breeders, details how InsP$_6$ and InsP$_7$ are able to complex with key proteins involved in Pi-sensing in plants [55]. Soils depleted in phosphorus lead plants to induce a suite of molecular and physiological mechanisms to enhance Pi scavenging, known as the Pi starvation response (PSR) [56,57]. The PSR is facilitated by an increase in transcription of a group of PSR genes, leading to increases in Pi transport and uptake. Upregulation of PSR gene expression is regulated by Phosphate Starvation Response Regulator 1 (PHR1), a transcription factor, along with its homologs [56,57]. PHR1 and homologous transcription factors have a high binding affinity for promoters containing PHR1 binding sequences (P1BS), which allows for the binding and up-regulation of PSR genes under low Pi conditions [56,57].

InsP$_6$ and PP-InsPs regulate Pi sensing via facilitating complex formations between the PHR1 transcription factor and the SPX domain-containing proteins (Figure 4) [58]. PHR and SPX proteins isolated from *Oryza sativa* (rice), known as OsPHR2 and OsSPX4, respectively, can complex with InsP$_6$ or InsP$_7$ in vitro [55]. InsP$_8$ has an even lower dissociation constant than InsP$_7$ in the OsPHR2–OsSPX4 complex formation [59]. Together, these data support the idea that InsP$_8$ is the main mediator of the PHR1:SPX complex formation in plants.

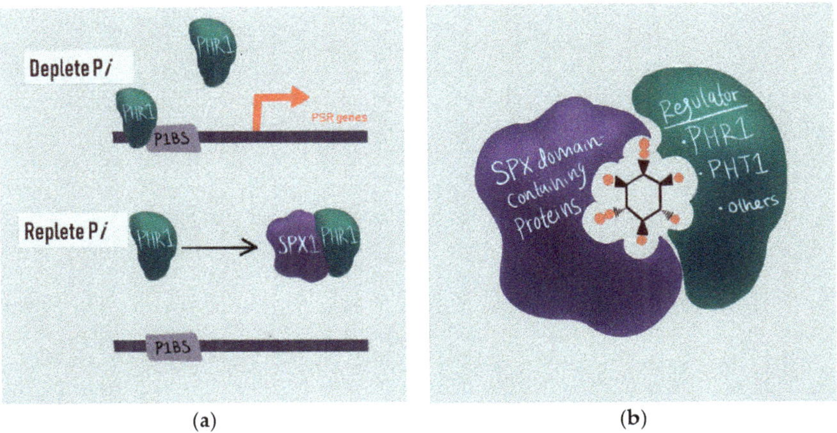

Figure 4. Model depicting how PP-InsPs regulate plant Pi sensing and the Pi starvation response (PSR). (**a**) PSR gene regulation under deplete (top) and replete Pi conditions (bottom). SPX1 binds to PHR1 under replete Pi, preventing SPX1 from binding to promoters containing P1BS. Under deplete Pi, PHR1 is uninhibited from binding to P1BS promoters. Adapted from [60]. (**b**) Model depicting a complex formation between SPX1, PHR1, and InsP$_8$. This model represents interactions between other SPX proteins and PSR regulators, such as PHR1 and its homolog, PHT1, along with others.

While InsP$_7$ has a ~7-fold stronger binding affinity than InsP$_6$ to OsSPX4, recent genetic analyses greatly support the idea that InsP$_8$ is the major signaling molecule, or proxy, that conveys information on the Pi status within the plant cell [37,61]. First, *atipk1, atitpk1, atitpk4* and *atvip1/vip2* mutants commonly show defects in Pi sensing, such that the PSR is turned on even when grown under Pi-replete conditions [30,33,37,61]. Additionally, all of these particular mutants have decreased InsP$_8$ levels (Table 1). In the case of *atvip1/vip2* double mutants, the PSR is likely so highly upregulated that growth becomes stunted and lethality occurs in double homozygotes [37,61]. In contrast, *atipk1* and *atitpk1* mutants can grow fairly normally under certain conditions, and can be stimulated to further increase the PSR under Pi-replete conditions [33,62]. All of this information suggests that InsP$_8$ functions to turn off the PSR.

We now know in greater detail that PP-InsPs likely function in binding to plant hormone receptors and transcription factor complexes involved in Pi sensing. One example where PP-InsPs potentially function as cofactors is in the case of auxin signaling [54]. Auxin is a phytohormone which regulates numerous plant developmental processes and responses to environmental stress [63,64]. Auxin modulates gene expression by binding to the auxin receptor TIR1, an F-box protein, and mediates the SCF ubiquitin–ligase-catalyzed proteolysis of AUX/IAA transcriptional repressors [65,66]. The crystallized Arabidopsis TIR1 protein complex has a tightly bound InsP$_6$ in the leucine-rich repeat (LRR) domain of TIR1, suggesting that InsP$_6$ is a cofactor for the auxin receptor [54].

A later study identified Ins(1,2,4,5,6)P$_5$ in the crystallized structure of a homologous plant hormone-JA co-receptor [67]. JA is a phytohormone critical for environmental and pathogen defense signaling as well as plant physiology [68]. Similar to auxin signaling with TIR1, JA is also perceived by interactions between F-box protein, coronatine-insensitive 1 (COI1), and the JA zim domain (JAZ) transcriptional repressors [69–71]. JA gene regulation is modulated by JA-hormone binding to COI1 and the degradation of JAZ repressors, freeing repressed transcription factors to upregulate JA genes [67]. Sheard et al. showed that Ins(1,2,4,5,6)P$_5$ binds to and stabilizes COI, suggesting that Ins(1,2,4,5,6)P$_5$ is a cofactor for the JA receptor [67,72]. A study using structural modeling predicted that the PP-InsPs have a much stronger binding affinity for COI1-JAZ1 than Ins(1,2,4,5,6)P$_5$ and InsP$_6$, suggesting that the PP-InsPs might be the true cofactor for the JA receptor [34].

6. Consequences for Targeting InsP$_6$ for Reduction

With our current understanding of PP-InsP synthesis, and analyses of genetic mutants in the pathway, it seems reasonable to question whether reducing InsP$_6$ in plants will also result in reduction in PP-InsPs. Most characterized Arabidopsis mutants showing alterations in InsP$_6$ also have impacted the intracellular levels of InsP$_7$ and InsP$_8$ (Table 1). Mutants with reduced PP-InsPs, such as *atipk1, atitpk1,* and *atvip1/vip2* mutants, show an upregulation of the PSR, which could impact engineered crop performance in the field [33,37,61]. Specifically, altered Pi sensing could negatively impact plant growth and development, reduce viability, and alter root architecture [56,57]. Alterations in Pi sensing can affect other signaling systems, such as the sensing and accumulation of nitrate and other micronutrients, along with the ABA signaling pathway, often linked to stress pathways [73,74].

Given the link between InsP$_8$ and JA signaling, decreasing PP-InsPs might result in crops that are more susceptible to pathogens and insects [34]. This impact is seen in Arabidopsis *atvip1* mutants, which have reduced InsP$_8$, and show increased susceptibility to insect herbivory as compared to WT plants [34]. Additionally, both transgenic potato plants and Arabidopsis mutants with reduced *myo-inositol phosphate synthase* (MIPS), the first committed step in inositol synthesis, have decreased InsP$_6$ and are more susceptible to pathogenic viruses [75]. While the PP-InsPs were not quantified in these mutants, it is possible that the PP-InsPs were also reduced and are a causative factor in the increased susceptibility to pathogens.

7. Concluding Remarks

Ultimately, understanding the genetic mechanisms for controlling PP-InsP synthesis and function is critical for developing novel low-phytate crops that are not compromised by changes in PP-InsP signaling. A recent review nicely details existing transgenic strategies used to reduce phytate in plants [76]. A common strategy is to use tissue-specific promoters to drive the overexpression of enzymes that break down $InsP_6$, such as Phytase, or to knock-down the expression of genes required for $InsP_6$ synthesis. By doing so, the idea is that only specific tissues will have reduced $InsP_6$. This strategy works well to reduce $InsP_6$ in seeds, for example, without negatively impacting vegetative tissues. One potential drawback of this approach is that the reduction in $InsP_6$ may also affect the precursor available for PP-InsP synthesis in these transgenic plants. It is also important to consider that plants store approximately 1% $InsP_7$ and $InsP_8$ in seeds [5]. We do not know much currently about whether and how PP-InsPs might regulate seed phosphate storage, however, we point out that existing data on *mrp5* $InsP_6$ transporter mutants indicate that $InsP_6$ modulation in seed can result in increases in PP-InsPs [5]. Given the emerging role of PP-InsPs in controlling critical plant sensing and signaling pathways, the future development of strategies for phytate reduction without compromising PP-InsP synthesis and function should be considered by plant breeders.

Author Contributions: Conceptualization, C.F., G.G. and O.A.; writing—original draft preparation, C.F., G.G. and O.A.; writing—review and editing, C.F. All authors have read and agreed to the published version of the manuscript.

Funding: We gratefully acknowledge the NSF for funding to GG (MCB 1616038). This work is supported in part by the USDA National Institute of Food and Agriculture, Hatch project VA-136334.

Conflicts of Interest: The authors declare no conflict of interest.

References

1. Livermore, T.; Azevedo, C.; Kolozsvari, B.; Wilson, M.; Saiardi, A. Phosphate, inositol and polyphosphates. *Biochem. Soc. Trans.* **2016**, *44*, 253–259. [CrossRef]
2. Chakraborty, A. The inositol pyrophosphate pathway in health and diseases. *Biol. Rev. Camb. Philos. Soc.* **2019**, *93*, 1203–1227. [CrossRef]
3. Shears, S.B. Intimate connections: Inositol pyrophosphates at the interface of metabolic regulation and cell signaling. *J. Cell Physiol.* **2018**, *233*, 1897–1912. [CrossRef]
4. Nagy, R.; Grob, H.; Weder, B.; Green, P.; Klein, M.; Frelet-Barrand, A.; Schjoerring, J.K.; Brearley, C.; Martinoia, E. The Arabidopsis ATP-binding cassette protein AtMRP5/AtABCC5 is a high affinity inositol hexakisphosphate transporter involved in guard cell signaling and phytate storage. *J. Biol. Chem.* **2009**, *284*, 33614–33622. [CrossRef]
5. Desai, M.; Rangarajan, P.; Donahue, J.L.; Williams, S.P.; Land, E.S.; Mandal, M.K.; Phillippy, B.Q.; Perera, I.Y.; Raboy, V.; Gillaspy, G.E.; et al. Two inositol hexakisphosphate kinases drive inositol pyrophosphate synthesis in plants. *Plant J.* **2014**, *80*, 642–653. [CrossRef]
6. Raboy, V. Seed Total Phosphate and Phytic Acid. In *Molecular Genetic Approaches to Maize Improvement*; Biotechnology in Agriculture and Forestry; Kriz, A.L., Larkins, B.A., Eds.; Springer: Berlin, Germany, 2009; pp. 41–53.
7. Otegui, M.S.; Capp, R.; Staehelin, L.A. Developing Seeds of Arabidopsis Store Different Minerals in Two Types of Vacuoles and in the Endoplasmic Reticulum. *Plant Cell* **2002**, *14*, 1311–1327. [CrossRef]
8. Yang, S.-Y.; Huang, T.-K.; Kuo, H.-F.; Chiou, T.-J. Role of vacuoles in phosphorus storage and remobilization. *J. Exp. Bot.* **2017**, *68*, 3045–3055. [CrossRef]
9. Bohn, L.; Meyer, A.S.; Rasmussen Søren, K. Phytate: Impact on environment and human nutrition. A challenge for molecular breeding. *J. Zhejiang Univ. Sci. B* **2008**, *9*, 165–191. [CrossRef]
10. Raboy, V. Seeds for a better future: 'Low phytate' grains help to overcome malnutrition and reduce pollution. *Trends Plant Sci.* **2001**, *6*, 458–462. [CrossRef]
11. Cowieson, A.J.; Acamovic, T.; Bedford, M.R. Phytic acid and phytase: Implications for protein utilization by poultry. *Poult. Sci.* **2006**, *85*, 878–885. [CrossRef]

12. Abelson, P.H. A Potential Phosphate Crisis. *Science* **1999**, *283*, 2015. [CrossRef]
13. Sharpley, A.N.; Withers, P.J.A. The environmentally-sound management of agricultural phosphorus. *Fertil. Res.* **1994**, *39*, 133–146. [CrossRef]
14. Raboy, V. Progress in Breeding Low Phytate Crops. *J. Nutr.* **2002**, *132*, 503S–505S. [CrossRef]
15. Isayenkov, S. Plant vacuoles: Physiological roles and mechanisms of vacuolar sorting and vesicular trafficking. *Cytol. Genet.* **2014**, *48*, 127–137. [CrossRef]
16. Perera, I.; Fukushima, A.; Tatsuki, A.; Horiguchi, G.; Saman, S.; Naoki, H. Expression regulation of myo-inositol 3-phosphate synthase 1 (INO1) in determination of phytic acid accumulation in rice grain. *Sci. Rep.* **2019**, *9*, 14866. [CrossRef]
17. O'Dell, B.L.; De Boland, A.R.; Koirtyohann, S.R. Distribution of phytate and nutritionally important elements among the morphological components of cereal grains. *J. Agric. Food Chem.* **1972**, *20*, 718–723. [CrossRef]
18. Erdman, J. Oilseed phytates: Nutritional implications. *J. Am. Oil Chem. Soc.* **1979**, *56*, 736–741. [CrossRef]
19. Shears, S.B. Inositol pyrophosphates: Why so many phosphates? *Adv. Biol. Regul.* **2015**, *57*, 203–216. [CrossRef]
20. Williams, S.P.; Gillaspy, G.E.; Perera, I.Y. Biosynthesis and possible functions of inositol pyrophosphates in plants. *Front. Plant Sci.* **2015**, *6*, 67. [CrossRef]
21. Odom, A.R.; Stahlberg, A.; Wente, S.R.; York, J.D. A Role for Nuclear Inositol 1,4,5-Trisphosphate Kinase in Transcriptional Control. *Science* **2000**, *287*, 2026–2029. [CrossRef]
22. Perera, N.M.; Michell, R.H.; Dove, S.K. Hypo-osmotic Stress Activates Plc1p-dependent Phosphatidylinositol 4,5-Bisphosphate Hydrolysis and Inositol Hexakisphosphate Accumulation in Yeast. *J. Biol. Chem.* **2004**, *279*, 5216–5226. [CrossRef]
23. York, J.; Odom, A.R.; Murphy, R.; Ives, E.; Wente, S.R. A Phospholipase C-Dependent Inositol Polyphosphate Kinase Pathway Required for Efficient Messenger RNA Export. *Science* **1999**, *285*, 96–100. [CrossRef]
24. Fujii, M.; York, J.D. A role for rat inositol polyphosphate kinases rIPK2 and rIPK1 in inositol pentakisphosphate and inositol hexakisphosphate production in rat-1 cells. *J. Biol. Chem.* **2005**, *280*, 1156–1164. [CrossRef]
25. Seeds, A.M.; Sandquist, J.C.; Spana, E.P.; York, J.D. A molecular basis for inositol polyphosphate synthesis in Drosophila melanogaster. *J. Biol. Chem.* **2004**, *279*, 47222–47232. [CrossRef]
26. Stephens, L.R.; Irvine, R.F. Stepwise phosphorylation of myo-inositol leading to myo-inositol hexakisphosphate in Dictyostelium. *Nature* **1990**, *346*, 580–583. [CrossRef]
27. Brearley, C.A.; Hanke, D.E. Metabolic evidence for the order of addition of individual phosphate esters in the myo-inositol moiety of inositol hexakisphosphate in the duckweed *Spirodela polyrhiza* L. *Biochem. J.* **1996**, *314*, 227–233. [CrossRef]
28. Desfougères, Y.; Wilson, M.S.C.; Laha, D.; Miller, G.J.; Saiardi, A. ITPK1 mediates the lipid-independent synthesis of inositol phosphates controlled by metabolism. *Proc. Natl. Acad. Sci. USA* **2019**, *116*, 24551–24561. [CrossRef]
29. Berridge, M.J. Inositol trisphosphate and calcium signalling. *Nature* **1993**, *361*, 315–325. [CrossRef]
30. Stevenson-Paulik, J.; Bastidas, R.; Chiou, S.-T.; Frye, R.; York, J. Generation of phytate-free seeds in Arabidopsis through disruption of inositol polyphosphate kinases. *Proc. Natl. Acad. Sci. USA* **2005**, *102*, 12612–12617. [CrossRef]
31. Stevenson-Paulik, J.; Odom, A.R.; York, J.D. Molecular and Biochemical Characterization of Two Plant Inositol Polyphosphate 6-/3-/5-Kinases. *J. Biol. Chem.* **2002**, *277*, 42711–42718. [CrossRef]
32. Zhang, Q.; van Wijk, R.; Shahbaz, M.; Roels, W.; van Schooten, B.; Vermeer, J.E.M.; Zarza, X.; Guardia, A.; Scuffi, D.; García-Mata, C.; et al. Arabidopsis Phospholipase C3 is Involved in Lateral Root Initiation and ABA Responses in Seed Germination and Stomatal Closure. *Plant Cell Physiol.* **2018**, *59*, 469–486. [CrossRef]
33. Kuo, H.-F.; Hsu, Y.-Y.; Lin, W.-C.; Chen, K.-Y.; Munnik, T.; Brearley, C. Arabidopsis inositol phosphate kinases IPK1 and ITPK1 constitute a metabolic pathway in maintaining phosphate homeostasis. *Plant J.* **2018**, *95*, 613–630. [CrossRef]
34. Laha, D.; Johnen, P.; Azevedo, C.; Dynowski, M.; Weiß, M.; Capolicchio, S.; Mao, H.; Iven, T.; Steenbergen, M.; Freyer, M.; et al. VIH2 Regulates the Synthesis of Inositol Pyrophosphate InsP8 and Jasmonate-Dependent Defenses in Arabidopsis. *Plant Cell* **2015**, *27*, 1082–1097. [CrossRef]
35. Kim, S.-I.; Tai, T.H. Identification of genes necessary for wild-type levels of seed phytic acid in Arabidopsis thaliana using a reverse genetics approach. *Mol. Genet. Genom.* **2011**, *286*, 119–133. [CrossRef]

36. Laha, D.; Parvin, N.; Hofer, A.; Giehl, R.F.H.; Fernandez-Rebollo, N.; von Wirén, N.; Saiardi, A.; Jessen, H.J.; Schaaf, G. Arabidopsis ITPK1 and ITPK2 Have an Evolutionarily Conserved Phytic Acid Kinase Activity. *ACS Chem. Biol.* **2019**, *14*, 2127–2133. [CrossRef]

37. Zhu, J.; Lau, K.; Puschmann, R.; Harmel, R.K.; Zhang, Y.; Pries, V.; Gaugler, P.; Broger, L.; Dutta, A.K.; Jessen, H.J.; et al. Two bifunctional inositol pyrophosphate kinases/phosphatases control plant phosphate homeostasis. *eLife* **2019**, *8*, e43582. [CrossRef]

38. Adepoju, O.; Williams, S.P.; Craige, B.; Cridland, C.A.; Sharpe, A.K.; Brown, A.M.; Land, E.; Perera, I.Y.; Mena, D.; Sobrado, P.; et al. Inositol Trisphosphate Kinase and Diphosphoinositol Pentakisphosphate Kinase Enzymes Constitute the Inositol Pyrophosphate Synthesis Pathway in Plants (pre-print). *bioRxiv* **2019**, 724914. [CrossRef]

39. Sweetman, D.; Johnson, S.; Caddick, S.E.K.; Hanke, D.E.; Brearley, C.A. Characterization of an Arabidopsis inositol 1,3,4,5,6-pentakisphosphate 2-kinase (AtIPK1). *Biochem. J.* **2006**, *394*, 95–103. [CrossRef]

40. Shi, J.; Wang, H.; Hazebroek, J.; Ertl, D.S.; Harp, T. The maize low-phytic acid 3 encodes a myo-inositol kinase that plays a role in phytic acid biosynthesis in developing seeds. *Plant J.* **2005**, *42*, 708–719. [CrossRef]

41. Kim, S.-I.; Tai, T.H. Genetic analysis of two OsLpa1-like genes in Arabidopsis reveals that only one is required for wild-type seed phytic acid levels. *Planta* **2010**, *232*, 1241–1250. [CrossRef]

42. Kishor, D.S.; Lee, C.; Lee, D.; Venkatesh, J.; Seo, J.; Chin, J.H.; Jin, Z.; Hong, S.-K.; Ham, J.-K.; Koh, H.J. Novel allelic variant of Lpa1 gene associated with a significant reduction in seed phytic acid content in rice (*Oryza sativa* L.). *PLoS ONE* **2019**, *14*, e0209636. [CrossRef]

43. Stiles, A.R.; Qian, X.; Shears, S.B.; Grabau, E.A. Metabolic and signaling properties of an Itpk gene family in Glycine max. *FEBS Lett.* **2008**, *582*, 1853–1858. [CrossRef]

44. Shi, J.; Wang, H.; Wu, Y.; Hazebroek, J.; Meeley, R.B.; Ertl, D.S. The Maize Low-Phytic Acid Mutant lpa2 Is Caused by Mutation in an Inositol Phosphate Kinase Gene. *Plant Physiol.* **2003**, *131*, 507–515. [CrossRef]

45. Sweetman, D.; Stavridou, I.; Johnson, S.; Green, P.; Caddick, S.E.K.; Brearley, C.A. Arabidopsis thaliana inositol 1,3,4-trisphosphate 5/6-kinase 4 (AtITPK4) is an outlier to a family of ATP-grasp fold proteins from Arabidopsis. *FEBS Lett.* **2007**, *581*, 4165–4171. [CrossRef]

46. Majerus, P.W.; Wilson, D.B.; Zhang, C.; Nicholas, P.J.; Wilson, M.P. Expression of inositol 1,3,4-trisphosphate 5/6-kinase (ITPK1) and its role in neural tube defects. *Adv. Enzyme Regul.* **2010**, *50*, 365–372. [CrossRef]

47. Couso, I.; Evans, B.S.; Li, J.; Liu, Y.; Ma, F.; Diamond, S.; Allen, D.K.; Umen, J.G. Synergism between Inositol Polyphosphates and TOR Kinase Signaling in Nutrient Sensing, Growth Control, and Lipid Metabolism in Chlamydomonas. *Plant Cell* **2016**, *28*, 2026–2042. [CrossRef]

48. Choi, J.; Williams, J.; Cho, J.; Falck, J.R.; Shears, S. Purification, Sequencing, and Molecular Identification of a Mammalian PP-InsP5 Kinase That Is Activated When Cells Are Exposed to Hyperosmotic Stress. *J. Biol. Chem.* **2007**, *282*, 30763–30775. [CrossRef]

49. Fridy, P.C.; Otto, J.C.; Dollins, D.E.; York, J.D. Cloning and characterization of two human VIP1-like inositol hexakisphosphate and diphosphoinositol pentakisphosphate kinases. *J. Biol. Chem.* **2007**, *282*, 30754–30762. [CrossRef]

50. Lin, H.; Fridy, P.C.; Ribeiro, A.A.; Choi, J.H.; Barma, D.K.; Vogel, G.; Falck, J.R.; Shears, S.B.; York, J.D.; Mayr, G.W. Structural Analysis and Detection of Biological Inositol Pyrophosphates Reveal That the Family of VIP/Diphosphoinositol Pentakisphosphate Kinases Are 1/3-Kinases. *J. Biol. Chem.* **2009**, *284*, 1863–1872. [CrossRef]

51. Mulugu, S.; Bai, W.; Fridy, P.C.; Bastidas, R.J.; Otto, J.C.; Dollins, D.E.; Haystead, T.A.; Ribeiro, A.A.; York, J.D. A conserved family of enzymes that phosphorylate inositol hexakisphosphate. *Science* **2007**, *316*, 106–109. [CrossRef]

52. Weaver, J.D.; Wang, H.; Shears, S.B. The kinetic properties of a human PPIP5K reveal that its kinase activities are protected against the consequences of a deteriorating cellular bioenergetic environment. *Biosci. Rep.* **2013**, *33*, e00022. [CrossRef]

53. Draškovič, P.; Saiardi, A.; Bhandari, R.; Burton, A.; Ilc, G.; Kovačevič, M.; Snyder, S.H.; Podobnik, M. Inositol Hexakisphosphate Kinase Products Contain Diphosphate and Triphosphate Groups. *Chem. Biol.* **2008**, *15*, 274–286. [CrossRef]

54. Tan, X.; Calderon-Villalobos, L.I.A.; Sharon, M.; Zheng, C.; Robinson, C.V.; Estelle, M.; Zheng, N. Mechanism of auxin perception by the TIR1 ubiquitin ligase. *Nature* **2007**, *446*, 640–645. [CrossRef]

55. Wild, R.; Gerasimaite, R.; Jung, J.-Y.; Truffault, V.; Pavlovic, I.; Schmidt, A.; Saiardi, A.; Jessen, H.J.; Poirier, Y.; Hothorn, M.; et al. Control of eukaryotic phosphate homeostasis by inositol polyphosphate sensor domains. *Science* **2016**, *352*, 986–990. [CrossRef]

56. Rouached, H.; Arpat, A.B.; Poirier, Y. Regulation of Phosphate Starvation Responses in Plants: Signaling Players and Cross-Talks. *Mol. Plant* **2010**, *3*, 288–299. [CrossRef]

57. Chien, P.-S.; Chiang, C.-P.; Leong, S.J.; Chiou, T.-J. Sensing and Signaling of Phosphate Starvation—From Local to Long Distance. *Plant Cell Physiol.* **2018**, *59*, 1714–1722. [CrossRef]

58. Azevedo, C.; Saiardi, A. Eukaryotic Phosphate Homeostasis: The Inositol Pyrophosphate Perspective. *Trends Biochem. Sci.* **2017**, *42*, 219–231. [CrossRef]

59. Ried, M. Inositol pyrophosphates promote the interaction of SPX domains with the coiled-coil motif of PHR transcription factors to regulate plant phosphate homeostasis (pre-print). *bioRxiv* **2019**. [CrossRef]

60. Puga, M.I.; Mateos, I.; Charukesi, R.; Wang, Z.; Franco-Zorrilla, J.M.; de Lorenzo, L.; Irigoyen, M.L.; Masiero, S.; Bustos, R.; Rodríguez, J.; et al. SPX1 is a phosphate-dependent inhibitor of PHOSPHATE STARVATION RESPONSE 1 in Arabidopsis. *Proc. Natl. Acad. Sci. USA* **2014**, *111*, 14947–14952. [CrossRef]

61. Dong, J.; Ma, G.; Sui, L.; Wei, M.; Satheesh, V.; Zhang, R.; Ge, S.; Li, J.; Zhang, T.-E.; Wittwer, C.; et al. Inositol Pyrophosphate InsP8 Acts as an Intracellular Phosphate Signal in Arabidopsis. *Mol. Plant* **2019**, *12*, 1463–1473. [CrossRef]

62. Kuo, H.-F.; Chang, T.-Y.; Chiang, S.-F.; Wang, W.-D.; Charng, Y.; Chiou, T.-J. Arabidopsis inositol pentakisphosphate 2-kinase, AtIPK1, is required for growth and modulates phosphate homeostasis at the transcriptional level. *Plant J.* **2014**, *80*, 503–515. [CrossRef]

63. Teale, W.D.; Paponov, I.A.; Palme, K. Auxin in action: Signalling, transport and the control of plant growth and development. *Nat. Rev. Mol. Cell Biol.* **2006**, *7*, 847–859. [CrossRef]

64. Woodward, A.W.; Bartel, B. Auxin: Regulation, Action, and Interaction. *Ann. Bot.* **2005**, *95*, 707–735. [CrossRef]

65. Gray, W.M.; Kepinski, S.; Rouse, D.; Leyser, O.; Estelle, M. Auxin regulates SCF TIR1-dependent degradation of AUX/IAA proteins. *Nature* **2001**, *414*, 271–276. [CrossRef]

66. Tan, X.; Zheng, N. Hormone signaling through protein destruction: A lesson from plants. *Am. J. Physiol. Endocrinol. Metab.* **2009**, *296*, E223–E227. [CrossRef]

67. Sheard, L.B.; Tan, X.; Mao, H.; Withers, J.; Ben-Nissan, G.; Hinds, T.R.; Kobayashi, Y.; Hsu, F.-F.; Sharon, M.; Browse, J.; et al. Jasmonate perception by inositol phosphate-potentiated COI1-JAZ co-receptor. *Nature* **2010**, *468*, 400–405. [CrossRef]

68. Browse, J. Jasmonate Passes Muster: A Receptor and Targets for the Defense Hormone. *Ann. Rev. Plant Biol.* **2009**, *60*, 183–205. [CrossRef]

69. Chini, A.; Fonseca, S.; Fernández, G.; Adie, B.; Chico, J.M.; Lorenzo, O.; García-Casado, G.; López-Vidriero, I.; Lozano, F.M.; Ponce, M.R.; et al. The JAZ family of repressors is the missing link in jasmonate signalling. *Nature* **2007**, *448*, 666–671. [CrossRef]

70. Cui, M.; Du, J.; Yao, X. The Binding Mechanism Between Inositol Phosphate (InsP) and the Jasmonate Receptor Complex: A Computational Study. *Front. Plant Sci.* **2018**, *9*, 963. [CrossRef]

71. Mosblech, A.; Thurow, C.; Gatz, C.; Feussner, I.; Heilmann, I. Jasmonic acid perception by COI1 involves inositol polyphosphates in Arabidopsis thaliana. *Plant J.* **2011**, *65*, 949–957. [CrossRef]

72. Yu, H.; Wu, J.; Xu, N.; Peng, M. Roles of F-box proteins in plant hormone responses. *Acta Biochim. Biophys. Sin. Shanghai* **2007**, *39*, 915–922. [CrossRef] [PubMed]

73. Hu, B.; Chu, C. Nitrogen-phosphorus interplay: Old story with molecular tale. *New Phytol.* **2019**. [CrossRef]

74. Baek, D.; Chun, H.J.; Yun, D.-J.; Kim, M.C. Cross-talk between Phosphate Starvation and Other Environmental Stress Signaling Pathways in Plants. *Mol. Cells* **2017**, *40*, 697–705.

75. Murphy, A.M.; Otto, B.; Brearley, C.A.; Carr, J.P.; Hanke, D.E. A role for inositol hexakisphosphate in the maintenance of basal resistance to plant pathogens. *Plant J.* **2008**, *56*, 638–652. [CrossRef] [PubMed]

76. Reddy, C.S.; Kim, S.-C.; Kaul, T. Genetically modified phytase crops role in sustainable plant and animal nutrition and ecological development: A review. *3 Biotech* **2017**, *7*, 195. [CrossRef] [PubMed]

 © 2020 by the authors. Licensee MDPI, Basel, Switzerland. This article is an open access article distributed under the terms and conditions of the Creative Commons Attribution (CC BY) license (http://creativecommons.org/licenses/by/4.0/).

MDPI

St. Alban-Anlage 66

4052 Basel

Switzerland

Tel. +41 61 683 77 34

Fax +41 61 302 89 18

www.mdpi.com

Plants Editorial Office

E-mail: plants@mdpi.com

www.mdpi.com/journal/plants

www.ingramcontent.com/pod-product-compliance
Lightning Source LLC
LaVergne TN
LVHW070949170726
843515LV00005B/948